PHENOMENOLOGICAL METHOD: THEORY AND PRACTICE

CONTRIBUTIONS TO PHENOMENOLOGY

IN COOPERATION WITH THE CENTER
FOR ADVANCED RESEARCH IN PHENOMENOLOGY

Scope

The purpose of this series is to foster the development of phenomenological philosophy through reactive research. Contemporary issues in philosophy, other disciplines and in culture generally, offer opportunities for the application of phenomenological methods that call for creative responses. Although the work of several generations of thinkers has provided phenomenology with many results with which to approach these challenges, a truly succesful response to them will require building on this work with new analysis and methodological innovations.

PHENOMENOLOGICAL METHOD: THEORY AND PRACTICE

FRED KERSTEN
University of Winsconsin-Green Bay, U.S.A.

KLUWER ACADEMIC PUBLISHERS
DORDRECHT / BOSTON / LONDON

Library of Congress Cataloging in Publication Data

Kersten, F. (Frederick)
Phenomenological method : theory and practice / F. Kersten.
p. cm. -- (Contributions to phenomenology)
Bibliography: p.
Includes index.
ISBN 0-7923-0094-7
1. Phenomenology. 2. Methodology. 3. Space and time. I. Title.
II. Series.
B829.5.K47 1989
142'.78--dc19 89-2838

ISBN 0-7923-0094-7

Published by Kluwer Academic Publishers,
P.O. Box 17, 3300 AA Dordrecht, The Netherlands.

Kluwer Academic Publishers incorporates
the publishing programmes of
D. Reidel, Martinus Nijhoff, Dr W. Junk and MTP Press.

Sold and distributed in the U.S.A. and Canada
by Kluwer Academic Publishers,
101 Philip Drive, Norwell, MA 02061, U.S.A.

In all other countries, sold and distributed
by Kluwer Academic Publishers Group,
P.O. Box 322, 3300 AH Dordrecht, The Netherlands.

printed on acid free paper

Printed in The Netherlands

CONTENTS

PREFACE

This book has two parts. The first part is chiefly concerned with critically establishing the universally necessary order of the various steps of transcendental phenomenological method; the second part provides specific cases of phenomenological analysis that illustrate and test the method established in the first part. More than this, and perhaps even more important in the long run, the phenomenological analyses reported in the second part purport a foundation for drawing phenomenological-*philosophical* conclusions about problems of space perception, "other minds," and time perception.

The non-analytical, that is, the literary, sources of this book are many. Principal among them are the writings of Husserl (which will be accorded a special methodological function) as well as the writings of his students of the Göttingen and Freiburg years. Of the latter especially important are the writings and, when memory serves, the lectures of Dorion Cairns and Aron Gurwitsch. Of the former especially significant are the writings of Heinrich Hofmann, Wilhelm Schapp, and Hedwig Conrad-Martius.

There have been and still are many (often profound) disagreements about phenomenological method and the epistemic and axiological foundations it ostensibly produces. If this book helps focus the disagreements, and if it contributes even in the most modest way to resolving them, then it will have served its purpose. But such purpose, after all, traditionally has been a chapter in the perennial will to philosophical theory. Indeed, what I find most appealing and significant of Husserl's phenomenology is that he himself anchors it, both historically and analytically, in the will to philosophical theory. To be sure, the will to philosophical theory is much more than the will to settle disagreements, to resolve controversies; it is rather the very prospect and promise of phenomenology. What makes the prospect interesting is that frequently it outruns the promise, and a will to criticism is required to restore balance and sight to theory. At other times the promise proves to be more than expected, and a will to explore largely uncharted territory is needed. Both situations seem to me characteristic of phenomenology from its inceptions and its inceptional slogan of the "will to return to the 'things themselves'." It is therefore appropriate to begin the book with an Introduction that examines the will to philosophical theory from which Husserl's phenomenology emerges.

The remainder of the first part of the book is devoted to the somewhat archeological task of ferreting out a method of phenomenological reductions. Chapter One assembles some of Husserl's chief insights into his phenomenology, defines it so that the spirit if not always the letter of Husserl's thought is preserved, and spells out what I believe is entailed by transcendental phenomenological epoché. Chapter Two examines the specific procedures for exercising transcendental phenomenological epoché with respect to the general positing in which the naturalness of the natural attitude consists; critical revisions of Husserl allow, here, for taking the necessary first step in uncovering the "primordial quasi-objective 'world'" of space and time. In Chapter Three further revisions of Husserl make possible the necessary second step in the exercise of transcendental phenomenological epoché whereby still other "quasi-objectivities" are uncovered and the outward boundaries of transcendental phenomenological epoché are reached. Chapter Four concludes the first part with a reflection on the right order of phenomenological inquiry (which is by no means the order of the transcendental phenomenological reductions) and on the promise of phenomenology to provide a basis for a critical-epistemological understanding of space, time and other as an example of the "will to return to the 'things themselves'." The name given to this basis, in the second part of the book, is "transcendental aesthetics."

The second part of the book begins with a survey of some ideas about space and time in modern philosophy that will serve both as point of reference for the phenomenological analyses to follow and to set the nature of the problems of space, time and other. Chapter Five responds to these problems by carrying out what we shall call with Husserl a phenomenological "unbuilding" to the constitution of foundational spatialities ("prespaces," we shall call them) that pertain to the tactually, visually and auditorily presented, while Chapter Six examines the corresponding "building up" of certain aspects of space included in our perception of the real, objective world. Chapter Seven considers still other aspects of space perception so as to formulate a "transcendental deduction" of space, and Chapter Eight seeks to do the same for the perception of time. The concluding chapter, Nine, seizes upon some of the phenomenological results of the previous chapters concerning the perception of space, time and other in order to consider the relationship and difference of phenomenology and philosophy to which those results lead.

In writing this book I have not tried to consider every polemic about phenomenology; nor has it been possible to review all the discussions about perception of space, time and other that legitimately claim attention. There are several reasons for this. The first is that I have limited myself to considering those discussions that have been of most immediate help in formulating a systematically feasible view of at least one, if not other, version of transcendental phenomenological method. The second reason is that to deal with many of the current discussions about phenomenology requires that the phenomenological analyses reported here be carried into areas of the perception of space, time and other that lie outside the problems on which this book focuses. To be sure, every attempt has been made to sufficiently ground the phenomenological analyses so that further investigation and further discussion is possible with respect to a broad range of questions. The third and final reason is that it is my belief that not every problem that arises in connection with the phenomenological analysis of the percep tion of space, time and other is amenable to phenomenological resolution. Some problems, such as metaphysical ones of the meaning of life, death and fate, may be purely philosophical in nature and not have an analogue in phenomenology (which need not imply that phenomenological analysis is irrelevant to them). In other words, there may well be limits to the promise and prospect of Husserl's and of Husserlian phenomenology, no matter how defined. Of course, in making these and other statements of that sort in the book that follows it may seem and perhaps it is true--that, as Hume says, I have forgotten "our scepticism, but even our modesty too; and make use of such terms as these, *'tis evident*, *'tis certain*, *'tis undeniable*; which a due deference to the public ought, perhaps, to prevent." And with Hume, at the end of the first book of *A Treatise of Human Nature*, I wish to add here that "I may have fallen into this fault after the example of others; but I here enter a *caveat* against any objections, which may be offer'd on that head; and declare that such expressions were extorted from me by the present view of the object, and imply no dogmatical spirit, nor conceited idea of my own judgment, which are sentiments that I am sensible can become no body, and a sceptic still less than any other."

I take pleasure in acknowledging my indebtedness to my colleague, Professor James Clifton, for his truly collegial generosity in making available to me the means for producing the text of this book. I am also indebted to Mr. David Griggs for his hard work and patience in producing a difficult text. Thanks are due to an anonymous reviewer whose comments were of inestimable value in improving various and difficult parts of the text. My gratitude is also extended to Professor William McKenna and Barbara Johnson at Miami University, Oxford, Ohio without whose generous, intelligent and skillful help this book would not have been published. Finally, I wish to thank the Frankenthal Foundation for financial support in producing this book, and Martinus Nijhoff, Publishers, for permission to quote extensively from their publications by and about Edmund Husserl, both in the original German and in English translation.

This book is dedicated to the memory of my wife, Raquel Kersten (1929–1988).

Fred Kersten
July, 1987

PART ONE

THE METHOD OF PHENOMENOLOGICAL REDUCTIONS

INTRODUCTION

"Wir wollen auf den 'Sachen selbst' zurückgehen."

1. THE LITERARY BACKGROUND OF HUSSERL'S PHENOMENOLOGY

Following the publication of the *Philosophy of Arithmetic* in 1891, the decade of the 1890's was a busy one for Husserl. For example, in 1894 he was busy reworking his lectures of 1891/92 on psychology,[1] the results of which he would publish in "Psychological Studies for Elementary Logic."[2] In turn, this essay formed the basis of the first chapter of the third Logical Investigation.[3] Two years later, and still at Halle, Husserl delivered lectures covering material on which all but the last chapter of the *Prolegomena* to the Logical Investigations is based[4] (the line of thought in that last chapter of the *Prolegomena* dates back to two years earlier[5]). It was also at this time (1896) that Husserl noted that it had taken him almost ten years to understand his own philosophical position and to make the "very hard break" with Brentano's philosophy. In the next year, 1897, Husserl published his "Review of German Writings on Logic of the Year 1894"[6] followed in 1898 by a series of lectures and essays that became the basis for dealing with the problems of a phenomenology of perception, phantasy, time, and physical thing (but only publicly presented in his lectures at Göttingen in 1904/05). Finally, in 1899, while writing out the rest of the *Logical Investigations*, Husserl achieved his first philosophical breakthrough in the formulation of the universal apriori correlation of experiential objects and manners of givenness or presentation. If we add to this the work in the next year, 1901/02, represented in the famous "double lecture" to the Göttingen Mathematical Society on which the first book of *Ideas* is based, it can be seen that in less than six years of intensively creative work Husserl produced the thought and view that would occupy him for the next two decades.

By November of 1899 all of the *Prolegomena*, except for the Preface, was printed; the Preface itself was published at the beginning of 1900. 1901 saw the publication of the second volume of the *Logical Investigations*, with an Introduction to all six of its investigations. At almost the same time, Winter Semester of 1901, Husserl moved from Halle to the University of Göttingen where he had been apointed by Altdorf over the objections of the faculties in psychology and philosophy. Coincident with the move, Husserl worked out extensive analyses of external perception with respect to its memory-modifications, phantasy-objectivations, and the like, all conceived as continuations of the *Logical Investigations*.[7]

It is also worth recalling that this was the period (extending to about 1912) of the formation of Husserl's greatest following, the period when "phenomenology" was a call to arms and even acquired a slogan. Coincidentally, it was the time when Husserl worked out his method of regress to the mentally productive life in all its domains; it was the time when the method of "*Wesenschau*," of clearly seeing essences, got extended to all realms of being. In addition to still other analyses of external perception, Husserl's students and others inspired (often critically) by his work carried out research in a wide variety of areas--Schapp, Conrad, Conrad-Martius on perception; Scheler on perception ǒf the alter ego, sympathy, feelings and ethics; Pfänder on sentiments; Reinach on civil law; Katz on manners of appearance of colors; Leyendecker on perceptual illusion; Hofmann on sensations.[8]

But no matter how far-flung the ramifications of Husserl's work and that of others at this time, what is striking is the fact that the period from the publication of the *Philosophy of Arithmetic* to the publication of *Ideas*, First Book, depends for its *philosophical* significance on his work during the last six years of the last decade of the nineteenth century. And it is even more important to realize, as did Husserl himself, that the "progress" of phenomenology in the later work is not so much the fulfillment of prospects and promises announced in the earlier work as it is the making explicit and deepening in meaning of what was already there at the turn of the century: the phenomenological method of reductions and large segments of the body of doctrine that will make up transcendental phenomenological philosophy.[9] What Husserl realized he had accomplished lies in summary in the Introduction to the first edition of the second volume of the *Logical Investigations* where the first literary expression of the theory of phenomenological reductions is found under the heading of the slogan, "Wir wollen auf die 'Sachen selbst' zurückgehen." That slogan (and its slightly different formulation in *Ideas*, First Book) will serve as the heading under which not only Husserl's, but more importantly, a Husserlian method of phenomenological reductions can be developed. This method, I shall also argue, is no mere method but instead an integral part of the will to philosophical theory.

2. THE WILL TO PHILOSOPHICAL THEORY

Expressed in full, Husserl's slogan reads:

> By no means do we wish to be satisfied with 'mere words,' that is, with a merely symbolic verbal understanding [such as we have at first blush in our reflections on the sense of the laws about 'concepts,' 'judgments,' 'truths,' etc., with their manifold specifications established in pure logic]. Significations that are inspired only by distant, confused, non-genuine intuitions--if by any of those <intuitions> at all--are not enough. We will to return to the 'things themselves' [*Wir wollen auf die "Sachen selbst" zurückgehen*]. In fully developed intuitions we will to make evident <the fact that> what is here given in actually performed abstraction is truly and really what is meant by the verbal significations in the expressions of laws; and [with respect to cognitive practice] we will to awaken in ourselves the disposition to retain the <verbal> significations in their unshakable identity by means of repeatedly and sufficiently measuring <them> against reproducible intuition (or against the intuitive performance of abstraction)....[10]

The will to return to the "things themselves" and what that entails are occasioned by the very specific task Husserl set himself at the beginning of the Introduction to the second volume of the *Logical Investigations*. There we learn that the basic concepts and the ideal laws of pure logic of the first volume, *Prolegomena to a Pure Logic*, arise out of "subjective sources." The idea is that by tracing the concepts and laws back to their "subjective sources" it will be possible for phenomenology then to bestow upon them a clarity and distinctness which, in turn, will be the basis for a critical-epistemological understanding of pure logic.[11] To achieve such understanding is the principal goal of the six investigations that follow in the second volume. In one way or another, they all express Husserl's belief at that time that all the objects and predicatively formed affair-complexes meant and intended to in thinking and knowing are ideal objects in the sense of synthetical unities presented relative to a multiplicity af actual and possible acts of thinking and knowing.[12] Pure phenomenology is made up, then, of descriptions of those acts of thinking and knowing in general and, in particular, of the concatenations of acts with respect to which the meant and intended-to objects of thinking and knowing are presented in their most original ways as ideal unities evidentially seized upon in further acts founded on those of thinking and knowing.[13]

Theory of knowledge is not theory, however, in the sense that genetic psychology is theory; theory of knowledge does not purport explanation or proof. It is rather theory in the sense of what Husserl calls "*Besinnung*" and "*Vertändigung*," of "reflection" and "understanding" which make clear and distinct any thinking and knowing whatever; it is theory based upon presented acts of thinking and knowing. In this phenomenological sense theory is reflection on the "constituting" of objectivities of all sorts for thinkings and knowings and, on that basis, the phenomenologist seeks to explicitly formulate the problem of clearly understanding how such objectivities are presented especially (i.e., "intuitively") as "things themselves," as ideal unities just as meant and intended to in and through a multiplicity of ostensible knowings and thinkings. Theory in this sense fully embodies the phenomenological slogan because it expresses the will to return to the "things themselves" departing from their mere verbal understanding in order to discover, in fully developed intuitions, that what is presented in actually performed abstraction is truly and indeed what the verbal understanding signifies.[14]

At issue for Husserl in the *Logical Investigations*, then, is not a naive but instead a critically conscious purpose. As the means to realize its end, the theory envisions not a thematizing of the "things" so much as *a thematizing of the acts of thinking and knowing* that mean and intend to and, in some cases, evidentially intend to (or: actually are presentive of) the "things as 'things themselves'." Expressed is not a will to purely "objective" knowledge gained by an exclusive preoccupation with objects. Instead, what the theory expresses is a *will to a critique of ostensible knowledge* by a reflective examination of the acts of consciousness in and through which what is ostensibly thought and known are reflectively seen to be actually presented as "they themselves." In short, phenomenological theory is that theory which expresses a will to knowledge as a foundation for a critically justified formal ontology (inclusive of formal ontology of apophansis as well as of predicatively formed affair-complexes). By no means is it a will to a straightforward ontology, based as it may well be in fact on (unexamined) intuition.

Unfortunately, and almost from the outset, the reflection and understanding that comprise phenomenological theory was frequently and even grossly misinterpreted. More often than not, the slogan was interpreted as expressing the simple purpose of being "objective." This interpretation turned up in the attempt to develop eidetic ontologies such as those of Hering, Reinach, Daubert,[15] Conrad-Martius, Schapp, Pfänder, Spiegelberg, and even Heidegger, just to mention a diverse few.[16] Undoubtedly there are a number of reasons for such misinterpretation, but the chief and perhaps most general one would seem to be that by the time of the First World War and afterwards German philosophers generally and readers of Husserl in particular were tired of theory of knowledge and wanted metaphysics instead.[17] Thus by taking the slogan out of context and taking stock of it only with respect to the *Prolegomena*, i.e., to the *first* volume of *Logical Investigations*, the slogan promised them relief from the "psychological" as well as the neo-Kantian theories of knowledge prevalent just prior to the First World War,[18] and of which they were tired. Thus one may conclude that the *historical* situation was more responsible for the naive ontologizing than was Husserl himself.

Indeed, this may be substantiated by the historical facts themselves. After all, Husserl himself never indulged in naive ontologizing; moreover, other naive metaphysical movements got underway at the same time, largely outside Husserl's influence (e.g., Meinong, Dilthey, Scheler, Bergson).[19] The other historical factor important in this connection is the well-known fact that by the time of the publication of "Philosophy as a Strict Science" (1911) and *Ideas*,

First Book (1912), that is, when the genuine sense of Husserl's phenomenology had become clear, Husserl had lost his influence on even those members of the generations before and during the First World War who had accounted themselves his followers.[20] In fairness it is necessary immediately to add that Husserl's one-time followers usually attributed their divergence from what Husserl meant by the slogan not to a progressive clarification of the pervasive tendency in Husserl's thought, or even to the historical situation itself, but instead to what they believed was a change in that tendency.[21] And because so many others since, not just in the German-speaking countries, but in the French- and English-speaking countries as well, have heeded the naive ontologists and the cult of the expressionist metaphysicians (called "existentialists" after the Second World War),[22] the influence of the real Husserl will not spread until people--a new generation--will become dissatisfied with the numerous dogmatisms (including "empiricism," "logistics," "skepticism," and "relativism") that still inspire confidence.

3. PHENOMENOLOGY AND EPISTEMOLOGY; LOGIC AND PSYCHOLOGY

The thematic objects of purely descriptive phenomenology and, later on (e.g., in the Introduction to the three books of *Ideas*), purely descriptive, eidetic and transcendentally pure phenomenology, are mental life-processes (*Erlebnisse*). In the case of the Introduction to the first edition of the second volume of *Logical Investigations*, they are overtly psychical mental life-processes and, as we shall suggest shortly, covertly transcendentally pure mental life-processes.[23] Let us also note that for Husserl the psychical mental life-processes are not unqualifiedly identical with those investigated by the empirical psychology of his time.[24] Nevertheless, some sort of identity is in fact presupposed by the relationship obtaining between phenomenology and empirical psychology: the one, phenomenology, analyzes and describes the identical mental life-processes that the other, empirical psychology, seeks to explain genetically. To express the situation another way, Husserl says that phenomenology describes the empirical objects the genetic concatenations of which empirical psychology traces.[25]

The distinction and relationship are not, however, so obvious as they might seem at first sight. Critical-epistemological understanding in particular, and theory of knowledge generally, are universal clarification of the ideal essence of psychical mental life-processes that are, quite specifically, knowings and thinkings; yet there is nothing in the Introduction to the second volume of *Logical Investigations* that states that phenomenology is not a description of psychical mental life-processes *as factual matter*. For example, when Husserl says that the logician has no immediate reason to make his own concrete, singular mental life-processes the goal of his logical interest, the suggestion is that his own mental life-processes, as concrete, singular factual mental life-processes, are the goal or theme of the phenomenologist.[26] Husserl's suggestion is important and worth examining not only because it is often overlooked, but because it also points to the contrast between phenomenology and logic in general and between phenomenological analysis and logical analysis as ideal. Indeed, the very fact of the contrast is significant because it would not obtain at all were phenomenology conceived simply as a descriptive analysis of ideal essences rather than of the instances of those essences, scl. individual, or singular, factual acts of consciousness.

Judging from Husserl's use of the word, "ideal," in this context, it would seem that the ideal lies on the side of the objective. In the Introduction to the second volume of *Logical Investigations* "ideal" signifies the ideality, we may say, of objects in general rather than just the peculiar ideality of essences. More particularly, "ideal" signifies the intentional unity of objects presented in and through the multiplicity of acts of consciousness of them. In addition, it is in reflection that acts are made into objects; that is, Husserl tells us,[27] it is necessary to consider them in new acts of intuition and thought, to analyze and describe them, to make them thematic objects of comparative and differentiative thinking so that, finally, they may be ordered into species.[28] Consequently a basis is established for ideation which, in turn, allows for the formulation of truths of unrestricted universality.

Thus phenomenology may be said to be *pure* phenomenology in a number of senses of the word, "pure." In the first place, phenomenology is pure in the sense that it is free from confusing the objective affairs of logic with its own proper theme: psychic mental life-processes, acts of consciousness in the broadest sense of the word, "act." In the second place, phenomenology is also pure in the sense that it is restricted to description without empirical-genetic explanation: its theory is *pure* theory, its description is *pure* description.[29] Moreover, it is in this second sense of the word, "pure," that the distinction between the two sorts of psychologism is made in the *Prolegomena*:[30]

> 1. The interpretation of the objects of pure logic as components of the logical acts, i.e., of judgments intended to as components of the judgings themselves;
> 2. the interpretation of the apriori laws of pure logic as universal empirical-genetic truths about mental life-processes.

The foregoing distinction points to an important fact, likewise frequently overlooked, namely that Husserl does not deny that general descriptive truths about judging and knowing are possible; nor does he deny that empirical-genetic explanation of those processes is possible. *What he does deny is that pure logic is at all about mental life-processes and that the laws of pure logic are empirical-explanatory laws or based on such laws or even based on phenomenological description of mental life-processes.* In addition, Husserl's distinction enables the further denial that the laws of pure logic are or could be based on even apriori description of mental life-processes.

Consequently phenomenology and pure logic deal with quite different regions of things, whereas, in contrast, phenomenology and psychology deal with the same region of things but in quite different ways. In other words, phenomenology is relevant to pure logic, not because it deals with the objects comprising the field of investigation of pure logic, but rather because it deals critically with those mental life-processes in which the ostensible thematic objects of pure logic are intended to and perhaps themselves presented as ideal unities. Phenomenology examines not the laws of pure logic, nor even their extension, but rather the acts, the intuitings of the laws of pure logic, the objects of pure logic (propositions, predicatively formed affair-complexes, and the like). Such critical-epistemological understanding depends chiefly on the ability of pure phenomenology to provide descriptive analyses of the acts of "empty" judicative intuitings of categorial objects (of affair-complexes themselves), and the intentive syntheses of fulfillment of the two--a case in point of the will to return to the "things themselves."

Providing such descriptions that make up critical-epistemological understanding is the phenomenological goal which bestows a single teleological meaning on all six of the subordinate analyses in the second volume of the *Logical Investigations*. The other side of the telic coin signifies that the *Logical Investigations* then do not merely cease with the sixth investigation; their descriptive-phenomenological results comprise the *terminus ad quem* so that in their light one ought to be able to submit straighforwardly developed pure logic (as apophantics and as formal ontology) to an epistemological criticism which would show the principles of pure logic to be genuine knowledge ("objective truths"). This criticism, of course, should not be confused with the reflexive criticism of logic by its own principles applied normatively; that is a wholly unphenomenological undertaking, such as the testing of the laws of logic themselves for formal consistency. And because the relevance of phenomenology to pure logic of this sort does not involve the two kinds of psychologism mentioned in the *Prolegomena*, Husserl's use of phrases such as "the phenomenological founding of logic,"[31] or the "analytic phenomenology, needed by the logician in his preparatory and foundation-laying concerns,"[32] should be understood only in the light of the goal of phenomenology as a consequence of which phenomenology is fundamental to an epistemologically self-critical logic rather than to a straightforward one. Indeed, the *Logical Investigations* themselves as a whole are said to present work preliminary to the epistemological clarification of a "future construction [*Aufbau*] of logic."[33]

The relevance of pure phenomenology to psychology (but not to psychology in Husserl's later sense of "phenomenological psychology") is quite different from that to pure logic. Pure phenomenology describes what psychology explains genetically. There is no corresponding distinction such as that between a phenomenological description and an explanation of the subject matter of pure logic. This is because acts of judging and knowing are parts of the concrete world-process; they come and go in time, are what they are in any factual instance as individual processes in the functionally interrelated nexus of individual events. Thus they can be investigated with respect to the empirical conditions determinative of their factual existence and nature as *hic et nunc*. In contrast, the subject matter of pure logic is not comprised by objects that come into and go out of being and exist *hic et nunc*. Pure logic's subject matter is made up of an objectivity quite distinct from individual things and mental life-processes.

Once Husserl has made the foregoing distinction and clarifications he can, at the end of the Introduction to the second volume of the *Logical Investigations*,[34] raise a series of questions concerning what, in his later writings (such as *Formal and Transcendental Logic* and *Cartesian Meditations*), would be called a "genetic" account of objects of a pure logic--an account for which the *Prolegomena* leaves room yet does not make explicit. Of course the account will be "genetic" in a sense quite different from the empirical psychology discussed by Husserl in the first edition of *Logical Investigations*. In a word, "genesis" concerns the production of an ideally identical object by "subjective sources."[35] Indeed, the multiplicities of evident intendings to objects are said to be "sources" from which the concepts of logic arise by means of intuitive abstractions. The unity of the object is then a "unity in multiplicity"--which is not to say, for Husserl, that intentive syntheses create the objects "in themselves," nor that the concepts of logic are even identical with objects in themselves of a higher order (as in Meinong or Benussi). Yet already in the Introduction to the second volume of the first edition of *Logical Investigations*, at the end of section 2 (a text almost wholly unchanged in the second edition), the phenomenological *problem* of genesis is felt by Husserl. That "feeling" points to a deeper meaning of the slogan expressing the "will to return to the 'things themselves'" as the will to philosophical theory.

4. KNOWING, OSTENSIBLE KNOWING, AND THE "IDEA OF PHENOMENOLOGY"

Husserl may be said to be open to what *others* have condemned as psychologism--namely, "transcendental idealism" in the sense of the view that (briefly expressed) all "objectivity" is of essential necessity the "dependent" ideal product ("*Leistung*") of a mental living ontically prior to all "objectivity." *And this is a view never denied by Husserl.* Indeed, it is a view largely implicit in the earlier writings,[36] yet securely anchored in Husserl's thought from the

1890's on. Of course it is clear, even for the most cursory reading, that in the *Logical Investigations* and elsewhere Husserl did not concern himself primarily with genetic-constitutive problems to the extent that he had in the *Philosophy of Arithmetic* and would again almost forty years later in *Formal and Transcendental Logic* and *Cartesian Meditations*. Perhaps it was because he realized that before genetic-constitutive analyses can be successfully carried out he would have to deal more thoroughly with "static" analyses (so the rest of the first edition of *Logical Investigations*, *Ideas*, First Book, along with the many lecture courses in between them[37]). But this de facto course of Husserl's thought ought not blind us to the crucial fact that the de jure context of "static" phenomenological analyses always remains the same in his work: "genetic" phenomenological analyses motivated by the will to return to the things themselves productive of critical-epistemological understanding.

From its outset, the "progress" of phenomenology has always rested on and been guided by a "feeling" for the problem of transcendental phenomenological "genesis." Later on (section 13) a conceptual framework will be established within which "static" and "genetic" phenomenology may be defined more precisely. Here and for the moment it is worth while to ferret out the "feelings" for the problem of "genesis" as it is at work within Husserl's thought, haunting and coloring it, by briefly reconstructing two different examples of the feeling from the literary history of the will to return to the things themselves.

It is in his lectures on the consciousness of "internal time" of 1905 that Husserl expresses for the first time in public the self-conscious exercise of the phenomenological epoché. Two years later, in a set of lectures posthumously published under the title of the *Idea of Phenomenology*,[38] Husserl develops the specifically *critical-epistemological* import of the 1905 lectures and carries to greater self-conscious exercise the will to a critique of ostensible knowledge as the context in which the will to return to the things themselves is set.[39] This set of lectures is especially appropriate, therefore, as an example by which to gauge Husserl's "feeling" for the problem of transcendental phenomenological genesis.

Husserl begins his 1907 lectures by asking whether any ostensible knowing that something is the case can be found to be a genuine knowing that something is the case.[40] To be sure, the ostensible knowing is always presupposed whenever such a question is asked. In so far as the ostensible knowing is all that is presupposed when reflectively seizing upon cognitive mental life, one may be said not only ostensibly but also genuinely to know the process in question as a case of ostensible knowing. At least prima facie, the presupposing is a justified presupposing. *Yet the genuineness* of the ostensible knowing can be questioned if the knowing is ostensibly a knowing of an object other than one's own immediately presented mental life. Thus the problem remains whether any nonreflective, ostensible knowing can also be genuine. Moreover, the same question must be raised with respect to the ostensible knowing of one's immediately presented mental life not only as having *its own* immediately presented, really inherent temporality (the topic of the 1905 lectures) but also as being related--temporally or otherwise--to an object cognitively meant as not its own immediately presented, really inherent self.

It is in this connection that, in the 1907 lectures, Husserl once again exercises the phenomenological epoché so that he can discriminate and correctly state the genuine problem of critical-epistemological understanding (more generally, the problem of "theory" in the phenomenological sense). The reason why phenomenological epoché is exercised at this juncture is because Husserl's line of thought, when made explicit, contains a remarkable conclusion, which may be expressed in the following way:

With Husserl I began with the analytic observation in which I find that the objects of my ostensible knowings are none other than objects I ostensibly mean in my ostensible knowings, hence as "dependent" on, "relative" to, my knowings. Yet at the same time and all along I see that I have been simply presupposing those objects by regarding them implicitly as objects of my ostensible knowings and my own ostensible knowings as on a par with other worldly processes and intraworldly events per se. Simply presupposing in this way I may proceed to elaborate, let us say, a dogmatic theory that their status as objects of my ostensible knowings is incidental to the world and its components. But if this theory is true, then the objects I consider to be truly objects of my ostensible knowings (and, by extension, thinkings) cannot also be the ostensibly known and thought-about world itself and its components themselves. As a result, I am forced to conclude that there must be a numerical distinction between something in itself and something as object of my ostensible knowings, as ostensible "idea." *And yet this distinction must be one I make between two objects of my own ostensible knowings and thinkings.*

Husserl's response to this situation is familiar: he takes an "officially neutral" attitude toward it which in no way alters, denigrates or interprets away the something in itself and something as ostensible "idea." The "feeling" for the problem of phenomenological genesis we may say with Husserl acts like the warning of a Socratic *daimonion* if we are to respond to the situations in a way that makes consistent sense of it rather than "logicizes" it away.[41] Several examples of Husserl may help make this clearer for there is yet another side to this situation marked by the will to return to the things themselves, and one moreover that leads us back to the first edition of the second volume of the *Logical Investigations* and ahead to the *Ideas*, First Book, along with the rewriting of the second volume of *Logical Investigations*. The rather tortured literary history of phenomenology allows of no escape! The 1907 set of lectures prove to be, in that history, a sort of fulcrum that enables one to see Husserl wrestling with a remarkable problem.

The second example is found in the Sixth Logical Investigations where, in the *first* edition, Husserl makes the following statement:

> The world in itself is a sensuous unity: For Being in the individual sense is equivalent to being sensuously perceivable. But the world in itself is not given *to us* in simple perception; for us the world is only a unity of theoretical inquiry, quite inadequately meant, partially through simple and categorial intuition, partially through signification....To doubt whether the world actually is as it appears to us, or as it is meant in theoretical science and in the latter's well-founded persuasion, has its legitimate sense; for inductive science can never shape the idea of the world adequately, no matter how far it may take us. But it is countersensical to wonder whether the actual course of the world, the real connectedness of the world, could not conflict with the forms of thinking. For that would imply that a determinate, significatively and hypothetically supposed, sensuousness, namely the one which would bring the world itself to adequate self-presentation for us, would in truth be capable of taking on the categorial forms; but these forms would demand unifications that are universally excluded by the universal essence of those same forms.[42]

In this passage the feeling for the problem of phenomenological genesis has already touched ground, something which Husserl himself sought to make explicit in the second edition of *Logical Investigations* written after the experience of the 1907 lectures and the publication of *Ideas*, First Book. For example, in the second edition, the first sentence reads, "The world is constituted as a sensuous unity," rather than "The world in itself is a sensuous unity" of the first edition.[43] Husserl then alters the second part of the first sentence to read in the second edition: "With respect to its sense, <the world> is a unity of actual and possible perceptions simpliciter." Correspondingly, the first clause of the second sentence, "But the world in itself is not given to us in simple perceptions," is altered to read: "With respect to its true Being, however, <the world> is neither given without qualification in self-contained perception-processes, nor is it given adequately." The second clause of the second sentence in the first edition is now made the third sentence in the second edition; the rest of the paragraph remains the same in both editions. Both the changes in the first sentences and the unchanged status of the rest of the paragraph express still more emphatically the same sort of phenomenological situation confronted head-on in the 1907 lectures: the things themselves to which Husserl has willed to return comprise, at the same time, the mutual inclusiveness of the "world itself" and the "world as constituted."

According to the gist of the whole paragraph as it stands in the first edition together with its changes in the second edition, to make "world in itself" and "world as constituted" mutually exclusive would be to ignore the beginning of reflection; or, if the beginning is acknowledged, it would be to refuse to grant the end, the result of reflection. Cast into the form of a question: Without falling into one or the other position of realism and idealism, and without mutual inconsistency and contradiction, how can one respond to what in fact one has entertained at the same time--at the beginning, the "world in itself" and at the end of reflection the "world as constituted"? In the light of the 1907 lectures (and certainly others from the first decade of this century) this is precisely what Husserl wrestled with in seeking to rewrite the paragraph just cited. The literary history of phenomenology is equally the history of the will to return to the things themselves that prove to be ambivalent as well as mutually inclusive opposites. The only way to make consistent sense out of this situation is to heed the Husserlian "daimonion," to exercise phenomenological epoché with respect to ostensible knowings and thinkings and their objects.

When taken together the different examples of the feeling for the problem of phenomenological genesis signal mutually exclusive ways of confronting, and having always already confronted, "world" in a two-fold way: *first*, "world for" my ostensible knowings and thinkings ("world as constituted") and "world in itself;" and, *second*, "world for" my ostensible knowings and thinkings as numerically distinct from "world in itself." Both confrontational and presentive alternatives are perfectly valid theoretically. And in each case the response is to hold in abeyance, to place a moratorium on, any decision to choose the one alternative over the other, or to interpret the one in terms of the other.

The phenomenological slogan has, then, a progressively profounder meaning that finds expression in *Ideas*, First Book:

> But to judge rationally or scientifically about things [*Sachen*] signifies to conform *to the things* themselves [*sich nach den Sachen selbst richten*] or to go from words and opinions back to the things themselves [*auf die Sachen selbst zurückgehen*], to consult them in their self-givenness and to set aside all prejudice alien to them.[44]

This expression of the slogan may be said to sum up the telos of two decades of philosphic thought. And, I suppose, the final formulation of the slogan can be found two decades later in *Cartesian Meditations*:

> ...a *first methodological principle.* It is plain that I, as someone begining philosophically, since I am striving toward the presumptive end, genuine science, must neither make nor go on accepting any judgment as scientific *that I have not derived from evidence*, from 'experiences' in which the affairs and affair-complexes in question are present to me as '*they themselves*'.[45]

Here the phenomenological problem of genesis fully surfaces in the slogan itself that gets its just due as a "first methodological principle."

Phenomenological epoché then turns out to be not only the means of securing access to the things themselves to which we will to return, and which we garner in reflection; it is also the "rule of law," the "justice," one might even say, done to the things themselves. To express the matter with respect to the context of the slogan's first formulation, we may say that the increasing self-conscious exercise of phenomenological epoché is nothing else but the quite concrete form of the will to philosophical theory, of the striving toward the "presumptive end, genuine science"--to borrow the phrase from the final formulation.

Phenomenological epoché is therefore not so much one among other methodic tools of the intellect as it is an intrinsic sight of the philosophic-theoretic will and striving required by the things themselves to produce critical-epistemological understanding.[46] Such is the "idea of phenomenology" to be developed in this book. The feeling for the problem of phenomenological genesis has emerged from its philosophical closet and has acquired a life of its own by the time of *Ideas*, First Book, with which, accordingly, this study begins.

CHAPTER ONE

THE TRANSCENDENTAL PHENOMENOLOGICAL REDUCTIONS

5. TRANSCENDENTAL PHENOMENOLOGY DEFINED

The full title of Husserl's book, *Ideas Pertaining to a Pure Phenomenology and to a Phenomenological Philosophy*, applies to three projected books of which only one, the first, was published during Husserl's lifetime and with the subtitle, *General Introduction to Pure Phenomenology* (1912).[1] Appended to the First Book was an "Introduction" to all three books which Husserl continued to reprint with the First Book despite his dissatisfaction with the manuscript of the Second and the continued non-existence of the Third Book. Yet the full sense of the fragmented whole of *Ideas* may be recovered from the "Introduction" in the light of which the title to all three Books may be expressed in these words:[2] *Ideas Pertaining to a Purely Descriptive, Eidetic, Transcendentally Pure or Transcendental Phenomenology and to a Transcendental Phenomenological Philosophy.*

When, at the time of *Ideas*, First Book, Husserl speaks of his phenomenology as "pure," his thought has reached the point where it is explicitly concerned with returning to the "things themselves" as phenomena that are "*transcendentally* pure." Such phenomena he contrasts now with what he calls "real phenomena," by which he means just essentially possible individual phenomena in an essentially possible spatio-temporal world. Moreover, although "transcendentally pure phenomena" may be said, with qualification, to be identically the same as the essentially possible individual phenomena, they are understood nevertheless as "purified" from "what confers upon them reality."[3] The phenomena in question, the transcendentally purified things themselves, are those mental lifeprocesses ("consciousness") comprising all that Descartes included under the heading of "cogito," such as thinkings, judgings, believings, doubtings, perceivings, willings, strivings, feelings (*Ideas*, First Book, sections 20, 28, 34). Accordingly, the "transcendentally pure phenomenology" of which Husserl now speaks and proposes to develop in all its ramifications concerns mental life-processes that have been purified from what confers upon them the characteristic of reality, from positing them as essentially possible processes of an essentially possible mental life in an essentially possible spatio-temporal world.

Equivalently but more broadly stated, "to be real," as Husserl understands it in *Ideas* signifies to be spatio-temporal in an essentially possible spatio-temporal world. Mental life-processes that have been "purified" from positing "real" in this sense must therefore be characterized in a radically different way. In the "Introduction" to *Ideas* Husserl settles on the word, "irreal," to designate their status as "transcendentally pure." But to designate transcendentally purified mental life-processes as "irreal" *does not* signify that they are "unreal" or even "real" in some other sense of the word, "real," let alone "ideal." However else "irreal" may eventually be defined (below, section 9f.), it must be noted at the begining that Husserl's use of the word does imply that, as in the fourth part of *Ideas*, First Book, for instance, the word, "real" ("*Real*") has a very broad *and non-genuine* signification so that the only genuine signification of "real" proves to be "irreal." But what, then, does "irreal" signify as a name for "transcendentally purified" mental life-processes?

The answer, I believe, lies in what Husserl understands by "transcendental reduction" and which is established in striving to return to the things themselves in the will to philosophical theory. But before it is possible to clarify this now from the perspective of *Ideas*, First Book and later writings, let alone and more importantly gain some insight into what "transcendental" and "transcendentally pure" signify, several other factors must be accounted for.

Transcendental phenomenology, Husserl says, is "in fact, a *purely descriptive* discipline, exploring the field of transcendentally pure consciousness by pure intuition." So understood, phenomenology is *also* "pure" in the additional sense of being *eidetic*: "*pure or transcendental phenomenology will be established, not as a science of matters of fact, but as a science of essences* (as an '*eidetic*' science); it will be established as a science that seeks to ascertain exclusively 'cognitions of essences' and *no 'matters of fact' whatever*."[4] Husserl adds the qualification that "Our phenomenology will be an eidetic doctrine, not of phenomena that are real, but of phenomena that are transcendentally reduced."[5] Rather than with "matters of fact," phenomenology deals with things themselves that are not only transcendentally pure (*scl.* "irreal") but also descriptively and, as well, eidetically pure.

As a first step toward explaining phenomenology as an eidetic science Husserl contrasts it with what he calls a "science of realities" and with the "experience of realities" (realities and experience then in non-genuine senses of "reality" and "experience"). The relevant example of such a science in the "Introduction" to *Ideas* is psychology, and the reality-phenomena of concern to psychology are real events having actual existence and with coposited real, percipient subjects who have their place in the one spatio-temporal world posited as the *omnitudis realitatis*.[6] Psychology is for Husserl a science of matters of fact, whereas in contrast transcendentally pure phenomenology is a science of essences. Yet the two sciences are not, as we have seen (above, section 3), unrelated. As Husserl indicates in the "Introduction" to *Ideas*, the eidetic phenomenological reduction proceeds from the psychological reality-phenomena, or matters of fact, to materially filled pure "essences;" or, in the case of judgemental thinking, from a matterof-fact (or: "empirical") universality to "eidetic" universality. Husserl then goes on to say that there are *other* phenomenological reductions to be developed in addition and in contradistinction to the eidetic reduction.[7] Phenomenology therefore is not an eidetic science of real phenomena but instead of transcendentally reduced phenomena.

In other words, Husserl has proposed, *first*, an eidetic transcendental reduction *and then* other, so-called "specific" transcendental reductions. When he goes on to say that the other specific transcendental reductions "purify" phenomena of their reality and being in the world, i.e., disclose them as "irreal," I take him to mean that world and its phenomena are comprehended just and only as correlates of transcendental "consciousness *of*" them.

A further phenomenological result of drawing the contrast between transcendental phenomenology on the one hand and, on the other hand, psychology in particular and natural science in general,[8] is the ascertaining of an *order of phenomenological reductions* in the light of which a specific set of descriptive methods will be established. In the "Introduction" to *Ideas* it is that of, first, the eidetic reduction and then, second, the "specific" phenomenological reductions. What is especially interesting about this order is that it is contrary to what Husserl actually (and, I believe, correctly) practices in the rest of *Ideas*, First Book, as well as in later publications such as *Formal and Transcendental Logic* and *Cartesian Meditations*. A careful look at the nature of these contrary, indeed contradictory, orders will lay a critical basis for formulating the right order and nature of phenomenological reductions, thus returning us to the things themselves[9] in "fact"[10] and "essence."

6. PHENOMENOLOGICAL EPOCHE AND REDUCTIONS

It is not an exaggeration to say that the innermost possibility of phenomenology depends on the right order and nature of phenomenological reduction. The point at issue here is the fact that the order sketched in the "Introduction" to *Ideas* is not that practiced in *Ideas*, First Book. This raises the questions of what it means to speak of a right or correct order of reductions and, correspondingly, of how such an order is established. But before we can directly address that issue we must ask what, indeed, should be understood by "reduction" and "epoché," *and in what sense are they "phenomenological"*?

We may begin, like Husserl, not in the "Introduction," but in the first chapter of the second part of *Ideas*, First Book: We begin with the real, objective world in the ostensibly customary sense of "real" and of "world" (section 27). World appears as one, intersubjective world existing at every moment for us--a real world (section 28) in which we find ourselves as egos living naturally and, for the most part, straightforwardly under the form "cogito" in which we are turned to the world in spontaneous advertence (section 29). (It is just this world and its positing which is the subject matter of what Husserl calls the "sciences belonging to the natural attitude" (sections 1, 30).)[11] Taking Husserl at his word, and practicing what each must do for himself: reflecting on my natural mental life-processes as they proceed straightforwardly in spontaneous advertence to the world and things in it, I seize upon the believedness of what I am actually and spontaneously busied with while also at the same time "officially dissociating" myself from all and any actual positions taken toward the world and things in it;[12] I now regard the world and things just as "what I believe them to be," as "what I see," as "what I think or feel about them." To express the situation wholly in the language of the second chapter of Part Two of *Ideas*, First Book, I exercise "phenomenological epoché" that consists of refraining from (*Sich-Enthalten*) accepting the actually existing, real world in the customary sense of being posited as an actually existing, real world. The exercise of phenomenological epoché then signifies a setting out of action ("*Ausser-Aktion-Setzen*") or excluding (*Ausschalten*) of the general positing belonging to the natural attitude--an excluding of the accepting ("*Hinnehmen*") of the actually existing, real world as an existing world. *Correlatively*, the exercise of phenomenological epoché signifies the parenthesizing ("*Einklammern*"--section 32) of the real objective world posited as actually existing--again in the customary senses of "actuality" and "existence."

Phenomenological epoché *is phenomenological* because it leaves *unparenthesized* and, *correlatively*, *still in acceptance*, transcendentally pure (or irreal) mental life-processes[13] as comprising the region of being that can become a field of inquiry for a meditating transcendental phenomenologist. Phenomenological epoché, Husserl says, "will deserve its name only in consequence of the insight that the fully conscious exercise of the epoché will show itself to be the operation necessary in order to make 'pure' <*scl.* irreal> consciousness, and subsequently the whole phenomenological realm, accessible to us" (section 32). But, I may now ask with Husserl (at the begining of section 33), "What can remain if the whole world, including ourselves with our cogitare, is excluded?" What can remain *as not parenthesized*, in other words, if the whole world is excluded? Further on in section 33 Husserl provides the still controversial answer that because mental life-processes still remain in acceptance for me even when the whole actually existing real world is parenthesized, they have, in themselves, a "being" of their own. Whatever else this assertion may mean, it does not mean that mental life-processes are concretely apart from the world, nor that they are in any way deprived of their mundane sense. To be sure, the "own-being" of mental life-processes still requires explanation toward which a beginning will be made in the next section. The point to be made at the moment is that with Husserl the phenomenological epoché is to be called in addition *transcendental* phenomenological epoché precisely because mental life-processes are made manifest by this limited and circumscribed operation of refraining. At the end of section 97 of *Ideas*, First Book, Husserl returns to what he has said in section 33:

> The characterization of the phenomenological reduction and, likewise, of the pure sphere of mental processes as 'transcendental,' rests precisely on the fact that we discover in this reduction an absolute sphere of stuffs and noetic forms whose determinately structured combinations possess according to *immanental eidetic necessity* the marvelous consciousness of something determinate and determinable, given thus and so, which is something over against consciousness itself, something fundamentally Other, non-really inherent [*Irreelles*], transcendent; and the characterization of 'transcendental' further rests on the fact that this is the primal source in which is found the only conceivable solution of those deepest problems of cognition that concern the essence and possibility of an objectively valid knowledge of something transcendent.

Husserl then proceeds to describe what he means by saying that mental life-processes are "irreal:"

> 'Transcendental' reduction exercises the epoché with respect to actuality: but what it retains of actuality includes the noemata with the noematic unity lying within themselves, and accordingly the manner in which something real is consciously intended to and, in particular, given, in consciousness itself.

The passages just cited also make it clear that for Husserl, and indeed for anyone who phenomenologizes, the operation of parenthesizing is therefore *not* a matter of excluding anything actual, real, so much as it is one of making explicit, manifestly clear and understandable the genuine sense of actual, real, which is already included and largely implicit in our customary experience. What remains, what is "untouched" as it were, by the parenthesizing is called the "phenomenological 'residuum'" or, as we may also say, what is "reduced." "Residuum" and "reduction" are therefore to be understood *relative to the limits or range placed on the phenomenological epoché.*[14]

It must be mentioned in this connection, finally, that the operation by which the phenomenological residuum is laid bare "will be divided into different steps," and that thus "our method will assume the characteristic of a step-by-step reduction" section 33). These various steps, to be formulated in the next chapters, make explicit the various hierarchical levels (or: "stories," strata, layers) of mental life-processes; the steps are to be carried out under the general heading of what Husserl calls "oriented constituting" in the *Cartesian Meditations* (see below, section 12).[15] The steps too have an order the principles of which must be explained.

7. PHENOMENOLOGICAL THEMATIZATION OF MENTAL LIFE-PROCESSES. THE CORRECT ORDER OF PHENOMENOLOGICAL REDUCTIONS

So far I have tried to indicate the sense in which, for Husserl, epoché is not only phenomenological but also transcendental in the rather novel meanings he gives to those terms. And, by extension, those adjectives apply mutatis mutandis to what is reduced and the residuum. To be sure, what Husserl says now in *Ideas* in response to the dilemma in his thought stemming from the work of the previous decade was and remains controversial, afflicted as well with literary and terminological difficulties that themselves still offer separate problems of interpretation. At this juncture in our discussion there seem to be two complimentary alternatives. Either one rehearses almost seventy years of criticism concerning what Husserl means and can mean by "transcendental phenomenological epoché and reductions"--assuming one can legitimately mean anything at all by the words; or one can proceed to systematically assemble according to certain themes a *Husserlian* view which follows the spirit if not always the letter of Husserl's own literary expression. Both alternatives have advantages and disadvantages; if the second alternative is chosen here it is only because it provides a more direct route for returning to the things themselves that can be tested critically by examples and concrete cases especially in the second part of this book.[16] More particularly, two themes will be formulated to express the nature of the phenomenological epoché and reductions, and to show their correct order.

Theme I: "My own mental life-processes as they proceed, some of which are spontaneously turned toward the world and things in it straightforwardly; my own mental life-processes presented as actually existing processes, functionally related to, even dependent upon, processes, events and things other than themselves in the world." On the basis of seizing upon and making explicit my mental life-processes, I now proceed to describe them apart from any other explanations (such as those of an empirical-genetic sort) only with respect to those intrinsic determinations and relations they are presented as having in their actual, current phase, and with respect to which, as such non-selfsufficient components of the world, they are believed in as actually existing in and with the co-presented spatio-temporal world. When I do this I am describing my mental life-processes in the "*psychological*" phenomenological attitude or epoché or refraining; and what is disclosed by exercising that epoché or refraining may be said to be "psychologically-phenomenologically reduced:" precisely my own positing of factual mental life-processes with respect to their posited and meant intrinsic determinations and their objects just as objects of those processes. With Husserl we shall call the psychologically-phenomenologically reduced mental life-processes "*psychic*" mental life-processes. Before saying anything more here it should be noted that by "psychology" and "psychic" something quite different is meant from at least some of what those words signified in the *Logical Investigations*.[17]

If, now, in reflection, I adopt a further attitude of refraining, and accept the actual current phase of my mental life-processes as only essentially possible exemplifications of mental life-processes as mental life-processes, I thereby refrain from believing in, or positing, the factuality of the current phase of my own mental life-processes as non-selfsufficient components in and of the co-presented world and things, events, and processes in it. Following Husserl's lead, we may call this attitude a psychological *eidetic* phenomenological attitude. What is then disclosed are my own essentially possible mental life-processes in and with an essentially possible world and essentially possible things, events, and processes in it.

According to the "Introduction" to *Ideas* the next step would be to carry out the "specific" transcendental phenomenological reductions. However, in the psychological eidetic phenomenological attitude my own essentially possible mental life-processes are in fact presented as having an essentially possible sense which they can have only in consequence of their basic characterization of meaning and positing the essentially possible world and essentially possible things, events, and processes in it. *Parenthesized correspondingly* is the essentially possible factual connection of my essentially possible mental life-processes with the essentially possible percipient organism which localized my mental life-processes in the essentially possible spatio-temporal world. This signifies that the psychological eidetic phenomenological refraining, and its corresponding parenthesizing, thus discovers (or: reduces) my essentially possible mental life-processes as going on in that of which they apperceive themselves as essentially possible non-selfsufficient components (functionally related to, and even dependent on, these and those essentially possible things in the essentially possible spatio-temporal world).[18] As a consequence, even though essentially possible localization is parenthesized, essentially possible mental life-processes *still retain*, as it were, *their essentially possible sense of being mundane proceses*, which is precisely what I mean in speaking of them as *psychic* processes made explicit in a *psychological* phenomenological attitude of refraining. And moreover the most that can be discovered in further refraining is the essentially possible mundaneness of my own essentially possible mental life-processes; they still have not been "purified" from what confers upon them the characteristic of essentially possible actual existence. Clearly a more radical refraining is required.

Theme II: "My own mental life-processes believing in themselves as actually existing processes in the actually existing world perhaps even if nothing else in the world exists" (the purport of section 49 of *Ideas*, First Book). I refrain from accepting the mundane sense, "being in the world of which my own mental life-processes apperceive themselves as non-selfsufficient components." This more radical refraining is necessary and must be expressed in the foregoing way because, in the psychological-phenomenological *eidetic* epoché and reduction, even though my own essentially possible mental life-processes are not presented as having the sense of localization or "embodiment" of an essentially possible real percipient subject with an essentially possible place in the one, essentially possible spatio-temporal order, mental life-processes are nevertheless still presented as having the sense, "essentially possible mundaneness." The phenomenological result, in other words, of the psychological-phenomenological eidetic epoché or attitude of refraining and reduction is presentation of my own mental life-processes as having an essentially possible sense which they can have only in consequence of their basic characterization of intending to and positing world and things in it.

Equivalently stated: the psychological-phenomenological eidetic attitude of refraining discovers my mental life-processes as going on in that of which they apperceive themselves as non-selfsufficient components. However, that also signifies that my own essentially possible mental life-processes are that *for which* the essentially possible world is and must be such that they apperceive themselves as non-selfsufficient components of the world. Even so, I still have not refrained, then, from believing in or positing my own mental life-processes *themselves as actual*; *they remain in acceptance*. Thus refraining from accepting the sense, "being in the world of which my mental life-processes apperceive themselves as non-selfsufficient components," *is not equivalent to* refraining from accepting the sense, "actuality of my mental life-processes for which the world is."[19] This is of the utmost importance: *because of the non-equivalence of these refrainings, it is impossible to proceed*, as Husserl would like to in the "Introduction" to *Ideas*, *from the psychological-phenomenological eidetic epoché or attitude and reduction to the* transcendental *phenomenological epoché or attitude and reduction* (the other "specific phenomenological reductions").[20]

Reformulated, the second theme is now: "My own *transcendental* mental life-processes believing in themselves as being in the world even if nothing else in the world exists." In consequence of what we may now call the transcendental attitude of refraining, my own mental life-processes are presented as that for which being in the world and the world is, as "prior" to all mundane actuality or being, thus to any actual *or mundane* sense whatever. By virtue of this more radical refraining, my own mental life-proceses are said to be "transcendentally purified," i.e., "purified" from that positing which bestows or confers upon them the characteristic of being real processes in a real world. In their transcendental "status," as we shall also say, then, my mental life-processes are "irreal" in the sense of being purified from a specific positing. In this respect we may also say that they are "prior to" all reality; transcendentally reduced, mental life-processes do not presuppose or beg the question of reality. In other words, to proceed from the psychological-phenomenological eidetic to the transcendental-phenomenological reductions would indeed beg the question of reality--which is what Husserl would seem to have forgotten when writing the "Introduction" to *Ideas* and which has occasioned so many difficulties in following his thought. Yet he correctly saw the right or correct order of phenomenological attitudes and reductions in the text of the First Book of *Ideas*, and he saw it again in contrast to what he wrote in the "Introduction" to all three Books of the *Ideas*.[21] I said what Husserl *correctly* saw. There is, then, still more to be considered concerning transcendental phenomenological method, i.e., that method which has as its theme transcendentally *purified* mental life-processes as the "things themselves" to which we will to return.

Indeed, only if refraining from accepting the sense, "being in the world" (to use a shorthand expression), were equivalent to refraining from accepting the sense, "actuality of my own mental life-processes" (to use another shorthand expression), would it then be possible to proceed from the psychological phenomenological eidetic epoché and reduction to the transcendental one. But because they are clearly not equivalent, because the one refraining does not ipso facto entail the other, it is necessary therefore to exercise another and further attitude of refraining, to adopt a new attitude in which I refrain from accepting the current, de facto phase of my mental life-processes for which there is, correlatively, an essentially possible example of world. As a consequence, it is only on the basis of seizing upon the *irreal* actuality of my mental life-processes as that for which the world is, regardless of those processes being non-selfsufficient components, that I can proceed in a further act of refraining from acceptance of that irreal actuality and effect a transcendental phenomenological *eidetic* reduction.[22]

8. THE POSITING AND ACCEPTANCE BELONGING TO THE NATURAL ATTITUDE

So far, now, four cases of refraining, four attitudes, have been distinguished: the psychological phenomenological attitude and its counterpart, the psychological phenomenological eidetic attitude; the transcendental phenomenological attitude and its counterpart, the transcendental phenomenological eidetic attitude. We can only distinguish these attitudes in a meaningful way, however, once the order of refrainings and reductions in the "Introduction" to *Ideas* has been revised in a way consistent with the text of the First Book and later writings. Of significance and, in fact, the crux of the whole question of correct order, is that there is a *fifth* attitude to be considered: what Husserl called the "*natural attitude*."

I understand Husserl to mean by the "natural attitude" that attitude in which there is continuous believing in and intending to the world along with self-apperceiving of mental life-processes as non-selfsufficient components of the world. Yet it would be better to speak here of the *natural transcendental* attitude. One of the reasons for this is not difficult to see: for mental life-processes to continuously believe in the world as actually existent and in themselves as components of the world is ipso facto to believe in themselves as that for which the world is. In other words, the transcendental "status" is my mental life-processes is already "there" beforehand and regardless of whether or not I refrain from accepting the "being in the world" and the "actuality" of my mental life-processes.[23] Even though it severely strains the already overburdened language, it does seem appropriate here to say that that attitude in which I reflectively seize upon my mental life-processes in this manner is an "*unnatural*" *transcendental* attitude, an "*unnatural*" *refraining*. Indeed, it would appear now that it only makes sense to speak of my mental life-processes as "natural" or "unnatural." In the language of the "Introduction" to *Ideas*, it only makes sense to speak of the phenomena with which phenomenology is concerned--mental life-processes--as "real" or else as "transcendentally purified," as "irreal" actualities or possibilities. Accordingly, as meditating phenomenologist it is possible for me to elaborate a body of cognitive doctrine on the basis of just what is presented: my mental life-processes as having a transcendental "status."

More precisely stated: transcendental phenomenological epoché discovers the sense, "being in the world pertaining to my mental life-processes as transcendental mental life-processes *in the natural attitude*." Differently expressed, that signifies that the general positing belonging to the natural attitude, the sense, "being in the world," is--contrary to Heidegger[24]--a sense conferred by my transcendental mental life-proceeses apperceiving themselves as non-selfsufficient components of the world, thus "masking" their unnatural "status." "This relationship," Eugen Fink said,

> can be briefly characterized as the relationship between constituted and constituting intentionality. The intentionality which is *given* in the first place is the psychical intentional

> life (as the manner in which human being is in the world) prior to all philosophical illumination. This life flows along now in determinate salientness, in an articulatedness into acts, position-takings, and so forth. By the reduction, however, the transcendental life which is continuously *masked* by the given mundane life is exposed--the transcendental life, more particularly, which first of all becomes visible in articulation according to and coinciding with the psychical <or: mundane> life. In this particular perspective, the reduction acquires the characteristic of a regressive method that inquires back beyond the mundane objectivizedness of the intentive stream of life (as an intra-mundane psyche). The reduction *de-objectifies*, *de-mundanizes* intentive life by abrogating [*durch die Aufhebung*] the mundanizing self-apperception orienting it in the world.[25]

The refraining, then, that characterizes the natural transcendental attitude is a refraining from accepting the "unnatural status" of mental life-processes: it is a "masking" of the "unnatural" (or transcendental).

Thus what is accomplished in the psychological phenomenological reduction is precisely the making visible of that conferred--and, therefore, abrogated--naturalness of mental life-processes. If that is true, then it follows that it only makes sense to speak of the psychologically-phenomenologically reduced ("psychical intentional life") in the light of the transcendentally-phenomenologically reduced. That signifies, in turn, that the transcendental "status" of my mental life-processes must already have been seized upon before the psychological phenomenological reduction to fact can be accomplished. Seeing this situation, i.e., the positing belonging to the natural transcendental attitude, the *conferred mundane* status of mental life-processes (hence a status which can be abrogated), warrants the order of phenomenological reductions, makes the order a *correct* one. As a result, the "eidetic and other specific reductions" presuppose in advance the transcendental phenomenological reduction to fact: the unmasking of natural transcendental mental life-processes.

9. THE QUESTION OF MOTIVES FOR SELF-CONSCIOUS EXERCISE OF THE TRANSCENDENTAL PHENOMENOLOGICAL EPOCHE

In the Introduction I sketched the dilemma that emerged in the course of Husserl's thought when trying to establish a critical-epistemological understanding of thinkings and knowings by striving to return to the "things themselves." When placed in the historical context of his publications and lectures, Husserl's response to the dilemma turned out to be similar if not the same since the early 1890's, and in the preceding section I have tried to express the response largely in the light and language of *Ideas*, First Book. Confronting the epistemic dilemma may, of course, be said to be a "motive" for exercising the phenomenological epoché or attitude of refraining in the first place. But discussing the dilemma in Husserl's actual course of thought raises the question of whether or not there are any "motives" for exercising the epoché in any actual or possible course of thought whatever.

The question is not so simple, however, and requires some precision in formulation. Granted that the transcendental phenomenological epoché occurs under circumstances, is motivated in some fashion, do the circumstances under which it occurs favor its exercise or not? Moreover, these circumstances, these motives, must, after all, take place in a wholly *natural* life if the epoché is performed for the first time. In other words: What are the motivating circumstances which trigger the ego to *refrain from those positings in which the naturalness of its life consists*? When expressed this way, the question about motives leads to a new facet of the dilemma: that there should be in the natural life of consciousness motivating circumstances for no longer living naturally.

A full exploration of this issue lies outside the immediate purposes of this and the following chapters. Nevertheless, a few preliminary observations may be made here.

In the first place, the *literary* background and *expression* of phenomenology that has been discussed so far may best be understood, indeed, *can only be understood*, *as an actual example of* a first *exercising of the transcendental phenomenological epoché* in the course of one actual, particular life--Husserl's--at various stages which led up to the decisive step in 1905/06 when the transcendental phenomenological epoché was self-consciously exercised. In other words, the question of motive has been treated, in the first place, as a *transcendental empirical* problem. In the second place, Husserl's own course of thought serves as a transcendental empirical example which reveals certain factors that perhaps are essential and others which are not for the self-conscious exercise of the transcendental phenomenological epoché. For our immediate purpose several such factors may be mentioned.

Probably the most obvious factor that lays claim to being essential as a motivating circumstance is that of reflection and which Husserl emphasized already in the Introduction to the first edition of *Logical Investigations.* To be sure, I may be engaged straightforwardly or reflectively in my mental life-processes in the natural attitude. To speak of being "straightforwardly engaged" in mental life-processes and, correspondingly, "straightforwardly busied" with the objects of those processes signifies that I am busied with objects and things per se; in contrast, when I am reflectively engaged in mental life-processes I am busied with those objects and things as meant and intended to by those specific processes rather than as per se. But reflection in this specific sense does not by itself make the exercise of the transcendental phenomenological epoché likely; reflection is never a sufficient but only at best a necessary motivating circumstance because there is nothing to prevent it turning back into straightforward engagement and busiedness in the natural attitude. Differently expressed, reflection of itself does not entail any theoretical interest in the exercise of the transcendental phenomenological epoché.

Still another factor that may lay claim as a motive is, in reflection, the seizing upon the intentionality of mental life-processes and, correspondingly, of the status of objects as intended-to objects of those processes. But it would also seem equally necessary as well to seize upon the positing peculiar to mental life-processes, upon their "belief-characteristic." Indeed, were I not to seize upon the positing of mental life-processes it would not be possible to proceed to adopt an attitude of refraining from engaging in the very positing of which the naturalness of mental life-processes consist. Both--seizing upon intentionality and upon the positionality of mental life-processes--would seem to be no more than necessary but not sufficient motives for exercising transcendental phenomenological epoché. In the one case, it would be possible to develop a "descriptive" psychology of mental life-processes without ever exercising the epoché (as did Brentano); in the other case, it would be possible to develop a "direct realism," again without ever exercising transcendental phenomenological epoché (as did Heidegger).

Taken separately, it is very doubtful that any of those factors would make transcendental phenomenological epoché likely.[26] This would not seem to be true, however, when all three factors are present; together, they are at least necessary motives for exercising transcendental phenomenological epoché as a response to the dilemma in Husserl's thought. Moreover, seeing the dilemma, and the discontent with this situation, may be a fourth factor, another motivating circumstance for exercising transcendental phenomenological epoché. Stated in terms of *Ideas*, First Book (see above, section 6), the dilemma may be expressed by saying that I begin with the assumption of a world, an assumption that is part of the natural attitude, in which there exists one minor item, myself, with my mental life, my mind. I now consider just my mind and its objects as clues to the "constituting" of them. I find that they are essentially nothing but objects intended to and posited by me, that as intended to and posited, they have certain structures and features for my mind. Indeed, these objects and world are the only ones I can mean in any way and that to speak of any other would be nonsense (*Ideas*, First Book, section 32f.). But how do I reconcile this with the starting point of myself as but one minor item in the world when, as a matter of fact, I am "constitutive" of the whole world? As it were, I am of "two minds" about myself and my world.

Perhaps more than the other factors in the transcendental natural attitude, it is dissatisfaction with that situation that favors the exercise of the transcendental phenomenological epoché because it makes prominent the question of how the analysis of "mind," of "constituting subjectivity," to use Husserl's expression, and its objects, can be relevant to the analysis of the nature of what is. But such a question is to engage directly the will to philosophical theory so that, as a matter of fact, while I live unnaturally my natural life still proceeds. In other words, simultaneously I live both in the transcendental unnatural and the transcendental natural attitudes. Only by exercising transcendental phenomenological epoché, however, is the simultaneity of unnatural and natural living preserved. Otherwise, in all likelihood, I can lapse back into the natural attitude. Thus the fourth motivating circumstance in the natural attitude is at best also only a necessary but not a sufficient one for exercising itself as a solution to the dilemma regardless of whether it turns out to be legitimate in the end.

Is there, then, any motive which is sufficient? This is a question that cannot yet be addressed here.[27]

10. TRANSCENDENTAL INTENTIONALITY. CLASSIFICATION OF TRANSCENDENTAL MENTAL LIFE-PROCESSES

The transcendental phenomenological epoché is a step-by-step epoché and reduction. Given the right order of reductions sketched in sections 7 and 8, the question arises as to what determines the different steps and their order. The answer lies, I shall suggest, in the very articulation of transcendental mental life-processes themselves. This answer was indicated earlier at the beginning of this chapter when, with Husserl, I said that the genuine meaning of "real" consists of the "irreality" of mental life-processes; or, consistent with Eugen Fink, it may also be said that the order of steps of the transcendental phenomenological epoché and reduction takes place in the abrogation of the non-genuine sense of "real," of the mundane sense of "real as real." It is therefore necessary to examine some aspects of the transcendentally purified mental life-processes themselves.

Exercising transcendental phenomenological epoché, I reflectively seize upon and observe my own transcendental mental life-processes in their current phase that make up my transcendental mental life in a concrete whole, and the objects of those processes. Most immediately I find that no matter how great the specific differences among transcendental mental life-processes might be (seeings, hearings, judgings, wishings, willings, and the like)--they are all alike in sharing a basic feature of "intentionality;" they are all alike in being awarenesses of something or, as I have also expressed it, in being *intendings to* something. And because transcendental phenomenological epoché is refraining from not only certain "active" and "passive" intendings, but also from those positings wherein my transcendental natural attitude and the transcendental naturalness of my whole transcendental mental life consist, it is also a refraining from those positings that are "active" and passive." To explain this it is necessary to refer to a "classification" of transcendental mental life-processes.

Husserl has provided a number of suggestions for ways of classifying transcendental mental life-processes, one of which is with respect to their being engaged in, or not engaged in, by the transcendental ego. This classification, implicit in Husserl's lectures after 1905, explicit by the time of *Ideas*, First Book, was fully developed in *Formal and Transcendental Logic* and *Cartesian Meditations*. By remaining with, and elaborating, this classification we can formulate a transcendental phenomenological law governing the "nature" of transcendental mental life-processes that will ground the method for describing and making judgments about them. At the same time, consideration of this classification will make it possible to bring up to the level of *Cartesian Meditations* the phenomenological results of the "Introduction" to *Ideas* and *Ideas*, First Book.[28]

The title of section 4 of *Formal and Transcendental Logic* reads: "The Problem of Ascertaining the Essential Limits of the 'Thinking' Capable of the Significational Function." Under this heading Husserl defines "thinkings" as those mental life-processes the objects of which are expressed or expressable by locutions, or which can be significations of mental life-processes. Such mental life-processes are said to be "*sinngebend*," that is, to bestow signification of locutions. Examples are: judgings, wishings, askings, commandings, and doubtings. These are distinguished from those processes that do not have the capacity to bestow signification, that is, the objects of which cannot be significations of expressions or expressable by locutions. Obvious cases would be objects intended to by sensuous perceivings; for instance, the touched as touched cannot itself be expressed by a word or a sentence. This is one possible classificatory scheme. But section 4 offers in addition a somewhat different one.

Judgings, wishings, askings, and the like, are distinguished from mental life-processes characterized by what Husserl calls "primary" or "original passivity;" examples he gives are associating processes, processes such as "retentions" and "protentions."[29] This distinction entails another classification that cuts across processes capable and processes incapable of bestowing signification. Schematically, the classification may be represented in the following way:

1. Actional mental life-processes, i.e., acts in the narrowest sense of *Ideas*, First Book: those mental life-processes in which I am engaged and busied with some object or other intended to (for example, a judging in which I judge step by step; or a collecting in which I collect this and that and something else).
2. Those mental life-processes in which I am not now engaged, but in which I might be engaged, or in which I have been engaged but am no longer engaged. Having judged and drawn a conclusion, or having formed a collection, I am no longer engaged in those mental life-processes; although, to be sure, I can always return to the judged and judge again, to the collection and collect again.
3. Mental life-processes which proceed in original or primary passivity, that is, those processes in which I cannot be engaged in principle (such as retentions and protentions).

In later chapters I shall explore the observation that, despite all such differences among mental life-processes as the foregoing, they are temporally extended and make up a continuous flow intentive both to themselves and to world and things in it. Husserl gives the name, "synthesis," to that union and unification peculiar to all mental life-processes as intentive, and the most fundamental kind of synthesis, he says, is that union by virtue of which, in a multiplicity of intendings, something is intended to as self-identical and non-identical with something else. In an example of perceiving a die in section 17 of *Cartesian Meditations*, Husserl describes the die presented as enduring and says that this "Objective temporality" of the die must be distinguished from the temporality of the *die-perceiving*, that is, from the temporality of its appearing which "flows away" with "its temporal extents and phrases which, for their part, are continually changing appearances *of* the one identical die." We may also say that throughout the course of intentive mental life there is a continuous identifying and distinguishing of earlier and later phases; a current phase of mental life is continually "identified" as self-identical through a multiplicity of changing "appearances" as it receded into the past, for example. Or, a physical thing, such as the die, may be perceived as identical throughout a multiplicity of changing appearances, now in near-appearances, now in far-appearances; a physical thing may also be perceived both as changing and as identical throughout changes in it.

Such identifying and distinguishing syntheses--central examples of which will be considered in the phenomenological analyses of Part Two--are basic to another sort of synthesis, to which Husserl gives the name, "associating synthesis,"[30] and which, following the practice of Dorion Cairns, will be given the name of "assimilative and differentiative synthesis" (chiefly to avoid the connotations of the term, "association," in modern philosophy and psychology). By way of a brief example, consider the case where two or more things are intended to, each as self-identical and as different from each other: they are intended to, however, not as identical with each other, but as more or less similar to each other, more or less different from each other in this or that respect, or, perhaps, even as perfectly alike or perfectly different (the limiting cases). This is a matter of assimilative and differentiative synthesis and, *like identifying and distinguishing synthesis, it may proceed either actionally or passively*. Indeed, the three-fold classification of mental life-processes is throughout synthetical. Husserl's well-known example of this second kind of synthesis is what he calls "pairing" in section 51 of *Cartesian Meditations*. Because it will play such a crucial role in determining the order of transcendental phenomenological refrainings and reductions as well in the phenomenological analyses of the second part of this book, his account of "pairing" is worth citing in part:

> ...two data are given intuitionally, and with prominence, in the untiy of a consciousness and that, on this basis...as data appearing with mutual distinctness, they *found phenomenologically a unity of similarity* and thus are always constituted precisely as a pair. If there are more than two such data, then a phenomenally unitary group, a plurality, becomes constituted...We find, more particularly, a living mutual awakening and an overlaying of each with the objective sense of the other. This overlaying can bring a total or partial coincidence, which in any particular instance has its degree, the limiting case being that of complete 'likeness.' As a result of this overlaying, there takes place in the paired data a mutual transfer of sense--that is to say: an apperception of each according to the sense of the other, so far as moments of

> sense actualized in what is experienced do not annul this transfer, with the consciousness of 'different'.

A brief but very useful example will help focus on what must be emphasized in the phenomenological situation Husserl refers to. Instead of a single die, a pair of dice may be considered in the simultaneity of their presentation. One die is itself presented and believed in as being here, and the other itself presented and believed in as being over there at the same time. Two things, not one, are perceptually intended to and posited, and this perception is a numerically distinguishing and identifying synthesis. More particularly, it may be said that the dice are perceptually presented and believed in as having certain determinations--size, shape, color, texture. These determinations are likewise numerically distinguished as well as the dice themselves: each is perceptually intended to and posited as "identical" with itself and non-identical with the other. It is on this phenomenological basis that a mutual transfer of sense takes place; the presented determinations of each die are "transferred to" or "overlay" the determinations of the other die, founding an assimilative and differentiative synthesis. For example, the transferred cubical shape and the presented cubical shape found a synthesis of likeness of shape; but the transferred size and the presented size may found a synthesis of unlikeness in size; the dice are perceived and believed in as like in shape, unlike in size. And Husserl's examples are instructive because the transfer of sense may proceed regardless of whether I am engaged in the perceiving; it may proceed in primary or original passivity.

11. ACTIONALITY AND PASSIVITY

The classification of transcendental mental-life processes considered with regard to their basic synthesizings expresses the fundamental transcendental *phenomenological datum.* It is that datum with which one begins and with which one ends. Some of those synthetical or unifying processes are presented as having the "quality" of being processes in which I am engaged and busied with something, and others do not present themselves as having such a "quality."[31] As I said before, the former are *actional* intendings and positings, "acts" in the pregnant sense of the term for Husserl. The latter are primary *passive* intendings or, again in the language of *Ideas*, First Book, *inactional* intendings and positings. But not all of what Husserl calls "inactional" modes in *Ideas* are what he later calls primary passive intendings and positings; some are what he calls "secondary" passive modes. For example, as Husserl notes in the Fourth of the *Cartesian Meditations* (section 31f.), there is continual "retention" (or, as I shall say later, retrotending to) all actional intendings; all the while *I* am seeing the dice, the seeings in which I am engaged or living do not then "disappear" when I am engaged in another mental life-process (e.g., counting the dice); they are rather "retained" (or: retrotended to) *passively*.

But the same also holds, *mutatis mutandis*, for all passive retentions; the retentions of retentions...proceed passively, that is, without my ever being engaged in them.[32] Indeed, all actional intendings and positings are "retained" (or, retrotended to) without my being engaged in the retentional mental-life processes. The same may be said also for all protentions. In the next chapters, but also especially in the concrete cases studied in the Second Part, phenomenologically significant situations will be examined where the sense of certain actional intendings, just as in case of any passive ones, are *passively* transferred to other intendings be they actional or inactional in such fashion that primary passive assimilative and differentiative syntheses are founded one upon another. Precisely this passive retention and protention of *actional intendings* and positings, and the passive assimilative and differentiative transfer or overlaying of their senses, is what Husserl includes under the heading of "secondary passivity"--the "habitualities" of mental living that characterize not only individual persons, but whole groups of persons and, ultimately, even whole "cultures."

Thus whatever else "passivity" may signify, phenomenologic ally it is not "pathos," a "suffering" of mental life-processes (which themselves may be actional or passive). Primary and secondary passive mental life-processes are rather processes that proceed but without an ego engaged in them. To express the situation in another way: the difference between secondary and primary passive mental life-processes consists in the fact that, in the case of the *former*, even though I may not now be engaged in such processes, they nevertheless "point back to" and "presuppose" actional intendings and positings; in contrast, in the case of the *latter*, they are mental life-processes that do not "point back to," nor "presuppose," actional intendings and positings.

In the next chapters this articulation of the transcendental phenomenological datum will be spelled out in greater detail with respect to the step by step transcendental phenomenological ref rainings and reductions. For the moment it is necessary to turn to the formulation of several tentative conclusions of a normative nature concerning the phenomenological method.

12. PRELIMINARY FORMULATION OF TWO TRANSCENDENTAL PHENOMENOLOGICAL LAWS. "ORIENTED CONSTITUTION" OF TRANSCENDENTAL PHENOMENOLOGICAL MENTAL LIFE-PROCESSES

There are several tentative conclusions which may be drawn now concerning method and which will be tested in the Second Part. More particularly, the conclusions concern the basis for establishing the correct order of transcendental phenomenological reductions. At the beginning of this chapter transcendental phenomenology included as part of its definition the science of essences (an eidetic science). The phenomenological "origin" of essences, I shall try to show, lies in primary passive assimilative and differentiative syntheses. The examination of such syntheses will then be of central importance which may be expressed in the preliminary formulation of two basic laws. Both laws purport nomological universality and express the intrinsic articulation of transcendental mental life-processes themselves.

The first law reads: All actional synthesizing mental life-processes, and all secondary passive synthesizing mental life-processes, of necessity belong to a constitutive stratum founded on primary passive mental life-processes. But no primary passive synthesizing mental life-processes (identifying and distinguishing, or founded assimilative and differentiating, syntheses) belong to or are founded on any actional mental-life processes. This law entails a second, which reads: Universally transcendental mental-life processes exhibit a *plane of demarcation* cutting across them, above which lie all actional and secondary passive intendings and positings and below which lie all primary passive intendings and positings: *correspondingly*, the line of demarcation cuts across objects intended to and posited by their respective processes. And it is precisely the plane of demarcation, this caesura, which provides the clue to the correct order and steps of the transcendental phenomenological epoché and reductions, that is, to the correct order of refrainings from various intendings and positings.

More particularly expressed, the clue to the order and steps is provided by the plane of demarcation that cuts across transcendental *natural* mental life-processes adverting to the world of which they apperceive themselves as non-selfsufficient, actually existing real components (to be sure, in a non-genuine sense of the word, "real"). Most immediately, this is the transcendental phenomenological datum with which we begin: transcendental natural mental life-processes which, by their intrinsic articulation, allow for discrimination of their various layers or strata of intending to and positing the world in which they apperceive themselves. Phenomenologically, it is always possible, as a result, to "abstract from," or discriminate, for instance, all actional and secondary passive intendings and positings and their corresponding senses and characteristics; the *phenomenological residuum* is then all primary passive intendings and positings and their corresponding senses and characteristics.

Thus, for example, in section 61 of *Cartesian Meditations* Husserl writes:

> If, within this primordial 'world,' we abstract from the reduced psychological being, 'I, the man,' primordial bare Nature remains, as the Nature pertaining to my own 'bare sensuousness.' As an initial problem concerning the psychological origin of the experiential world there emerges here the problem concerning the origin of the 'thing phantom,' or 'thing pertaining to the senses,' with its strata (sight thing, [touch thing,] and so forth) and their synthetical unity. The thing-phantom is given (always within the limits set by this primordial reduction) purely as a unity belonging to modes of sensuous appearance and their syntheses. The thing-phantom, in its variants as 'near thing' and 'far thing,' all of which belong together synthetically, is not yet the 'real thing' of the primordial psychic sphere. Even in this sphere the 'real thing' becomes constituted at a higher level, as a casual thing, an identical substrate of casual properties (a 'substance'). Ob-

viously substance and causality indicate constitutional problems of a higher level.[33]

In the forgoing passage Husserl not only sketches the content of the constituting of the actually existing real world with respect to the noetic-noematic multi-layered structure of consciousness of that world (to use the language of *Ideas*, First Book)--layers both above and below the plane of demarcation; Husserl is also talking about what he calls in *Cartesian Meditations* (sections 49, 58) an "oriented constituting" of the layers or strata of mental life-processes separated by the plane of demarcation:

> Constitution of 'worlds' of any kind whatever, beginning with one's own stream of mental processes, with its openly endless multiplicities, and continuing up through the Objective world with its various levels of Objectivation, is subject to the law of '*oriented*' *constitution*, a constitution that presupposes at various levels, but within the extension of a sense conceived with maximal breadth, something 'primordially' and something 'secondarily' constituted. At each of the levels in question, the primordial enters, with a new stratum of sense, into the secondarily constituted world; and this occurs in such a fashion that the primordial becomes the central member, in accordance with orientational modes of givenness. The secondarily constituted, as a 'world,' is necessarily given as a horizon of being that is accessible from the primordial and is discoverable in a particular order."

The various layers or strata of constituting the existing world, then, define the order of the step-by-step refrainings and reductions; what is manifest at each step, we may also say, is a "primarily" and "secondarily" constituted "world," a "quasi-objective world" noetically-noematically analyzable.

Once such an analysis is effected--the purpose of the Second Part of this book--of things pertaining to the senses, of thing-phantoms, of substantial and causal things, of predicatively and conceptually formed affair-complexes, the *task* of transcendental phenomenological method is to discover how what is primarily constituted enters into the secondarily constituted at the next higher layer or statum: for instance, how primordially constituted things pertaining to the senses enter into the secondarily constituted thing-phantoms; how, in other words, the primordial layer acquires the "appearance of" the next higher layer and is "necessarily given as a horizon of being that is accessible from the primordial and is discoverable *in a particular order*" determined by the articulation and memberment of mental life-processes themselves.[34]

13. GENETIC AND STATIC TRANSCENDENTAL PHENOMENOLOGY

In connection with Husserl's concept of "oriented constitution," and by way of conclusion to this chapter, it is necessary to add that all purely descriptive, eidetic, transcendental phenomenology is understood here as constitutive phenomenology in the specific sense of the critical-epistemic understanding of all actual and potential syntheses (actional and non-actional) in which objects are evidentially intended to and posited as self-identical and mutually distinct unities in and through a multiplicity of multi-layered similar and different transcendental natural mental life-processes. But the chief motivating circumstance for this critical-epistemic understanding that is transcendental phenomenology should not be lost sight of: the "feeling" for the problem of genesis and how it can be relevant for rendering an account of the world existing in itself. As already suggested in the Introduction (above, section 4), Husserl's answer is that the constituting of an object, of something intended to and posited as a self-identical unity (or: more particularly, as truly existent) is *broadly* speaking a "generating" of the object, is its "genesis."

Even though central to Husserl's thought from the beginning at the end of the last century, the idea of "genesis" becomes increasingly complex in writings after *Ideas*, First Book. One of the reasons why would seem to be that the term, and the concept it expresses, increasingly bears a heavier and heavier conceptual burden. Thus in *Cartesian Meditations*, section 37, it would seem that by "genesis" of something in the *strict* sense of the term, "genesis," Husserl means those features of the "monad" that are, as it were, "*einmalig*" and essentially irreversible in time. Moreover, "genesis" in section 37 is associated with "history" ("*Geschichte*") and is illustrated by the example developed in the previous section (section 36) of scientific theorizing. Thus the mental life-processes that make up such theorizing can only occur after things to theorize about have been constituted and when the corresponding habitualities (section 32) have been instituted. As a consequence, the rational constituting of a theory, a proposition, a physical thing "for the first time in one's mental life" is a unique experience. This would suggest that the founding relations in simultaneity point to "genetic" relations in succession. For example, that the relation of founding between perceiving a physical thing other than my own body, and perceiving a psychophysical thing other than my own psychophysical self points to an irreversible temporal genetic sequence. Although not self-evident, the example clearly shows the difference between "constitution" and "genesis" in the strict sense of the term, "genesis."

And, in general, it would seem that "genetic phenomenology" is concerned with "genesis" in the strict sense, while what is sometimes called "static phenomenology" is concerned with "genesis" (or: "constitution") in the broad sense of the term, "genesis." The names, "static phenomenology" and "genetic phenomenology," although used by Husserl, seem to me to be liable to confusion if they are not outright misnomers, and generally I prefer the distinction between "genesis" in the broad sense and the strict sense in order to bring *Ideas*, First Book, up to the level of analysis in *Cartesian Meditations*. Which meaning of "genesis" is analytically primary is a question left unresolved here. The assumption is, however, that "genesis" in the broad, and not the strict, sense is the basic meaning.

Husserl's concept of "oriented constitution" presupposes the distinction between the broad and strict meanings of "genesis" and, indeed, is "genetic phenomenology" in the broad meaning of "genesis." Although the term, "oriented constitution," would seem to be a late one in Husserl's writings, the concept it expresses clearly is not (just as the meanings of "genesis" are already present in the Introduction to the second volume of the first edition of *Logical Investigations*--specifically, in the last paragraph of section 2). It has already been emphasized several times that by 1905 Husserl had already explicitly introduced transcendental phenomenological epoché, so that it comes as no surprise that in the lecture of 1907 (now published as *Physical Thing and Space*) Husserl gives a number of examples of what he would later call "oriented constitution" when distinguishing the various strata of the constituting of space. In the case of visual space, for instance, the lowest stratum discriminated is an "oculormotoric space" in which "only the eyes are moveable and move;" the next highest stratum is a closed, two-dimensional space "constituted by moving the head on its axis while the otherwise normal position of the body remains fixed and in place." Husserl then states that the relationship between these two strata may be expressed by saying that the next highest stratum "takes on the appearance [*Apparenz*] of a new space, the oculomotoric body takes on the appearance of a body of a new kind."[35] In the Second Part of this book this as well as other similar cases will be considered. It is worth noting here, though, that Husserl's Göttingen student, Heinrich Hofmann, likewise used a similar idea in his dissertation;[36] Husserl's Freiburg student, Oskar Becker, develops even further Husserl's lectures on the constitution of space and also employs a concept similar to that expressed by the term, "oriented constitution."[37]

In the chapters that follow, transcendental phenomenological epoché will be exercised to spell out the basic features of the "oriented constitution" of the actually existing real world with respect to the problems of space perception, "other minds," and time perception to be considered in the Second Part.

CHAPTER TWO

SPECIFIC TRANSCENDENTAL PHENOMENOLOGICAL PROCEDURES

14. "UNBUILDING" AND "BUILDING-UP" PROCEDURES OF THE TRANSCENDENTAL PHENOMENOLOGICAL EPOCHÉ AND REDUCTION

Phenomenology was defined in the last chapter as a descriptive, eidetic science of transcendentally purified mental life-processes in the natural transcendental attitude. When seized upon in reflection those processes, moreover, are found to exhibit a noetic-noematic plane of demarcation according to which all actional and secondary passive mental life-processes are founded on and "presuppose" all primary passive mental life-processes (although the converse is not the case). This noetic-noematic plane of demarcation provided a clue for determining the correct methodic order in which the step-by-step transcendental phenomenological refrainings and reductions are to be carried out by means of what Husserl called "abstractive" or, perhaps better expressed, "discriminatory," procedures that will reveal and objectivate the variously reduced strata of oriented constitution. By making explicit these procedures for refraining from the general positing in which the naturalness of the natural transcendental attitude consists it will be possible at the same time to critically examine the specific context in which Husserl introduces the notion of oriented constitution in the Fifth Cartesian Meditation: objectivation of the constituting of the "primordial quasi-objective world."

The discriminatory procedures of the transcendental phenomenological epoché comprise what Husserl called an "*Abbau*," an "unbuilding" or "dismantling."[1] This set of procedures is complimented by another set whereby, after each constitutive stratum has been dismantled, isolated and explicated, the different ways are shown in which the strata are "built up;" these procedures comprise an "*Aufbau*," a "building up." The chief purpose of "building up" procedures is to show how each stratum of constituting acquires the "appearance" of the next higher one. Taken together, the unbuilding and building-up procedures are "abstractive" operations, although not in any empiricistic sense of the term, "abstraction." They are rather operations of objectivating *discrimination* and explication at each constitutive stratum of the noetic-noematic composition peculiar to the continuous concretion of the real, objective world posited in the natural transcendental attitude. These procedures ought not, of course, be confused with operations of generalization or construction, let alone inference, from sense data (as in Carnap, Price, Russell). Finally, because "abstraction" has such strong connotations of empiricistic psychology I shall, for the most part, speak instead of discrimination and discriminatory procedures to avoid those connotations. To be sure, "discrimination" is not without difficulties as a name for specific sorts of refrainings from the positings of the natural transcendental attitude.

With respect to the transcendental phenomenological datum expressed by the classification of mental life-processes according to their plane of demarcation (above, section 10), the *first step of unbuilding*, then, consists of discriminating (or: "abstracting from") those constitutive strata comprised by actional intendings and positings and their objects just as intended to and posited--thus, thinkings, judgings, conceptualizings, wishings, commandings and the like, and their subsequent habitualities. The *second step of unbuilding*, accordingly, consists of discriminating the highest noetic-noematic stratum that remains after the first step is taken. Like all those strata below it, this one consists of only primary passive intendings and positings and their objects only as objects of those intendings and positings. Taking his cue from the final sections of *Ideas*, First Book, in the *Cartesian Meditations* Husserl elaborates still further steps of transcendental phenomenological epoché in order to show how primary passive constituting of the real, objective world is built up on the primary passive constituting at various strata of quasi-objective "worlds," each of which has its own peculiar quasi-objective "world-form" that, in turn, is taken as a "transcendental clue" to the constituting of it.[2] But to express the phenomenological situation in this fashion already signifies that important and even radical alterations have been effected in the account Husserl renders of phenomenological method. A critical reflection will be required in the next sections to explain the need for the alterations.

15. THE CONTEXT AND MOTIVE FOR THE INTRODUCTION OF THE CONCEPT OF "ORIENTED CONSTITUTION"

Central to the method of transcendental phenomenological epoché is what Husserl called "oriented constitution" which he explicitly introduced in the Fifth Cartesian Meditation in the specific context of seeking to clarify the experience of someone else in the real, objective world.[3] The motive for Husserl is that, as transcendental phenomenological on-looker it is necessary to effect what he calls an "*eigenheitliche Reduzieren*," in other words, a refraining from intendings and positings such that all that is left in acceptance are those pertaining "to the sphere of the transcendental ego's ownness" in the natural attitude; such refraining ostensibly will disclose the phenomenological foundation of the experience of someone else. And it is just this transcendental phenomenological epoché and reduction that is critically at issue for establishing the right methodic order of unbuilding and building up the real, objective world.

Following Husserl to the letter, to effect the "*eigenheitliche Reduzieren*" signifies that I, as transcendental on-looker, take as my starting point fully concrete transcendental mental life-processes as they themselves are presented in the natural attitude, and discriminate nothing in the full noesis. What, in fact, I do discriminate are rather only the accomplishments, the "*Leistungen*," of *some* of the intendings and positings that belong to the fully concrete transcendental mental living in the natural attitude. In Husserl's words, "I screen off" ("*blendet ab*")[4] certain accomplishments of mental life-processes. As a result, ostensibly I will uncover an "ownness world," a noematically primordial quasi-objective world as the phenomenological residuum. In that quasi-objective world my mundanized transcendental mental living supposedly contains *all* the intendings and positings of the mundanized transcendental mental lives of others (and, still taking Husserl at his word, both actional and passive intendings and positings are included[5]). That is to say, *on the noetic side* there is no discrimination of (or "abstraction from") either the stratum of spontaneous ego-involvement in mental life-processes, or secondary passivity. As a result, the phenomenological situation Husserl sets himself to investigate consists of an *actional* mental living engaged in the *primordial* performance of all the acts performed by an ego when engaged in the mental life-processes in which the world proper, the real, objective world is constituted.

16. THE PHENOMENOLOGICAL PROBLEM AND PSEUDO PARADOX OF REDUCTION TO THE SPHERE OF THE TRANSCENDENTAL EGO'S OWNNESS

It is at this juncture that a difficult problem arises in connection with any attempt to carry through the methodic task Husserl proposed. Am I, then, as transcendental on-looker, to proceed to perform a *further* "*eigenheitliche Reduzieren*" on the mental life-processes given at the stratum of the primordial quasi-objective world, thereby disregarding the noematic sense, "mental life-processes of others in the world"? Or do I leave in acceptance the primordial mental life-processes in the inter-subjective world even though, by means of the "*eigenheitliche Reduzieren*," they are constituted in the primordial and what proves to be a "solipcistic" quasi-objective world?[6] Husserl does not resolve this problem. What he would seem to be saying may be expressed by noting that he does not effect the discriminatory refraining across the board, so to speak, of the fully concrete noetic-noematic strata of mental life-processes. If one rehearses Husserl's actual course of thought, it looks as though what one does is to keep the noetic side unaltered and in acceptance within the thematic sphere of phenomenologizing while, at the same time, setting out of action the correlative noematic side. It is as though the upshot of the methodic procedure were that, for instance, in the case of a sense perception, such as the seeing of the table in front of me, I should discriminatively examine the seeing of the table without taking into account the fact that this is a seeing of the table presented as in front of me.

Differently expressed, Husserl has violated his cardinal rule established in Chapter Two of the Second Part of *Ideas*, First Book, and followed in practice in the the first two Cartesian Meditations:[7] phenomenologically, an intending--any intending--cannot be objectivated without, at the same time, objectivating the corresponding noematic sense intended to; and the converse also holds. In these terms, Husserl, in the Fifth Cartesian Meditation, thus "screens off" only the noematic side of mental life-processes, *but not the noetic side*. As a result, just as in the "Introduction" to *Ideas*, First Book, Husserl tries to work out an intuition which, of essential necessity, cannot be carried out, so here, too, in the Fifth Cartesian Meditation, he cannot carry out his methodological intuition. Here, as there, instead of phenomenologizing there is a logicizing and, indeed, a logicizing that has a significant result for the metaphysical doctrine of what Husserl calls, with a Leibnizian nuance, a "monadology." The gist of this metaphysics is that the transcendental ego in its ownness includes all of the mental life-processes in the natural attitude, but not the intersubjective world as noematic correlate of the mental life-processes; just that inter-subjective world is "screened off." Accordingly, a plurality of transcendental monads of ownness are to be distinguished which do not overlap and which, notwithstanding, extend out beyond them-selves to constitute the intersubjective world which is not presen-ted as anyone's ownness. This leads to the weird metaphysical paradox that, on the one hand, monads are absolute with respect to which all else is relative; and, on the other hand, monads are relative with respect to which all else is absolute--a (pseudo) paradox already foreshadowed in *Ideas*, Second Book,[8] where the former monadic system is called "*Geist*" and the latter system is called "*Natur*."[9]

That Husserl's paradox is a pseudo paradox can best be seen by the absurdities to which it leads.[10] If one begins the discriminatory epoché at the level of intersubjectivity it then would seem possible to by-pass completely the first step mentioned above in connection with the plane of demarcation that cuts across the complexus of processes comprising mental life (any mental life). This is primarily what Husserl would seem to do in the Fifth Cartesian Meditation;[11] were I then to effect transcendental phenomenological epoché in the manner suggested by Husserl there would be disclosed a quasi-objective mental life in which the ego is engaged, actionally busied with the real, objective world--with the exception of actionally experiencing "other minds." Such a solitary, quasi-objective mental life would be aware of a cultural world, but it would not be pre-sented relative to society at large; the cultural world would be presented relative only to my own quasi-objective mental living in the natural attitude. Accordingly, there would be science, but only science of my own quasi-objective life and not *our* science. In other words, as Husserl has it, the noematic residuum refers only to my own quasi-objective life but not to anyone else's. The absurdities now multiply: I must now further assert that peculiar to my own quasi-objective life is my own literature and language-even my own literature and language as such--referred to my life as solely my own.

The more one spells out the absurdities, I believe, the more one is forced to return to *Ideas*, First Book (section 35ff.) where Husserl follows another (and, I believe, right) intuition, namely that the common, intersubjective world of self and others is already non-actionally there for me, where, to use Husserl's earlier expression, the ego "awakens" to a world in which the mental life-processes of others are presented as having already been actional. In other words, according to *Ideas*, First Book, I find the "sedimentations" of the actional mental life-processes of others "before" I myself, as transcendental ego in the natural attitude, am ever actionally engaged in perceiving and heeding the world around me and other transcendental egos in the natural attitude. To make this suggestion, however, requires further revisions and clarifications in the formulations of phenomenological method.

Note

In conversations and lectures Dorion Cairns suggested revisions of Husserl's phenomenological method on the basis of criticism of Husserl's Fifth Cartesian Meditation. Cairns' criticism has been of considerable benefit to my own account of the constituting of "other minds" which will be developed in the last chapter of the Second Part of this book.[12] In certain other respects I am in close agreement with the criticism of Husserl by José Ortega y Gasset in his posthumously published *El Hombre y la Gente* as well as that by Hans Reiner in his "Sinn und Recht der phänomenologischen Methode."[13] In this connection mention must be made of the important criticism of Husserl's Fifth Cartesian Meditation by Alfred Schütz in "Transcendental Intersubjectivity in Husserl,"[14] to which I shall turn shortly.

Briefly, on the one hand, I disagree with both Ortega and Schütz in holding that the methodological absurdities of the Fifth Cartesian Meditation signify that the phenomenological method is inapplicable to the problem of clarifying the experience of "other minds;" on the other hand, I agree with Cairns and Reiner that the method can be corrected to avoid the absurdities. Thus what follows will be a *Husserlian*, if not Husserl's, account of transcendental phenomenological epoché and reductions as I understand them based as much on the agreements as on the disagreements just mentioned.

17. THE FIRST STEP OF UNBUILDING THE GENERAL POSITING IN THE NATURAL ATTITUDE OF THE REAL, OBJECTIVE WORLD

a. Phenomenological Distinctions Reviewed

To avoid the absurdities mentioned in connection with following Husserl's method in the Fifth Cartesian Meditation, and to preserve the methodic intuition that emerges from the criticism of the "Introduction" to *Ideas*, First Book, it is clearly necessary that transcendental phenomenological epoché and reduction be carried out noetically *and noematically*, and that there is discrimination, in the first place, of all actional intendings to and positings of myself and others as well as the accomplishments of those intendings and positings. As a result, it is possible to consistently uncover a noetic-noematic stratum of constituting underlying the *actionally* constituted intersubjective world. Formulated in this fashion, the phenomenological datum evokes the distinction made earlier between those intendings and positings in which the ego is involved and those in which the ego is not engaged. Recall that in the latter case the further distinction was made between two sorts of passive intendings and positings: secondary and primary. Examples of the former were "ego habitualities," abiding convictions, cultural mores and beliefs; examples of the latter were not only certain perceivings but retentions and protentions. Recall, too, that not only all passive intendings and positings but also all actional intendings and positings are retrotended to and protended to in *primary* passivity so that the extension of secondary passivity includes primary passive retrotending and protending to all actional intendings and positings as well as to the primary passive assimilative transfer of their noematic senses. Universally, secondary passive intendings and their corresponding noematic object-senses point back to actional intendings and their noematic object-senses; in contrast, primary passive intendings are just those which do not point back to actional intendings and their object-senses.

Most immediately, then, what one has to consider at the outset with respect to the specific phenomenological datum are all noetic-noematic strata of actional mental life-processes and secondary passive mental life-processes, below which lie all noetic-noematic strata of primary passive mental life-processes. As will be evident later, the stratified hierarchy does not prove to be so simple. Nonetheless, a rough guide is provided for a beginning of carrying out the various steps of phenomenological unbuilding. To be sure, in making that beginning it must be borne in mind that transcendental mental life in the natural attitude always remains in full noetic-noematic concreteness, although now to be investigated in its basic structure which is the proper object of transcendentally pure phenomenology as an eidetic science.[15] Unbuilding, as noted before, is not abstraction in the empiricistic sense of removing, or "imagining away of something;" nothing whatever is removed or "imagined away."

b. The First Step of Unbuilding

We begin by discriminating all actional and secondary passive mental life-processes and their object-senses intended to and posited, just as intended to and posited. Consider, for example, the case of judgings and collectings. In such actional intendings predicatively formed affairs are constituted in an original way, to use Husserl's expression; supposed facts, supposed collections, and the like, are presented on the basis of actionally judging what is judged about or collected. In judging about, or in collecting things, those things acquire at the same time certain formal determinations as part of their noematic sense. For instance, something *A* acquires the determinational sense of being the subject of a predicate, while something *B* acquires the determinational sense of being the predicate of a subject; or the somethings *C*, *D*, *E*, *F* acquire the determinational sense of being elements of a collection.[16] As judged about predicatively, "This is round," the "This" is there for me now as the subject of the predicate "round;" or, in the case of collecting, "this and this and this," the "This" is there ready to be objectivated as the member of a collection by virtue of my having actionally collected it. Forthwith, these originally constituted formal determinational senses are carried over, transferred, in primary passivity to all other similar affairs so that even if some particular something has never been judged about before by me, or even if it has never been collected before by me, it is presented nonetheless as having the sense of being judgeable, collectable, of being the predicate of a subject whereof it has not yet been predicated, yet whereof it might be predicated. Not only do such affairs "call to me" to advert to them, become busied with them perceptually, for instance, but they also "call to me" to be judged about, collected, and so forth.[17] In short, the world is there for me to be judged about, collected, and it is there for me as such because of the fact that I have been engaged in actionally judging about it, in actionally collecting items in the world.

All of the foregoing is discriminated in the first step of unbuilding of mental life-processes: all of the mental life-processes and their noematic senses, including all formal senses, be they actional or sedimented and transferred in primary passivity. But the simplicity of the examples already betray a discriminatory complexity. The reason is that in carrying out this first step of unbuilding it is clear that judging, for instance, does not merely constitute syntactical forms; conceptualizing is also involved so that there is a framing of concepts and a seizing upon the things judged about in accordance with those concepts (specific and generic concepts under which the judged-about affairs are subsumed or arranged in some fashion). A concept is just that which can be signified by a verbal expression, and the phrase, "just that which," points to the fact that the full content of an actional perceiving, for instance, can never be expressed in words. Only as much of it as is conceptualized can be expressed in words; but in its full richness, a perceiving is ineffable because there always remains something unconceptualizable, something that cannot be seized upon. In other words, at the outset reflective exploration of this first step of unbuilding not only confronts again the plane of demarcation within mental life-processes; the plane of demarcation also exhibits a natural split, as it were, into the conceptual and subconceptual.

To express the phenomenological situation in a different way, I may say that judging involves conceptualization and, in particular, conceptualization of substrate objects judged about along with an apprehending of those objects as things of this or that kind. Now, not only are *formal* senses transferred or paired in primary passivity, but *material* (or: "non-formal") determinational senses are transferred as well (specific and detailed examples will be given in Part Two). Like formal determinational senses, material senses are "already there," as it were, to meet their "fate" of either being confirmed or cancelled by what is imputed to them. But every new object that I busy myself with calls for advertence on my part, and it already has the sense of being conceptualizable and subsumable under specific and generic concepts owing to primary passive transfer of sense founding syntheses of assimilation and differentiation. As a consequence, in discriminating actional and secondary passive mental life-processes and their noematic senses there is not only uncovered a subconceptual quasi-objective world, but also a material one as well, i.e., a spatio-temporal "causal" quasi-objective world. The phenomenological task is not simply to explore these constitutive processes but also to establish the very *limits* of conceptualization and subsumption of fully concrete mental life-processes in the natural transcendental attitude--a task to which a large segment of the Second Part of this book will be devoted.

It must be added here that, in contrast to Husserl, in so far as language is the accomplishment of ego-involvement in mental life-processes, and to the extent that language expresses concepts, it too is discriminated along with names and the nameable (so far as "naming" involves conceptualizing it is also discriminated even though it may not involve subsumption under specific and generic concepts). To be sure, if language is regarded in a much broader sense, so as to include gestures, expressive movements and forms of communication other than verbal expressions that signify, then something subconceptual is already present. And what has been said about language holds as well for works of art, cultural things and technics, mores and beliefs.

18. THE PHENOMENOLOGICAL RESIDUUM OF THE FIRST STEP OF UNBUILDING

So far what it is that is discriminated in the first step of unbuilding has been sketched. It is possible, now, to reflectively consider what it is that is uncovered (or: reduced to) by this first step--the phenomenological residuum. Most immediately it must be noted that the phenomenological residuum is comprised, not by a single substratum, but rather by a plurality of substrata constituted in primary passivity of which one is the "highest" (cf. *Cartesian Meditations*, section 58). More particularly, the transcendental phenomenological *theme* is made up of the *primary passive components* of those mental lives in the transcendental natural attitude constituted in primary passive syntheses of similarity and difference and, therefore, also as "kinds" (see below, Part II, Chapter Five). To be sure, the transcendental phenomenological theme now excludes the specific and generic differences among mental lives that are objectivated only actionally. The first question is, accordingly, what is the noematic correlate of the "highest" substratum of the "highest" primary passive intending which will serve as a "transcendental clue" to the constituting of it?

The answer would seem to be that it is the real, objective spatio-temporal and causal world presented and appresented as containing a plurality of quasi-objective physical things, at least some of which are intended to as presented with the sense of being "organs" related in a peculiar manner to processes comprising transcendental mental lives in the natural attitude. In addition, this quasi-objective world intended to and posited in primary passivity is intended to and posited with values, ends, means. In other words, no component of noematic sense of an object (a value, a perceptum, a volition) points back to an original constituting in actional mental life-processes. Expressed positively: The transcendental phenomenological theme is made up of a residuum of things and affairs mutually similar and different that belong to an all-embracing quasi-objective world constituted in primary-passive transcendental mental lives in the natural attitude. Expressed negatively: The transcendental phenomenological theme is made up of things and affairs that are non-objectivated, non-conceptualized and logically "formless."

Certainly the noetic-noematic quasi-objective substrata at issue here are but non-selfsufficient components of the concrete, real objective world constituted not only passively but also actionally. And just for this reason the various steps of the transcendental phenomenological epoché and reduction are necessary to acquire knowledge about such world-components. Here, however, caution must be observed about how the transcendental phenomenological situation is expressed because even though those components, such as those constituted purely in primary passivity, may be said to be quasi-objective quasi-worlds, *they are not, after all, themselves* "other worlds." It is rather the case that *what is so constituted in primary passivity is "the" real, objective world in so far, and only in so far, as it is constituted in primary passivity.* It is a quasi-objective real world. Only when the phenomenological situation is expressed in this way does it make any sense at all to speak of the oriented constituting of the real, objective world presented as it itself in transcendental mental life-processes in the natural attitude; only on this ground is it possible to clarify and attempt to make a "true speech" about the phenomenological genesis of the "things themselves" to which we will to return.

This first step of unbuilding uncovers as residuum, then, a spatio-temporal and causal quasi-objective world that is sub-conceptual, sub-linguistic. The possibility of a general descriptive account of these noetic-noematic substrata involves a unique problem raised by Alfred Schütz in his critique of Husserl's Fifth Cartesian Meditation.[18] It is a problem which might as well be faced now for its discussion will aid in revealing the nature of the first step in this unbuilding of the real, objective world.

19. SCHÜTZ'S CRITIQUE OF HUSSERL'S FIFTH CARTESIAN MEDITATION

In addition to (or, perhaps, aside from) all of the questions raised by Husserl's Fifth Cartesian Meditation, the truly "unsettling question" for Schütz is: "How can the isolated philosopher, the non-participating transcendental observer who performs the epoché, meditate with someone else? How can he perform the epoché in actual community with others? In short, how is '*symphilosophein*' possible?" The question is an unsettling one because unless an adequate answer is found to it what Husserl asserts at the end of the Fourth Cartesian Meditation is impossible. There Husserl says that "There belongs also, of course, the constitution of philosophy common to 'All of us' who meditate together--ideally a single *philosophia perennis*." Schütz raises his unsettling question about "philosophy common to 'All of us'" with reference to Eugen Fink's still important essay, "The Phenomenological Philosophy of Edmund Husserl in Contemporary Criticism," and specifically with reference to Fink's idea that there is a seeming paradox for the phenomenologist upon effecting the transcendental phenomenological epoché and reduction: indeed, Fink distinguishes three such ostensible paradoxes: 1) the paradox of what the phenomenologist says; 2) the paradox of the status of the transcendental phenomenological proposition or statement; and 3) the paradox of the phenomenologist's solitude. I believe that it is of the utmost importance to realize that Fink formulates these paradoxes in a polemical context which Schütz, like almost everyone else, ignores in referring to them: Fink argues that these paradoxes *arise*, *and can only arise*, precisely when the transcendental phenomenological epoché is *not* effected;[19] he then goes on to say that these paradoxes are what Husserl's critics, such as those of the Rickert school (*scl.* Zocher, Kreis), end up with when trying to produce philosophy in the natural attitude. The paradoxes are, then, *pseudo paradoxes* because *it is not seen* that mental life-processes in the world are *transcendental* mental life-processes in the natural attitude. Schütz, however, applies the critique of the three paradoxes even when the transcendental phenomenological epoché has been effected, and that application, *not Fink's*, touches directly upon what has been said about the first step of the unbuilding reduction.

Observe: When I, as transcendental on-looker, or observer, carry out the first step of the reduction I consider an intended- to and posited quasi-objective world of realities constituted only and just in primary passivity; it is presented in its own peculiar originality as a quasi-objective world of non-objectivated, non-conceptualized and therefore non-verbally expressed, non-named logically formless realities. But as transcendental on-looker I do indeed objectivate, conceptualize, bestow logical form and name that quasi-objective world and its presented components, but not of and to them as things and affairs in their own right. Rather I objectivate, conceptualize, give logical form to them as and only as intentional objects of primary passive mental life-processes. I make judgements about them, name them, but always only as just such intentional objects. As transcendental on-looker, I clarify and eventually communicate in language; but to do that I must conceptualize the non-conceptual, judge the unjudged, form the formless. As transcendental on-looker I cannot avoid doing that because I am committed to a purely descriptive discipline produced by exploring the field of transcendentally pure phenomena by means of pure intuition.[20] In other words, to describe what I explore by pure intuition necessarily involves that conceptualizing, judging, forming, naming. To be sure, my descriptions will seem inconsistent and paradoxical to my listeners and readers only if they understand me to be describing things and affairs pure and simple instead of as objects of certain, specific mental life-processes.

Thus as transcendental on-looker I will say that such and such is not objectivated, not conceptualized, not formed, not judged about, not named. But to say that it is necessary that I objectivate, conceptualize, judge about and name it. And *this* objectivating, conceptualizing, forming, and judging is only possible when I distinguish cleanly and sharply *between being busied with something pure and simple* on the one hand, *and*, on the other hand, *being busied with something as the* (synthetical) *determinational sense*, the intentional object, *of mental life-processes.*[21] Unless this distinction is made (and it was made implicitly from the very beginning of discussion of the exercise of the phenomenological epoché and its range) there is the *pseudo-problem* of whether or not a "science of man," or of sociology in the case of Schütz, is possible within the framework of transcendental phenomenology and philosophy.[22] Schütz's question is then unsettling primarily because it is a pseudo question, whereas the *genuine* problem of "symphilosophein" is just what is achieved by consistent and correct exercise of the transcendental phenomenological epoché.

20. SKETCH OF THE CONSTITUTING OF THE PRIMORDIAL QUASI-OBJECTIVE REAL WORLD

By taking the first step of unbuilding there is uncovered as residuum a sub-conceptual, sub-cultural quasi-objective world of logically formless things and affairs among which we count not only quasi-objective physical things but also some that prove to be related in a peculiar way to mental life-processes, themselves presented in equally original ways to primary passive intendings to and positings of them. Moreover, the quasi-objective temporality and spatio-temporality constituted in primary passivity are not presented with the same sort of determinational sense imputed to them in geometrizing or physico-scientific thinking. To be sure, even in primary passivity, as will be seen later, quasi-objective world-time and world-space are constituted as continua: quasi-objective time extending on and on, "forward" and "backward" from Now; quasi-objective space extending from Here or from any place "outward" in any direction. Yet it would be wrong to construe these as in any sense mathematical continua. Indeed, Oskar Becker persuasively argued in his dissertation[23] that the space of this quasi-objective world is neither "Euclidean" nor "Riemanian" but is rather non- or sub-geometrical. For him the question then arises: When geometrization (more basically: arithmetization) is considered at higher levels of constituting, how is it decided whether it is "Euclidean" or "Riemanian" geometry that most adequately "fits" space constituted in primary passivity? Formulated in terms of the concept of oriented constituting: How does the primarily constituted quasi-objective world-space acquire the appearance of the secondarily constituted Euclidian or Riemanian space? Perhaps one or the other "geometries" is more congruent with the sub-conceptual, logically formless data. The point is, however, that questions of geometrization--more generally, of idealization of whatever sort--can only be settled on the basis of a thorough examination of what is uncovered in unbuilding analysis (see below, Part II, Chapter Three).

From now on the quasi-objective world under consideration in the first step of unbuilding will be called the "*primordial quasi-objective world in the wide sense*."[24] And it is not only spatio-temporal in its own peculiar and original way, but it is also "causal" in a somewhat unusual but yet perhaps equally ordinary sense. Even though logically formless, the primordial quasi-objective world is not without sub-logical form: space, time, causality. At this stage of our discussion what "causality" signifies in primary passivity may be expressed by saying that when some quasi-objective physical reality of a particular sort is actual, then something else of a particular sort usually will be actual too. Or, to say that quasi-objective physical things are "causal" is to say that they are what they are under circumstances of one sort or another. One of Husserl's examples in *Ideas*, Second Book, is elasticity: to say that a thing is "elastic," for instance a steel spring, is to say that under certain circumstances it will be in a state of vibration; a piece of steel fastened at one end and loose at the other when bent down will not remain that way but rather, upon being released, will vibrate back and forth. As will be seen in Part Two, a reality in the primordial quasi-objective world is what it is under circumstances, that it has a "style" of circumstantial "behavior." And this "style" embraces not only the physical but the mental as well. Yet the "causality" is not rigorous; the typical correlations between "if this" and "then that" are always open to possible exception. For example, to say that something is red is to say that under specific circumstances, such as illumination of a certain sort, it will be seen as red; but under other circumstances it could be otherwise.[25]

Of equal importance is another feature of the phenomenological situation found here. The "causal" style peculiar to the primordial quasi-objective world makes possible the fulfillment of primary passive strivings by way of means to ends. An example would be (and one to be studied later in considerable detail) a striving to immediately effect a state of affairs of a certain sort presented as means for actualizing certain other affairs. In other words, determinations, qualities, relations that the primordial quasi-objective world is presented and appresented as having are not only mutually interconnected but also related to the percipient organism, the "lived body," in ways quite peculiar to it. Here it is sufficient to note that the primordial quasi-objective organism is most originally presented as a collection of organs, hence as a quasi-objective percipient organism for "action" in the quasi-objective world of quasi-objective physical things. Moreover, as Husserl says, "Among the bodies belonging to this <quasi-objective, primordial> 'Nature' and included in my peculiar ownness, I then find my *animate organism* as *uniquely* singled out--namely as the only one of them that is not just a body but precisely as an animate organism: the sole Object within my abstract <primordial> world-stratum to which, in accordance with experience, I ascribe *fields of sensations...*"[26]

Thus the transcendental on-looker exercising the transcendental phenomenological epoché is not left without any constitutive order. Regularities, synthetical and polythetical harmonies of experiences operate, to be sure, but with what might be called an "abstractive-reductive" co-efficient.

21. CONCLUDING REMARKS ABOUT THE FIRST STEP OF TRANSCENDENTAL UNBUILDING OF THE REAL, OBJECTIVE WORLD

The statement of Husserl just cited may serve here to express the transcendental phenomenological situation in which the primordial factors of oriented constituting are sought that acquire the "appearance" of the next highest level of constituting. The situation allows of being restated in the following way:

The one spatio-temporal form of Nature as constitutive product of sub-scientific thinking and experiencing is rooted in my perceptual field in which my own percipient organism is the central body among others of the field and in which the other bodies are presented as at rest or in motion relative to Here. But my own percipient organism, co-presented as at rest or in motion, is also presented as at rest or in motion relative to the earth in, on and over which I am in motion or at rest. But the earth is not, however, likewise presented as just another body of whatever sort. So far as the first step of unbuilding is concerned, this signifies that I refrain from positing the perceptually and apperceptually presented earth on which I and all other sorts of organisms and bodies find ourselves as having a place in the real space of the real, objective world; *and* I also refrain from positing *correlatively* my sensory-perceptual fields presented as having a place in and as being non-selfsufficient components of the real, objective world where those fields are constituted as located in and on the surface of, or inside, this real physical thing, my percipient organism "functionally related" to other events in the world.

The *correlative refraining* in the transcendental phenomenological attitude, however, in no manner signifies that, for example, earth and sensory-perceptual fields are therefore not transcendent to my own peculiar experiencing of them in primary passivity. Expressed positively: they are rather still posited and believed in as having *their own inherent quasi-objective spatial and temporal spreadoutness*. So abstractively reduced and co-presented with quasi-objective motion and rest from the primordial quasi-objective Here, the quasi-objective "earth" and my own quasi-objective "percipient organism" are constituted as "*anomalies*" and not as bodies in an indefinitely extended homogeneous, real objective space and time in which my organism is co-presented as a body like any other body, in which my organism can exchange places with any other body. "Earth" and my own quasi-objective "organism" are instead co-presented in an original way as each with its own peculiar quasi-objective spreadoutness, as each its own quasi-objective "place" that cannot be exchanged for any other quasi-objective place. Precisely these are some of the *primordial* factors of oriented constituting which will play a significant role at higher levels of constituting and to which we shall return in later chapters. But one transcendental phenomenological result may be mentioned here: While all bodies in and on and over and under the "earth" will acquire the objective "appearance" of being presented as having their own particular locations in an objective, universal system of places, "earth" and my own animate "organism" will not; they are presented originally as anomalies even at the highest levels of sub-scientific constituting.

But the foregoing is not the only discernable ingredient of order and regularity at the logically formless primordial level of oriented constituting. Certain fields of "sensation"--tactual, visual, auditory, thermal, visceral, and the like--are uncovered in this step of unbuilding as located in and on the quasi-objective organism. In the second part it will be necessary to distinguish the various ways by which these noetic-noematic determinational senses are transferred in primary passivity to other bodies on the basis of which, in some cases, the transfer is partially confirmed and, in other cases, partially cancelled. Very broadly stated, in so far as there is partial confirmation, the primordial quasi-objective world is passively constituted in my own quasi-objective mental living as a quasi-objective world containing not only many quasi-objective psychophysical things but some of which are presented in other passively constituted quasi-objective mental lives. In other words, as transcendental on-looker I find a primordial quasi-objective world in which primary passive experiencings and strivings are characteristic of a plurality of quasi-objective mental lives.

Thus a further phenomenological result of clarification of the quasi-objective primordial world is also clarification of the constituting, in primary passivity, of the root of "experiencing of someone else." In this regard we find ourselves in agreement with Husserl that the primordial quasi-objective world is where original insight is obtained into the overt as well as covert intentionality in which the quasi-objective psychophysical being of someone else becomes evinced and verified in one's own quasi-objective mental life; it is here that there must be carried out discrimination of "what intentionalities, syntheses, motivations, the sense 'other ego' becomes fashioned in me and, under the title, harmonious experience of someone else, becomes verified as existing and even as itself there in its own manner."[27]

In short, the purport of transcendental phenomenological discrimination and analysis is that by methodologically disregarding the fully constituted, real objective world those noetic-noematic substrata are revealed that are quasi-objective worlds, through which the fully constituted real, objective world appears and is experienced as fully constituted and at all. And it is at the substratum of the primordial quasi-objective world in the *specific sense*, Husserl says, that "myself" is constituted in primary passivity as *solus ipse*--though *not*, he emphasizes, in *a privative sense*: Where else, except in the stream of my own mental living can I ask about and expose the sense of another mental life, indeed of a plurality of mental lives, existing as they themselves there?[28]

Before it will be possible to consider *philosophically* this transcendental phenomenological situation, however, there is yet a *second* step of the unbuilding of mental living which must be carried out.

CHAPTER THREE

FURTHER TRANSCENDENTAL PROCEDURES

22. THE SECOND STEP OF UNBUILDING THE GENERAL POSITING IN THE NATURAL ATTITUDE OF THE REAL OBJECTIVE WORLD

As transcendental on-looker attending to the essential demarcations of mental life in the transcendental natural attitude, I have now taken the first step in unbuilding reduction of the fully real, objective world to the primordial quasi-objective world (in the wide sense of the term "primordial"). At the same time, I also indicated the sorts of phenomenological affairs uncovered as peculiar to this "upper story" of oriented constituting.

To further advance these methodological reflections another and *second step* of unbuilding the real, objective world must be taken. Effecting the second step consists of discriminating the highest noetic-noematic stratum of constituting that remains thematic after having discriminated those noetic-noematic strata belonging to the ego and its activity as well as those making up secondary passivity. In consequence of the first step of unbuilding was a quasi-objective noetic-noematic substratum characterized in primary passivity by "causality" and a plurality of other mental lives. The further step to be taken now consists of a still further refraining from positing the actually existing mental life-processes of others along with the positing of a quasi-objective world intendable to in and by mental life-processes other than my own. The second step of unbuilding reduction is, therefore, the exercise of a transcendental phenomenological epoché that excludes the accepting of my own actual mental life-processes in the quasi-objective primordial world intended and intendable to by mental life-processes of others in primary passivity.[1]

The residual noetic-noematic stratum which now remains thematic is likewise a "world"--indeed, a quasi-objective world--because it too has a quite specific and phenomenologically discernible space-time form; it is a quasi-objective world, moreover, in which my own mental life, in primary passivity, intends to itself, to be sure, as in the quasi-objective world but to no other mental life as posited or positable in this quasi-objective world. This substratum of oriented constitution now uncovered Husserl calls the "primordial <quasi-objective> world in the *specific sense*."[2]

What, now, is to be accomplished by this second step, and what are the phenomenological factors that define the outward boundaries of the range of this further exercise of the epoché?

The answer lies in the fact that only by having first effected the second step of unbuilding reduction, and by isolating all the substrata underlying the noetic-noematic stratum in which an intersubjective, quasi-objective world is constituted in primary passivity, can one inquire into the "motivations" in consequence of which a mental life proceeding in primary passivity does not rest with the experiencing of a quasi-objective world in which it is aware of itself as the only mental life and of its organism as the only animate organism, but instead necessarily "goes beyond" them to constitute a quasi-objective world containing a plurality of psychophysical beings at higher constitutive strata. To express the matter paradoxically, effecting the second step demonstrates its own impossibility by making it clear that it will not do to start out by presupposing a world with a plurality of "subjects" in it and then asking how it is that we come to mean others in the world.[3] It was noted before, moreover, that the primordial quasi-objective world (in the wide sense of "primordial") is a "causal" quasi-objective world characterized by an all-pervading and dominant "if-then" style of primary passivity. In this enlarged scope of the epoché, the second step of unbuilding reduction also consists of discriminating those meanings and intendings to the primordial quasi-objective world as "causal" and, correspondingly, the "causal" characteristics of that world. One finds then what Husserl sometimes also called a "phantom quasi-objective world" of "causally unrelated" things and events. But the *purpose*, again, of this further step of unbuilding is to reveal the "motivations" for confronting a quasi-objective world of "causally" interrelated things and events as an appearance of a phantom subcausal quasi-objective world. Not only is it possible now to inquire into why there is "causal" apperception by virtue of experiencing the manner whereby one phantom changes when another changes or undergoes "distortion" as we shall say later, but it also enables the examination of the anomalies of the fully constituted real objective world such as "after images," "mirror-images," and the like--that is to say, those events and things for which there is no "motivation" at higher levels of oriented constitution.

The highest noetic-noematic stratum disclosed by the second step of unbuilding is presented and posited as a sub-causal quasi-objective world. Yet just as "causally" related things are sensuously perceived through multiplicities of quasi-objective appearances, so phantoms are perceived through the "same" quasi-objective appearances; the phantom, too, is presented and posited as a unity over against a multiplicity of appearances, and each quasi-objective phantom appearance is itself presented and posited as a unity over against a multiplicity of variations of that quasi-objective appearance. The full phantom quasi-objective world is constituted tactually, visually, auditorily and, correlatively, each full phantom can itself be further analyzed into "separate" tactual, visual, auditory phantoms. In a manuscript dating from about 1920,[4] Husserl makes the point that "the phantom is then defined as the intuitional unity which prevails when we abstract from any causal, therefore substantive, connection, and we consequently abstract from the strata of causal intendings." Moreover, each of the components of the full phantom may be considered apart from all the others and taken as clues pointing back to their own peculiar constitutings: "It must be explicitly emphasized," Husserl says, "that the concept of schema (or phantom) is by no means limited purely to a single sphere of sense. A quasi-objective perceived physical thing also has its *tactual schema* which comes to light in the tactual seizing upon it; in general, there are *in the full schema* or full phantom *just as many strata to be distinguished as we can find genera of sensuous data.*"[5]

23. THE PHENOMENOLOGICAL RESIDUUM OF THE SECOND STEP OF UNBUILDING

This second step of unbuilding reduction may be further clarified by referring once more to a specific phenomenological situation of perception mentioned before: the fact that perspectival appearances in and through which objective as well as quasi-objective physical things are perceived are functionally dependent on, among other factors, the co-perceived states of motion and rest of my objective or quasi-objective animate organism within one or another orientation, under these or those aspects. The perspectival, one-sided appearances through which I see and touch the table vary in functional correlation with the motion of my organism as a whole and of parts when, for instance, I walk around the table or when I move my head from side to side perhaps squinting my eyes. In turn, these co-perceived states are themselves presented as dependent on the various sequences of flows of kinaesthesias: if there is a flow of kinaesthesias in one manner, my organism is co-perceived as in motion; if in another manner, my organism is co-perceived as at rest (thus, phenomenologically, kinaesthesias are said to both found and determine the order of motion/rest). There is also an additional correlation between sequences of flows of kinaesthesias and the perspectival appearances of quasi-objective or objective perceived physical things. Accordingly, any unbuilding reductions and corresponding extensions of the range of the phenomenological epoché exercised necessarily entail discriminations of the co-perceived organism and the perceived, real objective world as well as the mental life-processes in and through which they are constituted; and below this, the quasi-objective unities (primordial, in the wide sense, and phantom) which serve as perspectival appearances of the co-perceived quasi-objective organism along with those primary passive mental life-processes in which they are constituted.

In part, what the foregoing situation signifies for the *second* step of unbuilding the real, objective world is expressed by Husserl in the *Cartesian Meditations*.[6] Speaking specifically of the spatiality peculiar to the primordial quasi-objective world, Husserl makes the familiar observation that my quasi-objective co-perceived organism is presented in the mode of the central Here, while every other thing is always presented in the mode of There. Moreover, the orientation, There, can be freely altered by virtue of my kinaesthesias: By freely modifying my kinaesthesias (e.g., those of locomotion) my position can be changed such that the presentational mode, There, is converted into the presentational mode, Here. The implication is, Husserl continues, that the "same" quasi-objective physical things would be perceived from There, except in correspondingly different modes of appearance such as belong to my being There. Constitutively, then, the appearance-systems belonging to my being Here and perceiving from Here but also the appearance-systems, corresponding to the change of position whereby I am There and perceive from There, belong to each quasi-objective primordial thing: "if" there is perceiving from Here, "then" this appearance-system comes into play; or "if" there is perceiving from There, "then" that appearance-system comes into play; and so forth.

Among other things, what Husserl has done is to make thematic the noetic-noematic substratum of a spatially (and temporally) *centerless* quasi-objective world in which, as we may now say, systems of quasi-objective perceptival appearances are in play correlative to free modifications of kinaesthetic flow-sequences, hence a quasi-objective world in which any There may and can be converted into any Here and, correspondingly, any Here into any There. In this specific sense, we may speak of a *centerless* primordial and, by extension, *centerless phantom* (or schematic) quasi-objective spatiality (and temporality). *It is this phenomenological residuum upon which, now, further unbuilding discrimination is exercised.*

24. POSSIBLE FURTHER STEPS IN UNBUILDING NOT CARRIED OUT BY HUSSERL. THE QUESTION OF ESSENTIAL LIMITS OF UNBUILDING REDUCTIONS AND THE REFORMULATION OF "ORIENTED CONSTITUTION."

Precisely at this juncture a further possible step of unbuilding reduction suggests itself and which was not explicitly developed by either Husserl or Becker: an unbuilding reduction to a quasi-objective world of phantoms and their quasi-objective appearances intended to and meant not merely from Here *but instead only from Here.*[7] By extending still further the limits placed on the previous epoché, I now find, at the first level directly below a quasi-objective phantom world of inherently centerless spatial spreadoutness a unified quasi-objective phantom world intended to and meant only from Here; and then, second, at still lower levels, a whole series of quasi-objective phantom "worlds" such as those of purely tangible phantoms, of purely visual phantoms, of purely auditory phantoms--all, to be sure, presented originaliter as inherently spreadout only from Here and as transcendent to the primary passive meaning of them. This further discrimination is accomplished by heeding solely that quasi-objective sense which the here-centered quasi-objective world of full phantoms has, for example, as noematic object of past, present, and future tactual intendings and, concomitantly, by further and specifically refraining from positings or accepting any sense it has or might have as noematic correlate of intendings of any other sorts (e.g., hearings, seeings, etc.). The same procedure can be repeated for the visual, auditory and olfactory phantoms. In each case, a configuration of purely tangible phantoms, purely visible phantoms, and so forth, can be discovered and which, in each case, is co-perceived with the quasi-objective tangible, or visual, or auditory, organism-phantom presented as at the center of the quasi-objective phantom world that extends from Here as far as any part belonging to the tangible organism-phantom can or might reach, or that extends from here as far as can be seen, or heard. These cases of oriented constitution will be the subject of investigation in Chapters Six and Seven of Part Two.

Bizarre and capricious though at first it may seem, unbuilding reduction to the noetic-noematic *here-centered* quasi-objective phantom world nonetheless leads to the crux of these methodological considerations. This is because it raises the question of whether any further unbuilding of consequence is possible, of whether, indeed, phenomenologically it is possible to discover and discriminate still more fundamental founding substrata which *of essential necessity* figure into the oriented constitution of the fully concrete, real objective world. It is, of course, entirely possible now to turn from the here-centered quasi-objective phantom unities presented in and through multiplicities of quasi-objective phantom appearances to just those appearances themselves, to just those unities of quasi-objective visual appearances, or tactual appearances, or auditory appearances; and so forth. Yet the question persists: Are there any essential limits to the unbuilding reductions? Is there a point at which the transcendental-phenomenological methodological intuition is exhausted? Is there a moment when I, as transcendental phenomenological on-looker, have completely refrained from positing the naturalness of which my transcendental natural attitude consists? And if any further noetic-noematic substrata of oriented constituting are discoverable and to be discovered, what, if any, new intuitions and further methodological precisions are required to uncover them?

By exercising transcendental phenomenological epoché I refrain from positing the here-centered quasi-objective phantom spatial spread as having a place in not only the centerless quasi-objective phantom spatial spread but also ultimately in the real spatial spread of the real, objective world; correlatively, I refrain from positing those spreads as components of the real, objective world.[8] I am, of course, still positing them as *transcendent to* my quasi-objective experiencing of them in primary passivity. In primary passivity each quasi-objective noematic correlate is posited, believed in, as having its own transcendently inherent characteristics. And in each case it is to the *own transcendently-inherent*, to coin a ponderous phrase, that I turn and have in phenomenological view when I now ask if further unbuilding is possible of primary passive mental life-processes in the transcendental natural attitude.

It is important to bear in mind the foregoing not only to properly set the question of whether any further unbuilding reductions are possible. It is also important because the whole methodological intuition of unbuilding reductions is *in no shape or form a "reductionism."* Differently expressed, the *philosophical* significance of what has been said is that the methodological intuition consistently follows the principle that the phenomenological residua of one level of the multi-storied oriented constituting of the real, objective world cannot be automatically extrapolated as functioning in precisely the same way at the next highest level. Thus the way in which something "primordial" enters into something "secondarily" constituted with a new stratum of sense is in each case different.[9]

As a result, the question about further unbuilding, about the limits of range or scope of transcendental phenomenological reductions, will have a different meaning for each constitutional level. The specific case in question, here-centered quasi-objective phantom unities presented in a here-centered spread, is no exception. Unexceptional as it is, however, the question about further unbuilding raised in this connection does have exceptional consequences as much for the exposition of Husserl's own thought on these matters as for a certain set of problems in modern philosophy and for establishing the validity of the phenomenological-methodological intuition itself. The substance of these consequences may be made clear at this stage of our discussion by developing three critical reflections. The first is historical in nature, and concerns Husserl's theory of "hyletic data;" the second, somewhat broader in scope, deals with a set of philosophical problems of perception to which the theory of hyletic data is addressed; and the third establishes a Husserlian alternative to the theory of hyletic data, thus allowing further methodological precisions in unbuilding the real, objective world.

25. FIRST CRITICAL REFLECTION. "HYLETIC DATA."

The concept of "hyletic data" is basic to Husserl's account of sense perception, subtending his analysis of the way in which the quasi-objective phantom world is constituted. Although concepts somewhat similar to that of "hyletic data" had been employed by Husserl before *Ideas*, First Book, it is there that its clearest formulation is found a propos an analysis of perceptual fields within which unities of sensuous data are salient with respect to a background and each other. Mindful of his critique of psychologism according to which such data would be only elements of the stream of consciousness (as in Hume),[10] and at the same time wishing to avoid Brentano's over-ripe "objectivism" according to which sense data are only apparent spatio-temporal events forming the basis for well-founded hypotheses about a real (but non-apparent) spatio-temporal world,[11] Husserl introduces hyletic data to circumvent both (and similar) views:

In any concrete case of sense perception it is possible to discern a highly stratified structure, each lower stratum of which founds that immediately above it, and where the lowest founding stratum has, as a component "really inherent" to it, hyletic data.[12] In addition, reflective observation of any concrete perception of something physical reveals still another "really inherent" component, an intentive component. Together, these components make up certain mental life-processes. Hyletic data, Husserl says, designate the "phenomenological residuum of what is mediated in normal perception of something external by the 'senses'."[13] In other words, hyletic data mediate the founding-founded structure characterizing the perception of something physical.[14] And it is precisely this role played in perception of something physical that distinguishes hyletic from other sorts of sensuous data such as kinaesthesias, somatic data (muscle tensions, "feelings" of rest or motion of the organism), and the like.[15] To say, then, that perception of a physical thing is "mediated by the senses" signifies that between the stratum of hyletic data on the one hand, and the founded stratum of the perceiving of something physical on the other hand, there lies *still another* stratum *founded* on that of the hyletic data and which, in turn, founds the stratum of the perceiving of the physical thing.

An example of the intermediate stratum are the appearances in and through which concrete physical things are perceived. To express the ostensible phenomenological situation in the language of *Ideas*, the intentive component is both an "animating" construing of the founded stratum of hyletic data as adumbrations of a multiplicity of, e.g., visual appearances of the physical thing, and a perceiving (e.g., a seeing) of the physical thing meant as self-identical throughout those appearances. Thus the "sense data" in question are not merely subjective, merely components of the stream of consciousness, of mental living, since they allow of being construed as adumbrations of something objective (the physical thing) transcendent to the processes intending to them. Nor are they unreal, however, even though appearing: they are instead construed as real appearances of non-hypothetical real physical things. At one stroke Husserl confutes views as diverse as Hume's and Brentano's.

Two further observations must be made here concerning hyletic data. In the first place, hyletic data belong to the province of original or primary passivity--as early as *Ideas*, First Book, Husserl refers to them as a case of "intentionality" that does not bestow signification.[16] In the second place, Husserl believed that the *temporal form* of hyletic data is the same as the temporal form of the intentive component of mental life-processes. There is no quasi-temporality of hyletic data; they are "really" temporal, which means that their temporal form is to be distinguished from that of perceived physical things.

However persuasive Husserl's theory of sense perception might be owing to the concept of hyletic data, he nevertheless pays a heavy price for it. Of the several important criticisms of hyletic data, perhaps the most telling is that of Aron Gurwitsch.[17] If, for the sake of argument, we grant Husserl the distinction between determinations of the physical thing perceived, presented in and through appearances adumbrated by "animated" hyletic data, and the hyletic data themselves endowed with adumbration-functions,[18] then, Gurwitsch notes, we must also assert--as Husserl would certainly seem to do--that there is after all nothing peculiar to the nature of hyletic data themselves which *unambiguously* determines the adumbrations of this rather than that perceived physical thing. This difficulty is found in a variety of places in Husserl's writings in addition to *Ideas*, First Book. For instance, in the 1912 manuscript published as the fifth volume of *Husserliana*,[19] in a characteristic passage Husserl says that "sensation stands as something common at the limit, so to speak, of the second and third levels <of oriented constitution>. At the second level it is the making known of the feeling-state [*Empfindsamkeit*] of the organism. On the other hand, it is, at the third level, the material foundation of perceptive construings, e.g., of perception of something physical [*materielle Wahrnehmung*]. As a result, <sensation> consists in the dual functions of construing mentioned above: as kinaesthetic sensation in function of what is motivating, as presentive sensation in function of what is motivated...." Hyletic and kinaesthetic data are distinguished and distinguishable here only in virtue of their motivating and motivated functions. Without developing the special meaning Husserl gives to "motivation," it is still clear that in and of themselves such "data of sensation" remain identical whether construed as appearances through which a real physical thing is presented (the "presentive sensations"), or as kinaesthetic functions on which somatic states of rest and motion depend. In a similar vein Husserl says, in the posthumously published lectures on phenomenological psychology (1925), that it is possible to "abstract" from the appearance-and adumbration-functions and consider the "pure data of sensation" in and of themselves.[20]

Thus although Husserl first developed his concept of hyletic data solely with respect to the role they play in sense perceptions, to account for that role he must then consider hyletic data as indifferent to the perception as well as to other data. At best, then, his account would seem to be logical rather than phenomenological. Moreover, if it is the case that hyletic data remain the same throughout the different noetic construings, then it would seem to be impossible to account for the distinctions between "data of sensation" functioning to present perceived physical things and data functioning as kinaesthesias, as somatic "feelings" and the like. An example adduced by Gurwitsch to suggest the implausibility of this view is the following: Suppose that, in the mountains, I see at an altitude bluish-grey formations appearing now as clouds, now as mountainous sky-line; so long as the sky-line of the mountains appears, the bluish-grey color is attached to a bodily surface. But when clouds are perceived, the color is detached, spatially indeterminate, and would seem to "float" in the air. Thus Gurwitsch concludes that "variations ascribed by Husserl solely to noetic factors do not leave unmodified what is given in direct sense-experience."[21] Were those factors left unmodified, there would be no perceptual difference between sky-line of mountains and clouds; Husserl's theory cannot explain the simple perceptual difference between cloud and sky-line and their alterations. It would seem that the affairs of which Husserl speaks, such as hyletic data, rather than being discovered in the various phenomenological reductions are instead construed in the name of the reductions. Even so, Husserl nevertheless also would seem to have himself to have provided a much more satisfactory account. A clue to that account may be found in the *Lectures on the Phenomenology Internal Time Consciousness.*

At the beginning of the *Lectures* Husserl states that he is interested in the constitution of a pure datum of sensation, that is a hyletic datum, and in the primary passive constitution of "phenomenological time" underlying actional mental life-processes.[22] Thus Heidegger, in his 1928 Introduction to the *Lectures*, states the "official" view that hyletic data and the noetic components have the *same temporal form.*[23] However, what is of interest in this connection is that the de facto course of analysis carried out by Husserl, and which Heidegger ignores, is at variance with the "official" view. Just this variance provides a significant clue to the correct phenomenological procedures for further unbuilding reductions. The variance consists of the fact that in section 8 of *Lectures*, where the actual course of phenomenological analysis begins, Husserl speaks of the temporal form of an auditory sensation--e.g., the sound or tone he describes is said to have "its own time-form." Earlier in the *Lectures*, in fact, Husserl had referred to the "phenomenologically given" as "the abstract parts of mental life-processes which specifically found temporal construings as such and, therefore, found...the specific temporal content."[24] The temporal content, the datum's own temporal form, is, correspondingly, a *founded noematic stratum* and therefore is not identical with the temporal form of the noesis in question (*scl.* the hearing of the sound or tone).

If the results of Husserl's de facto course of analyses is now carried over into his theory of perception, then there is a very different picture of perception from the one usually found in his published writings and posthumously published lectures.[25] Certainly part of Gurwitsch's criticism is met because hyletic data no longer figure as really inherent components of a concrete perceiving; in fact, it is unnecessary to call them "hyletic data" at all because, as their different time-form signifies, they are instead constituted as transcendent to the noetic processes (although, as noted before, still meant and intended to as having their own inherent spatio-temporal spreadoutness).

As a further consequence, instead of distinguishing between the construing of sensuous data as adumbrative of appearances of a physical thing on the one hand and, on the other hand, of perceiving a physical thing itself in and through those appearances, what now emerges into phenomenological view are perceivings of objective, or quasi-objective, physical things noetically-noematically "founded" on sensings of sensuous data. But "founding" and "founded" now have a very different meaning--that of oriented constitution, of how the "primordial" enters with a new stratum of sense into the "secondarily" constituted (how the sensing of sensuous data enter with a new stratum of sense into the perceiving of physical things). There is no question of "mediation" by "animated" data, thus, in that respect, no construing of such "animated" data as adumbrations of appearances.

Terminologically, then, instead of speaking of one kind of sensuous process as "hyletic data," I shall speak of sensa or sensuous processes and, correlatively, of the sensings of sensa of sensuous processes. When it is necessary to distinguish them verbally from other sorts of sensuous processes, specific names will be given them--for example, kinaesthesias, or kinaesthetic patterns of sensa; feeling-sensa; and tactual, or visual, or auditory sensa. Further terminological clarifications will be made as we proceed. However, once it is seen that at issue are noetic-noematic perceptual strata or layers, and not the "animating" of non-intentive components of the noesis, then it is possible to objectivate and seize upon the various sensings of sequences of sensa presented as changing and non-changing throughout temporal extents. At best, the theory of hyletic data is a superfluous construction.

Returning to the specific unbuilding situation which occasioned this first critical reflection, it may be said that a negative consequence is the phenomenological finding that implicit in the intending to and positing of quasi-objective here-centered phantom unities are co-intendings to and co-positings of many sorts of sensa-sequences with respect to which further phenomenological epoché and discriminations may be exercised. At least some discriminable sensa-sequences intended to may still be called "kinaesthetic" because, at higher levels of oriented constitution, intending to them necessarily entails perceptual intending to quasi-objective or objective motion and rest of the quasi-objective or objective organism (or some part of it). At least some other sensa-sequences intended to necessarily entail, at higher levels of oriented constitution, perceptual intending to quasi-objective or objective physical things (or some part of them).[26] However, the primary reason for calling some sensa-sequences "kinaesthetic" is that, no matter what perception-"functions" might be legitimately imputed to them, changes and non-changes of particular sorts of these sensa-sequences are actualizable *immediately* in consequence of primary passive strivings *and* regularly accompany changes and non-changes in certain other sensa-sequences actualizable *mediately* in consequence of those primary passive strivings to actualize kinaesthetic sensa-sequences immediately.

The foregoing formulations will serve for the moment to suggest the direction to be taken by further unbuilding beyond the quasi-objective, here-centered phantom world. Further critical reflections are necessary, however, to establish the boundaries of this new phenomenological focus.

26. TERMINOLOGICAL NOTE

Immediately actualizable sensa-sequences regularly accompanied by mediately actualizable ones of whatever other sorts are examples of sensuous changes and non-changes of what will be designated as *prespace* in the *broad* meaning of the term, "prespace." Sensuous changes and non-changes not so correlated will be designated as prespace in the *narrow* or strict sense. The term, "prespace," is not the most satisfactory or unambiguous one; it does not signify "unspace" or "non-space." However, the term is consistent with earlier phenomenological literature, and was used, for example, in this context by Oskar Becker (although a somewhat different meaning is attached to the term here). Despite this, it still seems better than speaking of a sub-phantom quasi-objective space or of sub-phantom unities. Moreover, prespace in either the broad or narrow sense does not signify changes or non-changes that are not constituted in primary passive identifying and distinguishing syntheses: unusual as the idea of prespace may seem at first, it is still a case of the oriented constituting of the real, objective world in space and time.

Another term that has and will be used frequently is the term, "*field*." When used without qualification, "field" signifies a qualitied spatio-temporal spread. Each field, such as the visual, the tactual, the auditory field, is presented as having a continuously spreadout whole the "parts" of which are usually differentiated from one another (below, Part Two, Chapters Five and Six). In addition, it will be necessary also to discriminate visual, tactual, auditory perceptual fields of quasi-objective or objective physical things and qualities. Finally, I shall discriminate primary passive intendings to and positings of *sensum-fields* (visual, tactual, auditory sensumfields of sensa) which enter into, or take on the "appearance of" quasi-objective perceptual fields of quasi-objective physical things with a new stratum of sense at higher levels of oriented constitution. A number of different sensum-fields will be distinguished, some of which, I shall argue, are presented as the most basic cases of oriented constituting of the real, objective world.

27. SECOND CRITICAL REFLECTION: KINAESTHETIC DATA

The story of oriented constitution designated as prespace in the broad sense of the term, "prespace," can be regarded as a sort of phenomenological analogue to a wide variety of nineteenth and twentieth century accounts of sensation and space perception.[27] Almost without exception those accounts base themselves on the idea of space formulated by Kant which holds that only if space and time attach to things in relation to our sensibility can it be explained how it is possible to know a priori that whatever is presented to us must have the spatio-temporal determinations prescribed by mathematics. In other words, only if physical things give rise to "sensations" can their spatio-temporal properties appear as necessary and subject to the laws of geometry and kinematics. This position is basic not only to the theories of space perception developed by psychologists and philosophers such as Lotze, Helmholtz, Lipps, J.S. Mill and Bain, to mention but a few,[28] but it also has its roots in and bears kinship with theories developed by Locke, Berkeley and Hume.[29] In contrast, the phenomenological analogue of prespace would seem to be closely allied with the ideas of space perception developed by Katz[30] and Merleau-Ponty.[31]

However, despite all differences with the Kantian position, all of the writers mentioned (including Katz and Merleau-Ponty) hold, first, that the propositions of geometry and kinematics will be true of intuited spatio-temporal properties of things; and that, second, space is that in which all things are, hence that space is, basically, a "room for movement; which its German name, *Raum*, distinctly confirms."[32] In this light the propositions of geometry, which render the intelligibility of space, are reducible to those of kinematics.[33] To state the matter plainly, basic to the Kantian position is the idea that the "sensation of movement" is fundamental to space perception. Phenomenologically expressed, this signifies that the constitutive substratum of the correlation between immediately actualizable kinaesthetic sensa-sequences and mediately actualizable sensa-sequences (more specifically, between immediately actualizable *locomotor* kinaesthetic sequences or processes and mediately actualizable sensa-sequences) is indispensable to the oriented constituting of the spatial form of the real, objective spatio-temporal world.[34]

To be sure, the late nineteenth-century accounts of space perception stemming from Kant argue, in addition, that what I designate as mediately actualizable sensa-sequences are *aspatial* in contrast to physical things and their properties.[35] Indeed, the assertion of the aspatiality of such data proves to be the solution to the Kantian paradox where the spatial characteristics of things only can be explained if space defines the real; but if space defines the real, how can space itself be explained? Here the answer is that what defines space is what is by itself aspatial, that is, the "sensations" of physical things. But this is tantamount to introducing a supervenient factor to resolve the paradox.[36] Yet it remains to be seen if what has been held to be aspatial is not instead "intrinsically spatial"[37] (to employ a late nineteenth century term). And by following clues given by Husserl I shall try to show how the "intrinsically spatial" figures into the oriented constituting of the real, objective world.[38]

That is to say, although I find no reason to deny or abrogate the phenomenologically significant role played by the correlation of kinaesthetic processes and other sensa-sequences in the oriented constituting of the spatial aspect of the real, objective world, nevertheless I will suggest that this correlation plays this role only by virtue of a more fundamental stratum of constitution. It will be necessary, then, to examine this substratum with respect to two problems that arise in this connection:

1. If it is held that there is a substratum of oriented constitution more basic than the correlation between kinaesthetic processes and mediately actualizable sensa-sequences, then it is necessary to establish the foundation for differentiating between them.

2. Accordingly, it will be necessary to see how, in the first place, the primarily constituted prespatial substratum enters into and acquires the "appearance" of the secondarily constituted prespatial spread at the next higher level; how, in turn, that higher prespatial spread acquires the "appearance" of the phantom quasi-objective spatial spread and eventually that of the real, objective world. In the second place, it will be necessary to explain the role played by the regular accompaniment of kinaesthetic processes by other sorts of sensa-sequences for higher levels of oriented constitution of the real, objective world.

The foregoing brief formulation of some phenomenological problems in their historical setting serves to return us to the methodological question about further unbuilding reductions of the here-centered quasi-objective phantom world. The third critical reflection is therefore a methodological one.

28. THIRD CRITICAL REFLECTION. THE "SETTING OF KINAESTHESIA AT ZERO"

To resolve the two phenomenological problems of the last section and, thereby, to transform the largely negative results of the first two critical reflections into positive ones, further methodological precision is required. The precision concerns the elaboration of the phenomenological situation where in any perceiving of something physical the perceiver's organism is itself always co-perceived, where perspectival appearances in and through which something physical is perceived are, in part, functionally dependent on the co-perceived states of motion and rest of the percipient organism, where, in turn, these co-perceived states are presented as themselves functionally dependent on kinaesthetic processes.[39] This correlation provides a clue to further unbuilding:

I discriminate intending to and positing of the percipient organism along with co-intending and co-positing of the phantom quasi-objective world. More particularly, I discriminate those intendings to and positings of the hierarchical quasi-objective unities that function as quasi-objective appearances of the phantom organism, phantom tactual appearances, phantom visual appearances, and the like. My phenomenological residuum then consists of primary passive intendings to and positings of fields of sensa or sensuous unities in those fields which are either immediately or mediately actualizable. Still more particularly, I have not yet discriminatively refrained from primary passive intendings to and positings of those sensa-fields and to those unities, nor from those primary passive expective intendings to certain correlations between immediately and other mediately actualizable sensa-sequences; nor have I discriminatively refrained from primary passive likings and dislikings of sensa-sequences or expective sensa-sequences, nor from primary passive strivings "motivated" by such likings or dislikings of current or expective sensa-sequences.[40]

Is there any further unbuilding now possible? As indicated before, the question can only be answered if it can be shown that what has not yet been methodically discriminated is stratified, storied, and if, therefore, it is possible to discriminate between various strata of sensings and positings of sensa and their sequences. It is precisely here that further methodological precision can be introduced, specifically the procedure of "*setting kinaesthesia at zero*." Like the other steps in unbuilding reductions, this procedure is nothing else than the making explicit of "moments on the basis of the completed constituted sense in which what is abstracted from [*das Abstrahierte*] is already constitutively implied."[41] So Husserl characterizes the constitutive situation where there is, so to speak, "not yet" the "idea" of space and place, nor even the experience of somatic motion or rest. It is, rather, the phenomenological situation where kinaesthetic processes or flows are "held still," are "at zero" [*Null- oder Stillekinaesthese*] and where appearances persist just as they are.[42] In short, as phenomenological on-looker I consider the "zero-orientation" [*Nullorientierung*] of all immediately actualizable or kinaesthetic processes and their systems.[43]

Husserl had introduced this methodological procedure in various places to deal with rather different sets of problems at different levels of constitution. In contrast the use of the procedure here is restricted to just the lowest levels of the oriented constitution of the real, objective world for reasons to be explained shortly. It is necessary first to distinguish two basic meanings of the phrase, "setting kinaesthesia at zero."

In the *narrow* sense of the phrase, "setting the kinaesthesia at zero" signifies instituting either actually or in phantasy a course of kinaesthetic processes as are accompanied by quasi-objective perceivings of some part or parts of a quasi-objective organism as not moving relatively to the rest of the organism, or by a quasi-objective perceiving of all parts of an organism as not moving relatively to one another. In a *broader* sense of the phrase, "setting kinaesthesia at zero" signifies, in addition, discriminatively refraining from, first, the possibility of any different ("non-zeroed") past, present, future course of the "zeroed" kinaesthetic processes; and, second, discriminatively refraining from everything that functionally depends on the past, present, or future possibility of some different ("non-zeroed") course of the "zeroed" kinaesthetic processes.

For present purposes, these senses of "setting kinaesthesia at zero" may be illustrated with several brief examples. In the *narrow* sense, "setting *oculomotor* kinaesthesias at zero" signifies instituting actually or in phantasy of a course of kinaesthetic processes accompanied by quasi-objective perceivings of one's eyes as not moving in their sockets relatively to one's torso, and not moving as they would if a course of kinaesthetic processes were underway accompanied by quasi-objective perceivings of one's organism as moved by, e.g., one's legs relatively to one's torso. In the *broader* sense, "setting oculomotor kinaesthesias at zero" would consist, in addition, of discriminatively refraining from the *possibility* of a "non-zeroed" course of "zeroed" kinaesthetic processes and every sort of change in appearance functionally dependent on "non-zeroed" kinaesthetic processes.[44]

The *visual correlate*, then, of all oculomotor kinaesthetic processes "set at zero" is a visual intending to an enduring or persisting spread the "spatio-temporal" parts of which differ qualitatively but in which there can be nothing moving relatively to the "spatially" distinct parts or patches of the spread. The *purpose* of "setting oculomotor kinaesthesias at zero" is precisely to uncover this visual correlate pertaining to the "zeroed" processes and the "spatial" spread posited and seen. Moreover, at this stage of unbuilding, in addition to oculomotor kinaesthetic processes I "set at zero" all other such systems of processes correlated with sensings of certain specific kinds (such as visual, tactual, auditory sensings and positings). This is also the reason for restricting the procedure of "setting kinaesthesias at zero" to just this stage of unbuilding for, unlike Husserl, the phenomenological theme is also "reduced: to the lowest reaches of oriented constitution of the real, objective world. As a result, it is possible to distinguish, in the various cases, between sensing an enduring spread in which there are sensa "at rest" or "in motion" relatively to prespatially distinct parts of the spread, on the one hand, and on the other hand, sensing an enduring spread the prespatial parts of which may differ qualitatively but in which there can be no sensum "at rest" or "in motion." This latter constitutive stratum, examined in detail in Chapters Five and Six, comprises the narrowest meaning of prespace and *marks the limit of essentially possible refraining from positing the naturalness of which my transcendental natural attitude consists* (below, sections 97-99).

29. CONCLUDING CRITICAL NOTE

Earlier, in Chapter One (section 12), a passage from the *Cartesian Meditations* was cited in which Husserl spelled out the "problem concerning the origin of the 'thing phantom,' or 'thing pertaining to the senses,'" which, he says, "is not yet the 'real thing' of the primordial psychic sphere. Even in this sphere the 'real thing' becomes constituted at a higher level...." Similarly, with respect to the phenomenological result of the last step of unbuilding of the oriented constituting of the real, objective world it is necessary to speak of a constitutive sphere in which, in consequence of "setting kinaesthesia at zero," the organism is "not yet" constituted and which, like the "real thing," only becomes constituted at a "higher level." It must be emphasized that this conclusion is not Husserl's. It is rather a Husserlian one for, according to Husserl, the organism is presented as *res extensa* only in so far as it is constituted in and through kinaesthetic processes regularly accompanied by other mediately actualizable sensa-sequences; such kinaesthetic processes (or integrated systems of them) are that upon which the motion or rest of the organism are dependent. Kinaesthetic processes therefore "presuppose" the constitution of what is dependent on them in the first place.

In his book on Husserl's theory of space perception, Claesges succinctly summarizes Husserl's position: "The constitution of something spatio-temporal is only possible on the ground of the pregivenness of hyletic data. On the other hand, however, to the extent that it is determined essentially as *sensation*, the pregivenness of hyletic data presupposes the sensing organism."[45] The mutual presupposition and pregivenness of hyletic and kinaesthetic data are essential to the possibility of the constitution of something spatio-temporal. In contrast, the purport of this chapter has been to suggest that intentive sensings of different sorts of sensa-sequences or processes also involve positing of a segment or dimension of the naturalness of the transcendental natural attitude which is not to be confused with the positing of the quasi-objective or objective organism.

What then is essential to the possibility of the constitution of something spatio-temporal is the way in which what may be called "prespatial" naturalness enters into, with a new stratum of sense, and takes on the "appearance" of the "secondarily" constituted, i.e., the quasi-objective here-centered spatio-temporality. The constitution of something spatio-temporal is then not the problem of clarifying how the "sensing organism" is presupposed by the postulation of hyletic data which, as Gurwitsch has shown,[46] is a variation of late-nineteenth century theories of "sense data." Nor is the problem that of clarifying how "kinaesthetic data" are correlated with the mediating and animating functions of hyletic data.[47] The building up analysis of prespatial, "pre-objective" spreads will take on a very different shape and form (Part Two, Chapter Six).

In concluding this chapter reviewing further procedures of transcendental phenomenological method it is worth mentioning something that is not always obvious. It seems to me clear from many of Husserl's published texts as well as from presentation of unpublished manuscripts that Husserl has not carried out the phenomenological reductions far enough, that he has not, therefore, always reduced the theme, "concrete real, objective world," to the lower levels of constitution that he considers. This would seem to be the case especially with the important manuscripts such as "Die Welt der lebendigen Gegenwart und die Konstitution der ausserleiblich Umwelt" and "Notizen zur Raumkonstitution." While some of the difficulties that result will be considered later,[48] a note of caution must be entered regarding the reading of Husserl in this respect as well as with respect to any attempt to carry through the phenomenological reductions. Self-conscious exercise of phenomenological epoché yields no "mechanics" of application. The only assertion I make here is that the will to philosophical theory in giving form to the feeling for the problem of phenomenological genesis must try to square with the most "primitive" naturalness of which the transcendental natural attitude consists and to which this chapter has sought to find a way to return.[49]

CHAPTER FOUR

THE ORDER OF TRANSCENDENTAL PHENOMENOLOGICAL INQUIRY THAT WILLS TO RETURN TO THE "THINGS THEMSELVES"

30. METHOD AND SCIENCE

The preceding chapters have sought to identify the various transcendental phenomenological refrainings from those positings in which the naturalness of my mental life consists so that, step by step, there are uncovered the layers of actional and passive oriented constituting of the real, objective world and of my real, objective mental living in the world. Even though a diversity of Husserl's texts were consulted as a guide for establishing the phenomenological discriminations and their correlative reductive unbuildings (and, correspondingly, by implication the various stages of reductive building-up to be elaborated in the second part of this book), the basic literary expression and philosophic touchstone has remained the definition of transcendental phenomenology expressed in the Introduction to the three books that comprise the uncompleted *Ideas Pertaining to a Purely Descriptive, Eidetic Transcendentally Pure or Transcendental Phenomenology and to a Transcendental Phenomenological Philosophy*. There, in the "Introduction," Husserl defined transcendentally pure phenomenology as a purely descriptive discipline that seeks to render a cognitive account of the "irreality" of mental life-processes with respect to their conferred status as "being in the world," and that, as such a discipline, transcendentally pure phenomenology is a "science of essences," an eidetic science.[1]

By making thematic the sense, "being in the world pertaining to my mental life-processes as any mental life-processes you please in the transcendental natural attitude," transcendental phenomenological method makes possible a self-critical basis for elaborating a body of knowledge. In short, the method purports the prospect of a science. The specific task that remains, then, in this final chapter on the method of phenomenological reductions is the statement of its principal characteristics as a science of essences which I, as meditating phenomenological on-looker, produce. The statement hinges on the order of inquiry entailed by the "things themselves," namely mental life-processes in the transcendental natural attitude.

31. THE RIGHT ORDER OF SCIENTIFIC EIDETIC INQUIRY

For the purposes of this chapter a Husserlian account of seeing and seizing upon essences may be summarized as follows:[2]

1. Seeing and seizing upon essences constitutively presupposes and is based upon actual or feigned intendings to individual exemplifications of the essences in question; for instance, actual or feigned perceivings of individual cases of red colors exemplifying the essence or Eidos, "Red." This signifies that, for Husserl, I do not have to see any individual red colors but must at least feign the seeing of them.
2. There is no analogue in perception for the seeing and seizing upon essences; there is, therefore, no "empirical, scientific description" of seen and seized upon essences.
3. In the case of perception, what is perceived is believed in, posited as real, as individuated in space and time. In contrast, individuation in space and time are utterly unimportant for seeing and seizing upon essences. The clear recollecting of an essence, for instance, as previously seen and seized upon, is forthwith a clear seeing and seizing upon that essence with as much originality as before. Thus even though I may speak of "seeing and seizing upon essences," that need not signify that I must also speak as well of an originary acquisition of essences--anymore than the generic likeness and specific difference of seeing essences and of perceiving real individual things signifies that I must also speak of an analogy between that seeing on the one hand, and that perceiving on the other hand.
4. So far as perception is concerned, there is a clear-cut difference between actually perceiving something and feigning the perceiving of that same something. In the case of perception I may therefore speak of possible actual and feigned real individuals. In contrast, the distinction is unimportant for seeing and seizing upon essences.

Indeed, phenomenologically all the differences important for perception are not only unimportant for, but are also irrelevant to, seeing and seizing upon essences. The only important difference among seeing and seizings upon essences is that which obtains among those that have, and those that do not have, actual and possible individual exemplifications.

Some of the factors of this summary will be reconsidered in a moment; implicit in them is a right order of inquiry into essences which must now be made explicit.[3]

The narrative of seeing and seizing upon essences makes up a chapter in the phenomenology of those mental life-processes in and through which (predicatively) formed affair-complexes of whatever sort are constituted in an original way. Two sorts of intendings to such affair-complexes may be distinguished: 1) intending to senses of verbal expressions, and 2) intending to "universals." Despite the fact that the terms here are not entirely satisfactory, and despite the fact that this rough and overly simple classification requires extensive revision, the distinction is servicable at the moment because it immediately yields two broadly conceived sets of disciplines. The first set is comprised of theories such as theory of language, theories of possible senses of verbal expression (such as a theory of propositions, or arguments, even a theory of theories). The *second set* is comprised of a theory of "universals," the basic division being that between material and formal "universals." For the purpose of this chapter and indeed the rest of this book, I shall restrict (rather arbitrarily) the term, "essence," to just this second set of objectivities.[4] Now, just as in the case of delineating the various steps of the transcendental phenomenological epoché, so here the clue for the right order of inquiry is drawn from the very structure of mental life-processes themselves.[5]

In the first place, because mental life-processes are highly stratified, with a clearly discriminable plane of demarcation running through them (see above, section 10), it is necessary to consider the primarily constituted ("founding") before the secondarily constituted ("founded") levels or strata.[6] In the second place, in each case of the primarily constituted and the secondarily constituted it is requisite to develop an account of *direct* before *indirect* mental life-processes--for instance, perception is considered before so-called "picture consciousness." Likewise, to take a further step, a theory of perception is developed before a theory of memory.

Finally, and in the third place, it is necessary to consider mental life-processes in which real things are presented before considering those in which things and objects of other sorts are presented, e.g., affairs such as senses of verbal expressions and essences. The reason for this is not difficult to find: while seeing and even seizing upon senses of verbal expressions and essences presupposes, for instance, the perceiving of real things, the perceiving of real things need not presuppose seeing, let alone seizing upon, verbal expressions or essences. Similarly, the positing of real things does not entail the positing of senses of verbal expressions or of essences; and conversely positing of essences or senses of verbal expressions does not entail the posting of "individual existence" (see *Ideas*, First Book, sections 4, 5).

Accordingly it seems reasonable that one inquires into the seeing and seizing upon the senses of verbal expressions, and of essences by examining, to start with, those whose exemplifications are constituted as real things. In addition, and more narrowly, it follows that seeing and seizing upon material essences must be considered before seeing and seizing upon formal essences because, as will become clear in later chapters, the former are posited at lower and different levels of oriented constituting than are formal essences.[7] And because intending to real things proves to be more basic than intending to other sorts of things, those material essences that have real things as exemplifications should be examined before those that have any other sort of things as exemplifications. Finally, everyday exemplifications should more likely be examined before any other sort that are less immediate and familiar.

32. THREE THEMES OF INQUIRY

In short, seeing and seizing upon essences is based upon seeing and seizing upon their possible exemplifications. Because the exemplifications may vary widely, there are at least three themes of inquiry at the outset and to which the chapters of the second part will be largely confined:

1. The first theme is made up of direct seeing of material essences, presented an "they, themselves," on the basis of the perceiving of real things.

2. The second theme is comprised of seeing material essences as exemplified by straightforward perceivings of real things. A secondary theme, in this connection, is then reflective awareness of the seeing of essences exemplified by the reflective awareness of the perceiving of real things.
3. Finally, a third theme is that of seeing essences exemplified by not just the perceiving of real things, but also by the clear perceiving of real things.

There are, of course, still further and secondary themes as well as other themes of inquiry (such as those belonging to inquiry into formal essences and senses of verbal expressions). What has been said, however, is sufficient for the second part of this essay; it is also sufficient for the immediate purpose of formulating the principal characteristics of the phenomenological method in so far as it purports the prospect of a science of essences consistent with the idea of oriented constitution.

In section 34 of the *Cartesian Meditations*, Husserl points out the following:

> Eidetic phenomenology <or: phenomenology as a science of essences>, accordingly, explores the universal Apriori without which neither I nor any transcendental Ego whatever is 'imaginable;' or, since every eidetic universality has the value of an unbreakable law, eidetic phenomenology explores the all-embracing laws that prescribe for every factual statement about something transcendental the possible sense (as opposed to the absurdity or inconsistency) of that statement.... 'In itself,' then, the science of pure possibilities precedes the science of actualities and alone makes it possible as a science. With this we attain the methodological insight that, *along with phenomenological reduction[s], eidetic intuition is the fundamental form of all particular transcendental methods* (that both of them determine, through and through, the legitimate sense of a transcendental phenomenology).[8]

To this statement may be added one cited earlier from section 59 of *Ideas*, First Book,[9] that transcendentally pure phenomenology is "in fact, a purely descriptive discipline, exploring the field of transcendentally pure consciousness by pure intuition."

Accordingly, *at each and every level* of the oriented constituting of the real, objective world the "methodological insight that...eidetic <or: pure> intuition is the fundamental form of all transcendental methods" is followed for each case of unbuilding and, correspondingly, for building up as a consequence of which all-embracing laws prescribe the possible sense for "every factual <transcendental> statement about something transcendental." Another name for this eidetic or pure intuition is "ideation" or, at times, "abstractive ideation" (in conformity with the term, "abstractive reductions"). In producing a "science of pure possibilities," eidetic intuition involves what Husserl called "free variation in phantasy." It is to this further aspect of phenomenological method that we must now turn.

33. IDEATION

In an earlier chapter it was noted that salient things and events pair themselves as similar and different in primary passivity according to the law of "associative" or, as it was also called there, "assimilative," synthesis.[10] More particularly, they pair themselves as similar or different in some respect or other that remains the same, invariant. But seizing upon the fact that two things salient in the field of experience are paired is not equivalent to seizing upon, let alone seeing, the same or invariant respect as exemplified in them, nor is it equivalent to seizing upon their similar, individual moments. For example, I see two faces on the barroom floor, and I may be aware of the two faces as similar; yet I still may be at a loss to see in which respect the two faces are similar and consequently may still be at a loss to seize upon the invariant respect exemplified in and by the individual moments of similarity. But what does happen, nevertheless, is that pairing and mutual transfer of sense increases the "affectivity" of similarly constituted, non-individuated invariants so that there is a "tendency" to turn to them and to seize upon them in a further act.[11] This further act is called "*ideation.*" To be sure, even though it may be a direct and "founded" seeing, intuiting, of an exemplified essence, it may not be a clear one. It may be clarified, however, and converted into a clear as well as a direct intuiting by the active process of "free variation in phantasy"--which, however, must not be confused with the act of ideation upon which it is based. A brief example will serve to explain this difference.

Consider the material essence, Physical Thing. I have in mind golden mountains and winged horses, having just read Meinong and James. In so far as they serve to exemplify the essence, Physical Thing, it really makes no difference whatever whether or how they are more or less valid possibilities than the trees and stones I see from my window. That is to say, because essences are exemplified by products of feigning or phantasy no less than by the facts of actual perception, the operation of "free variation" need not begin with real things; feigned or phantasied things or real data of perception may serve equally well as starting points. The act of variation whereby a foundation is established so that an essence comes into original presentation is not confined to actual existence which forms but one sub-class among many of possible exemplifications. Thus to consider something, even a real thing, as a possible exemplification does not therefore divest that real thing of its actuality-determination. Nonetheless, in acts of free variation in phantasy it is possible to survey a domain of possible exemplifications without having to consider, or only consider, realization in actual existence.

For example: Beginning with a musical sound, be it actually experienced or simply feigned or phantasied, I vary that sound through a series of auditory phenomena; the sound is varied in this way, transformed in that--such that a whole series of related variations is produced. All members of the series are congruent in a certain respect, all exhibit an invariant moment in spite of differences in, e.g., pitch, resonance, and the like. They all exhibit the "same content." Now, instead of regarding the varieties as exhibiting the same invariant, exemplifying the "same content," I can rather seize upon that invariant itself no matter what its actual or possible exemplifications. This is an act of ideation. It is a direct and clear seeing of the invariant or essence intended to and posited in the process of variation and, in the seeing of it, is presented in contrast, *thus as necessarily related*, to the manifold of actual and possible varieties which, no matter what their differences, are all realizations of the same essence. Differently expressed: The essence, Sound, is seen and seized upon as an "identity" in contradistinction, but always with reference, to an indefinite manifold of possible varieties *within a limited range* beyond which the essence is no longer necessarily related to possible varieties, some of which may be actual.

Yet this is not because all varieties would have been surveyed actually or in phantasy. Indeed, it would be impossible to survey them all in the example of the musical sound. As a consequence, the process of free variation is broken off after a certain number of steps; nonetheless, it could be carried out *ad libitum*. And it is the possible *ad libitum* continuation that makes possible the reference of the essence not only to all those examples that have been, or may have been, or will be, considered, but also to more varieties *ad libitum*. In short, the essence has an extension not exhausted by the varieties; the extension of the essence, Physical Thing, for example, does not comprise simply real things but rather pure possibilities, purely possible varieties. Real things in actuality are, so to speak, "set aside," "neglected" (though not divested of their actuality); yet they can be returned to and, as it were, actualized by virtue of the continual reference to the particulars in question as *purely possible* exemplifications of an essence.

But if that is the case then it is necessary to speak of the "empirical extension" of an essence, such as Physical Thing, or Sound, and if the relations among essences hold even apriori for matters of fact, that is because the real things and their properties must be considered as purely possible exemplificatory varieties *which happen to be actualized*. And so for each and every level of unbuilding and building up to be examined in the second part of this book.

The conferred status of mental life-processes in the real, objective world, and the real objective world itself, are to be regarded as purely possible exemplifications of the essences, Mind and Nature, which happen to be actualized. For this reason "the sciences of pure possibilities precedes the science of actualities and alone makes it possible as a science." And on the basis of acts of ideation clearly seen and seized upon essences and their concatenations of relationships in turn provide the foundation for exploring the "all-embracing laws that prescribe every factual statement about something transcendental the possible sense...of that statement." But by no means does that eliminate the actuality of the real, objective world nor of mental life-processes in the world.

There is, of course, much more to be said about the phenomenological science of essences, much more to be clarified conceptually (especially the distinction between material and formal essences). Some further clarification will be necessary in the course of later chapters. The purpose here was to indicate the eidetic bones, as it were, of the phenomenology of essences to express the shape of the critical-epistemological understanding of the philosophic-theoretic will required by the things themselves.[12]

34. SPACE, TIME, OTHER

The second part of this book will consist of the application of the transcendental phenomenological method to the analysis of the oriented constituting of the purely possible conferred or bestowed sense, "being in the world pertaining to my mental life-processes as transcendental mental life-processes in the natural attitude." Taking a clue from the writings of Husserl as well as from other phenomenologists, this conferred sense of "being in the world" will be examined under the heading of the material essence of Space along with the material essences of Time and of Other. In short, Space, Time, and Other comprise the eidetic complexus of Being-in-the-World.

The material essences, Space, Time, and Other are not arbitrary topics for the task of analysis chosen for the second part. To be sure, space and the essence, Space, made up a set of important philosophical problems in their own right to which Husserl always seemed to have been drawn again and again during his lifetime. But in addition to that, the phenomenological investigation of space often served as a central paradigm for formulating his truly revolutionary study of the "origin" of the constitution of the "consciousness of internal time" and the "exposition of the problem of experiencing someone else." More broadly conceived, the phenomenological investigation of space was basic to excavating the bedrock for a vast scheme of a "transcendental phenomenology of reason" proposed at the end of *Ideas*, First Book. Yet even so vast an undertaking hardly exhausts the substance of a large number of Husserl's manuscripts dealing with space perception, some of which are now available in the posthumously published volumes of *Husserliana*.

Still other facets of Husserl's ideas about space appeared in Oskar Becker's dissertation on space and geometry, published in

1923. Becker's study was and still is valuable. In addition to working out in some detail Husserl's investigations of the constituting of space, Becker took considerable pains to develop a critique of the then-current scientific cognition of space in geometry and physics. The critique consisted of isolating and examining the various sub-scientific strata of importance for the constituting of the spatial form of the real, objective world as conceived by physics. Concluding that the spatial and temporal forms of those strata are neither "Euclidean" nor "non-Euclidean," that is, of necessity entail no system-forms of Euclidean geometry, Becker then examined the various substrata of visual "prespace" and of "schematic" or phantom quasi-objective space. The former proved for him to be a two-dimensional manifold founding the three-dimensional manifold of the latter.[13] This phenomenological result, and the novel "transcendental deduction" supplied to demonstrate it,[14] still does not encompass all of the significant changes that occur in Husserl's thought about space after he moved from Göttingen to Freiburg.

One of the reasons for this is that by the late 1920's and early 1930's Husserl had already enlarged and deepened not only the transcendental setting within which he located the problems of space constitution of interest to him. He had, in addition, refurbished and radicalized his critique of natural scientific cognition. By 1929 Husserl had already, it would seem, prepared much of what would later (1931) be published as the *Cartesian Meditations* in which he introduced a "transcendental monadology" the space-time of which originates in "primordial Nature." In this light the "genetic transcendental phenomenological problems of birth and death" as well as the "psychological origin" of the "idea of space" are recognized in their proper sense "by means of a change of the natural into the transcendental attitude...open to a 'Copernican conversion'" so that they have a well-defined place "within a transcendental 'metaphysics'."[15] Moreover, 1929 saw the publication of the *Formal and Transcendental Logic* in which Husserl tried to root out the "*original grounding of all the sciences*, and of the formal ontologies...exercising in their behalf the function of a theory of science, the normative function," which "*gives them all unity*, as branches of a constituted production from the one transcendental subjectivity."[16]

As a result, there is but one actual and genuine philosophy and science, according to Husserl, comprising a formal ontology in the quite specific sense of having the "Form:" the "allness of realities, with the allness-'forms' space and time...."[17] In turn, the result of the 1929 study is the foundation for the central ideas of *The Crisis of Western Sciences* of 1934 in which the historical form of modern science that makes a certain idea of space go bail for the "allness of reality" is subjected to a critique that explicates the space-time of the life-world as the source of science in the style of Copernicus and Galileo.[18] On the positive side, the results here, if consistently analyzed, are quite different from the earlier work on space by Husserl in his Göttingen lectures of 1907, to which reference has been made a number of times (and which were published as *Physical Thing and Space*).

Certainly the spatial form of the real, objective world, for instance, is still described as set into quasi-objective and prespaces, but these are now not only three-dimensional rather than two-dimensional manifolds; the very nature of the constitutive setting is also differently conceived. In many cases the results are different in still other ways to be examined now in the second part, although it must be emphasized that Husserl's transcendental phenomenological framework remains the same from the beginning to the end of his work on space.

35. CONCLUSION TO PART I: INTERPRETATIONAL PROBLEMS

The method of phenomenological reduction sketched in this and in the preceding chapters takes issue with any number of pronouncements of Husserl on the subject in addition to a wide variety of interpretations of Husserl's pronouncements--interpretations not just by Husserl scholars but also by philosophers as diverse as Sartre and Merleau-Ponty at one end of the spectrum, to Pace and Tran Duc Thao at the other end. In short, the history of Husserl interpretation may well seem an interpretational nightmare the outlines of which cannot be traced here.[19] But "history" is not the only means by which interpretational problems can be approached.[20] By way of a conclusion to Part I we may approach the problem of Husserl interpretation from both a broader and narrower optics than that of "history."

This first part began by setting the question of phenomenological method in a certain context, namely that of two different yet intimately related philosophical problems that arise in the "feeling" for the problem of the genesis of the real, objective world: *How is it that consciousness is, as it were, of two minds about itself as in and of the world*? And *how is it that the world in itself is at the same time the world for me*? Together these problems trigger the "will to return to the 'things themselves'" in conformity with which phenomenology is defined and the method of phenomenological reductions and discriminatory procedures formulated. The definition and formulation have, in part, a literary basis in Husserl's writings, although their legitimation must be sought finally in the critical-epistemological understanding phenomenological investigation produces.

Even so, the interpretational problem remains. Its recognition fostered the attempt to ensure that the present account of the method of phenomenological reductions does not exclude de jure other actual and possible phenomenological methods. Such an attempt is only meaningful if the *limits* of the method formulated find some warrant in the will to philosophical theory. In other words, a reasonable conclusion, and one explicitly drawn in Chapter Nine, is that other and perhaps non-phenomenological methods are necessary to advance the understanding of the problems of Space, Time, and Other. It further follows then that not all factors of Space, Time, and Other that figure into our presentive awareness of them allow of being clarified phenomenologically (or even philosophically, for that matter). Although other required methods cannot be developed within the scope of this book, their possibility must be allowed, the only caveat being that they not be inconsistent with transcendental phenomenological results by excluding them a priori.[21]

Nevertheless, it is clear that the definition of transcendental phenomenology and the formulation of the method of phenomenological reductions are distinct in many ways from other accounts, including at times Husserl's own. For example, the distinction is not made here between "descriptive phenomenology" and "transcendental phenomenology;"[22] such a distinction is rendered superfluous by the "feeling" for the problem of genesis of the real, objective world that motivates a refraining from that positing in which the naturalness of the natural attitude consists.[23]

But that is not the only distinction at issue in Husserl interpretation. Husserl himself spoke of "ways" to transcendental phenomenology.[24] The very choice of words is ambiguous: does "way" signify a separate method or only one among several approaches to the transcendental phenomenological method? With respect to the latter half of the question, Iso Kern and others have found at least three ways distinguished by Husserl in his writings. They are, first, the "Cartesian way" which can be found throughout the early as well as the later writings;[25] second, the "way by means of intentional psychology," a way perhaps more confined to Husserl's work in the 1920's;[26] and, third, the "ontological" way which, like the first, can be found throughout Husserl's writings but which is nonetheless to be explicitly dissociated from the first way.[27] There is also, perhaps, a fourth way to be distinguished, accomplished by means of a critique of positive scientific cognition; Kern believes this way to be essentially like the third, while Boehm reserves judgement as to whether this is still another way separate from the others.[28] The issue need not be decided here, and a brief sketch of the three ways will be sufficient to place, so far as that is possible, the formulation of the method of phenomenological reductions in this and the preceding chapters. It must be borne in mind that the various ways mentioned often co-exist in Husserl's writings, that they therefore do not necessarily mark off "stages" in the development of his philosophical thought.

The "Cartesian way" would seem to purport an absolute beginning for the grounding of science by seeking to realize absolute knowledge once all ostensible knowledge of a transcendent world is set out of validity so that "pure subjectivity" is uncovered.[29] A number of problems arise for Husserl in following this way, not the least of which is that consciousness turns out to be something left over from the phenomenological reduction,[30] that consciousness of "other minds" cannot be adequately accounted for.[31] Nor would it seem possible to deal persuasively with the ostensible apodicticity of consciousness even though Husserl's "Cartesianism" hardly resembles that of its eponymous origin.[32] In contrast, the way by means of intentional psychology begins by "abstracting" the purely psychical from the physical, and inevitably leads to transcendental philosophy.[33] The chief claim of this way is that the psychological phenomenological reduction is implicitly the transcendental phenomenological reduction. Be that as it may, the question arises whether by "abstraction" (in a distinct Berkeleyian sense) it is possible to arrive at the establishment of a concrete science:

> Through abstraction and one-sided interest, do we not incur the loss of a part of total reality, namely, of full concretion? In other words, can the way through psychology lead further than to psychology, namely to the *partial* science of the purely psychic? Does not Husserl's thesis--that the world is nothing other than the 'phenomenon' for this subjectivity, achieved through abstraction--occur as an unfounded claim of the natural scientist who suddenly declares this abstraction to be the whole reality, and thereby slides into materialism?[34]

To avoid the slide into materialism on the slippery slope of abstraction we are left, according to Kern, with the third, the ontological, way. This is a much more satisfactory way because it entails a change in philosophical attitude that is a gain and not a loss on the credit/debit sheet of phenomenology.[35] To be sharply distinguished from the Cartesian way, the ontological way is found as early (if not earlier) as the lectures posthumously published as *The Idea of Phenomenology*, is present in *Ideas*, First Book, and especially in *Formal and Transcendental Logic*.[36] The chief virtue of the ontological way is that it avoids sliding back into natural scientific cognition and produces an adequate account of the intersubjective world of experience.[37]

The three--or four, or however many--ways at least show Husserl continually reevaluating his methodological intuition in the light as much of old as of new problems. To be sure, Husserl's dissatisfaction and those of others with any or all of the ways need not mean that the ways are therefore wrong or inadmissable: at least some of the dissatisfactions of Husserl and others can be critically resolved, especially as concerns the problem of the experience of "other minds."[38] Or, again, the sort of confusion concerning the order of phenomenological refrainings noted in *Ideas*, First Book, certainly allows of correction.[39] Likewise the whole issue of "intentional psychology" can be critically resolved; it need not be abandoned as a phenomenological debit as Aron Gurwitsch and others have shown.[40]

To state the matter differently: I see no reason for the "dialectics" of ways such as proposed by Kern. The right measure of the method of phenomenological reductions, it seems to me, are the philosophical problems mentioned earlier and on which Husserl ceaselessly worked. And the method takes its initiative, so to speak, from consciousness or mental life itself.[41] Nevertheless, the method of phenomenological reductions as developed in the present study is best called "Husserlian" rather than Husserl's which remains but a transcendental empirical example of the self-conscious exercise of transcendental phenomenological epoché.[42] Thus while the formulation of the method of phenomenological reductions here may run counter to any number of Husserl's reflections on, and literary expressions of, the method, it squares with the philosophical problems that charged Husserl to develop the method in the first place.[43] Accordingly, my formulation, while not one or the other way, confirms in some cases and disconfirms in other cases what is evinced by Husserl's thought as a transcendental empirical example. The various ways distinguished by Husserl himself, elaborated by Kern, Boehm, and others, then prove to be *ways* precisely because 1) of themselves they do not yet make explicit in their own fashion refraining from positing the naturalness of which the natural attitude consists, 2) they do not yet adequately exercise the refraining *noetically and* noematically, and 3) they therefore misconstrue the order of refrainings. In short, none of the ways adequately seizes upon the general positing of the naturalness of the natural attitude as the general positing of transcendental naturalness.[44] As a consequence, the formulation of the phenomenological reductions in this first part does not so much replace anything Husserl developed as it makes explicit what Husserl too often left implicit as essential for any phenomenological method whatever evinced by his own transcendental empirical example.

There is, of course, no more a royal road to transcendental phenomenology than there is to anything else. But I find that a virtue of phenomenology for it allows, in the spirit of William James, a "phenomenological pluralism." And one effect of reviewing Husserl interpretation, however briefly, is that it returns one to the "things themselves."

PART TWO

TRANSCENDENTAL PHENOMENOLOGY OF SPACE, TIME, OTHER

INTRODUCTION

THE PROBLEM, PLAN AND HISTORICAL SETTING OF THE CONSTITUTION OF SPACE AND TIME

36. THE PROBLEM AND PLAN OF INVESTIGATION OF SPACE CONSTITUTION.

Drawing on literary, critical and systematic reflections, the previous chapters examined some of the elements necessary for a phenomenological investigation of the real, objective world and of transcendental natural mental life-processes in and of that world. That examination was also the occasion for stating some of the results of a Husserlian account of transcendental mental life-processes and the principal features of the method of transcendental phenomenological epoché and reductions employed to achieve those results. Central to the results was Husserl's idea of constitution, especially of what he called "oriented constitution." By way of laying the groundwork for an investigation of the oriented constituting of the spatial and temporal aspects of the real, objective world, in step-by-step reductions we discriminated the various substrata in and through which that world is presented in primary passivity--the multi-storied noetic-noematic "building" without which there would be no world at all: the primordial quasi-objective world presented as containing a plurality of mental lives and psychophysical bodies, things, events "under circumstances;" beneath this a substratum at which a quasi-objective world is presented with but one primary passive mental life and just one psychophysical organism; below this a full phantom quasi-objective world founded in turn on tactual, visual, auditory and other phantom quasi-objective worlds; and, at still lower levels, quasi-objective fields the events and things of which are presented as quasi-objective appearances of phantoms; then, at still lower and archontic levels, spreads of sensuous quale. That, in brief, is the picture that has emerged so far.

Mostly for the sake of terminological consistency with Husserl's writings as well as those of Becker and others, I called the spatio-temporal form of the primordial and phantom quasi-objective worlds *quasi-objective spatio-temporalities* (or some variation thereof, such as quasi-objective spaces or times). The spatio-temporal forms of fields and spreads of sensuous quale of various sorts I called *pre-spatio-temporalities* (or some variation thereof, such as prespaces or pretimes). To be sure, to speak of "prespace" and "pretime" does not signify no space, or no time. There are, of course, other than literary reasons for the terms. For example, with regard to sense perception, an essential law of perception is its one-sidedness and presumptively evident intentionality. As I shall try to show, while this law holds for the primordial and phantom perception of spatio-temporal things, it does not hold for the lowest substratum of oriented constitution discriminable where one can only speak of a "plenum" and not a one-sidedness of perception. Thus some sort of sharp terminological distinction must be made (see below, sections 73, 84).

In any case, however one may resolve the terminological difficulties the *problem* is to discover how each pre-spatiotemporality acquires, with a new sense, the "appearance" of each quasi-objective spatio-temporality at the next higher levels; and how each quasi-objective spatio-temporality acquires, with a new sense, the "appearance" of the spatio-temporality of the real, objective world at still higher levels of oriented constitution. Lest the impression be given, however, of a rigid, hierarchical structure of "being in the world," it may be worth while to risk a metaphorical expression of the phenomenological problem of the investigation of space constitution. The oriented constitution of the spatio-temporality of the real, objective world may be expressed metaphorically by the image of the spaces between the pickets of a fence.[1] With respect to the formulation of the problem it may be said that just as the intelligibility and the being itself of the fence as a whole depend on the spaces between the pickets, so the intelligibility as well as the being of the real, objective world depend on the constituting of picket-like "worlds" the various sorts of "spaces" of which yield a real, objective spatio-temporal world in the first place. Without the "spaces," and those "times," the real, objective world, like the fence, would remain wholly without intelligibility.

To be sure, it is always possible to "build" a world out of the many constitutive "spaces" and "times" just as an architect may build a house by borrowing the spaces between the pickets of the fence. There is, then, a danger posed by the setting of the problem, namely of ending up with pickets having no space between them, with oriented constitutings of "worlds" having no particular order among them so that we end up with a Humean mind "as a kind of theatre, where several perceptions successively make their appearance" with "properly no simplicity in it at one time, nor identity in different" (*Treatise*, IV, 1, section 6). As it were, to avoid illicit borrowing of "spaces" and "times" considerable methodological caution must be exercised. Most immediately, one way to exercise that caution is by taking note of the plan, the blueprint, peculiar to a phenomenological investigation of space and time constitution for, like the picket fence, the very phenomenological problem also marks off the boundary lines of the "genesis" of the "idea of space and time."

Space and time, after all, have been the subject of many investigations in the history of science and philosophy. But regardless of what differences subsist, or have subsisted, among the results of those investigations, the following would seem to hold true (see above, section 20):

All de facto natural scientific-philosophic thinking assumes as a matter of course that the spatio-temporal form of the real, objective world encountered in daily life exists. Thus an investigation of the spatio-temporal form of the real, objective world as determined and further determinable by any natural scientific-philosophic thinking whatever involves an investigation of the spatio-temporal form of the real, objective world encountered in daily life as the "product" of subscientific-philosophic thinking. More particularly, an investigation of space and time as determined and further determinable by natural scientific-philosophic thinking also involves an investigation of space and time just as "products" of subscientific-philosophic thinking and, still more particularly, of subscientific-philosophic experiencing in primary passivity. But the plan or blueprint is also a transcendental phenomenological one, sharply distinct from even a psychological phenomenological plan. That is to say, the focus of the plan is that of the optics of the transcendental dimension by which is uncovered the positing of the naturalness belonging to the natural attitude within which natural scientific-philosophic thinking and, additionally, subscientific-philosophic thinking and experiencing operate. This focus of the plan of phenomenological investigation of space and time needs to be made explicit before any attempt can be made to gauge the historical setting of the investigation and before an actual investigation can be carried out.

37. THE FOCUS AND ARGUMENT OF A TRANSCENDENTAL PHENOMENOLOGY OF SPACE

Earlier, in Part One (sections 21, 27) it was noted that central to a transcendental phenomenological investigation of space constitution is the clarification of the oriented constituting of the correlative "anomalies" Earth and Own Animate Organism in their own peculiar intrinsic "spatiality." Equally important, the clarification consists of showing how, at higher levels of oriented constituting, they acquire, with a new sense, the "appearance" of bodies among other bodies, of occupying places among other places and, in addition, of how the constituting of "anomalies" plays a necessary and not an incidental role in determining the meaning of space as a product of natural scientific-philosophic thinking. This latter, it may now be said, "presupposes" a primordially constituted Earth and Animate Organism that enters into the secondarily constituted space "produced" by natural scientific-philosophic thinking such that it is necessarily presented as a horizon of being "accessible from the primordial and discoverable in a particular order."

Accordingly, the transcendental phenomenological "origin" of the "idea of space" is to be found in the basic traits of the constitutive substrata of the "anomalies," Earth and Animate Organism. If this be the case, then a radical departure has been taken from the prevailing ideas of space in modern science and philosophy. The phenomenological problem of space constitution is then equally an argument that the "origin and genesis of the idea of space" lies not only in the more primitive layers of the oriented constituting of Earth and Animate Organism, but also in a constitutive layer "presupposed" by motion and rest of bodies of whatever kind, "anomalies" or not, in space. Thus, as Husserl noted in a late manuscript,[2] one need not, therefore, perpetuate the absurdity of "tacitly presupposing beforehand the naturalistic...conception of the world...and of then seeing human history...anthropologically and psychologically...as an obviously accidental event on the earth which might just as well have occurred on Venus or Mars."

The "naturalistic conception of the world" has been formulated before (section 27) in connection with the unbuilding procedure of "setting kinaesthesia at zero." Briefly in review, it states that only if space attaches to things in relation to sensibility can it be explained how it is possible to know a priori that whatever is presented to us must have the spatio-temporal determinations prescribed by natural scientific thinking. Only if, therefore, physical things give rise to sensations, specifically sensations of motion, can their spatio-temporal properties appear as necessary and subject to the laws of geometry and kinematics. The two assumptions that underlie this view, to be considered in more detail shortly, are, first, that the propositions of geometry and kinematics are true of intuited spatio-temporal properties of physical things and, second, that space is that in which physical things are. As a result, the propositions of geometry, traditionally conceived as rendering the intelligibility of space, are reduced to those of kinematics. Expressed phenomenologically, the naturalistic conception of world and world-space asserts that indispensable to constituting the spatial aspect of the real, objective world as determined and further determinable by natural scientific thinking are more primitive experiential substrata of various sorts comprising in different ways correlations between changes in somatic states and the perspectival appearances in and through which perceived things on and in and over the earth are presented. *In contradistinction to this phenomenologically expressed naturalistic conception, the transcendental phenomenological conception to be developed in the next chapters is that those more primitive substrata are indispensable only because of a still more basic and primitively intrinsic "spatiality" or "spreadoutness" of what is neither in motion nor at rest and upon which the perceptions "Earth" and "Animate Organism" are built.*

The problem and plan of a transcendental phenomenological investigation of the "genesis" of the "idea of space" acquire, as a consequence, a critical dimension. The ontic and epistemic claims at the heart of modern scientific-philosophic thinking may be expressed by saying that if "geometry" and "kinematics" define space by rendering it intelligible, and if space defines the reality, let us say, of Nature and Earth, it then follows that "geometry" and "kinematics" define reality. Just these claims, I shall argue, do not find confirmation--indeed, are disconfirmed--in a transcendental phenomenological investigation of space. Instead, I shall try to show how the definition of reality is rather to be found in transcendental mental life in the natural attitude apperceiving itself as on and in and over the "earth" as well as that for which Earth and World are in the first place.

Taken together, these two sides of the transcendental phenomenological problem can be formulated as a transcendental phenomenological "deduction" of space (below, Chapter Seven). If the words "spatial" and "space" are now used to include not only space as conceived by natural scientific thinking, but also to include the transcendentally reduced "intrinsic spreadoutness" presented originaliter at various constitutive levels, then this signifies that transcendental mental life in the natural attitude is not only temporal and aware of spatio-temporal Nature; it is also aware of itself as temporal and apperceives itself as within the spatio-temporality of Nature (as "animate organism") at various levels of oriented constituting. But, as I will also try to show (below, Chapter Eight), one and the same transcendental mental life constitutes itself as intrinsically temporal, as a "product" of the "retentions" and "protentions" comprising the "consciousness of internal time"--a "product" meant as identical at each of the levels of oriented constituting which is the "appearance,"in a particular order and with a new sense, of the level immediately beneath it. The phenomenological result, accordingly, may be expressed by saying that at each level of oriented constituting, transcendental mental life in the natural attitude is spatio-temporalized with the identical "consciousness of internal time." If that is the case, then it is also true that it is an eidetic necessity that I find in fact my transcendental mental life and my worldy living in possesion of a like "content"--a necessary, though not a sufficient, condition for being of "two minds" about myself.

This conclusion provides a still more precise formulation and focus of the transcendental phenomenological problem and plan of investigation of space: To critically-epistemologically understand how and that my transcendental mental life constitutes "in itself" all that which belongs to me as my own peculiar essence and, at the same time, in my worldly living, that and how my own peculiar essence is psychophysically objectivated as among psychophysical others and things, as thus apperceptively included as a component at different constitutive levels of the one spatio-temporal real, objective world that is itself constituted in my transcendental mental life in the natural attitude.

38. THE HISTORICAL SETTING OF THE INVESTIGATION OF SPACE CONSTITUTION. THE TWO PRINCIPAL LINES OF THOUGHT ABOUT SPACE IN MODERN SCIENCE AND PHILOSOPHY

It may be true, as Merleau-Ponty says in "L'Oeil et l'esprit," that Descartes was right in setting space free by daring to idealize and construct it. The idealization and construction were accomplished by conceiving the external world as identical with space, by asserting that geometry deals with space and that therefore space may be said to have mathematical intelligibility, and by insisting that all things which are real are in space.[3] Setting space free in this fashion creates a difficulty of profound consequences for the naturalistic conception of space. The gist of the difficulty is that if positional relations change, then there is motion; if they do not change then there is rest.[4] Real things then turn out to be nothing but delimited, defined positions in space so that nothing belongs to real things except their geometrical and spatial properties.[5] Such a view immediately gives rise to two questions: 1. How and by what means are positional, geometrical and spatial characteristics connected with each other? 2. If "real thing" is defined as that which is in space, what, then, about space itself? Is space itself contained in space? Space to be sure, may be so idealized and constructed that it defines the real. But what, then, is space itself?

The problem acquires further ramifications when the Cartesian "geometrization" of space is taken over by Spinoza in his transformation of Descartes's notion of substance an attribute of which is space (*extensio*). This "geometrical rationality," it may be said, turns out to be the very "divinity" of substance for Spinoza, the systematic interconnectedness developing in accordance with different dimensions two of which are known: *extensio* and *cogito*. Thus space, set free by Descartes, is more real than things in space. Whatever elaborations we may wish to make in this connection, the fact remains that together Descartes and Spinoza represent but one line of thought about space in modern science and philosophy.[6]

Another and equally important line of thought is also present at this time. In his *Elements of Philosophy* of 1656, Hobbes, departing from Descartes, raises the question of whether or not space comprises or encompasses all things. In a famous passage in the second part of the *Elements*, he says that we may begin by supposing the whole world to be annihilated except for the thinking subject.[7] Hobbes's question is, then: What remains in our experience even if the experienced is annihilated? Although there are no longer presentations of sense data, there is, nonetheless, a meaning in which space remains, for whenever I try to conceive of sense data as external to me, the thinking subject, I cannot conceive of them except as spatial. This signifies, Hobbes says, that space is a "phantasm." I cannot help but have the conception of space whenever I try to think of something as real and external to me. Yet, is space real? According to Hobbes, space is not real because it is a creation of my conceiving of things external to me. Time, likewise, is a "phantasm" because in conceiving external things I cannot avoid conceiving of them as enduring in time.

Thus at the very beginning of modern science and philosophy two distinct lines of thought about space have arisen. According to the one, space is more real than the things in space (Descartes, Spinoza); according to the other, space is the product of feigning the world annihilated, something wholly "subjective" and therefore deserving of the epithet, "phantasm" (Hobbes). And in both these lines of thought there is present the paradox that if space defines what is real, it seems impossible to render a satisfactory account of what space itself is.

It might be said that this dilemma is the "motor force" behind the further development of the problem of space. If space defines the real, what then is space? Is it Unreal? Suprareal? Surreal? Perhaps the first attempt to resolve the dilemma is found in the Platonism of Henry More in his correspondence with Descartes.[8] By strongly opposing the Cartesian identification of matter and extension, with the result that space is materialized, More worked out a possible synthesis of the Spinozistic and Hobbesian views. If one can demonstrate that endless extension is essentially distinct from things in space, then one has, in effect, demonstrated the presence of spiritual rather than a material existence. In short, one has demonstrated the existence of God. At the same time, More rejected the Hobbesian view that space is a "phantasm" because space proves to be an all-encompassing extendedness existing apart from anyone actually or possibly experiencing it. What is perceived in the world are limited things, delimited portions of space, and from such fragments an endless space can never be constructed. For More God is the substratum of space and, for this reason, space symbolizes God's omnipresence. And if things in space presuppose space which precedes all things in space, then those things too presuppose the omnipresence of God. If that is so, then not only is the Cartesian construction of space impossible but also the idealization of space because, as we experience it, space is at best the image of real space; as it were, geometry deals only with images, with shadows in the Platonic world-cave, although they do point beyond themselves to the original, to the ultimate *sensorium dei* or absolute space.

The final significance of More's idea of space is perhaps best revealed in Newton. Although Newton does not use the rather remarkable term, "*sensorium dei*," in his physical writings, he does speak of absolute space (and time) in the Introduction to his *Principles of Natural Philosophy*. There he draws a distinction between motion in space and motion with respect to space. In the text of the *Principles* proper, however, he confines himself to the former and it would seem that his physics does not require the idea of absolute space and time once the foregoing distinction has been made. However, if the Leibniz-Clarke correspondence is read as expressing the implications of Newtonian if not Newton's physics, then the distinction between absolute and relative space becomes central. Clarke states that copies of objects are engendered in the mind owing to sense impressions; in contrast, God seizes upon things in their originality without the intervention of copies. The implication is, then, that there is a real thing behind our copy of it, that there is a real space behind our copy of space, and that that real space is the *sensorium dei*. This would seem to mean that space is understood as an all-encompassing medium of container in which things are; space is both one and unique but yet completely intelligible because of the geometry from which all possible spatial figures are deduced. In this respect, absolute space can become the central subject matter of a rational science.

Yet despite this further refinement by Clarke of the Newtonian view, which shifts, with More, the accent of reality to space itself, the two approaches to the problem of space remain: the Cartesian-Spinozistic approach, where space is one of the attributes of Deity and where, with Newton-Clarke, space is a correlate of geometry; and the Hobbesian approach, where space is a "phantasm" of things so far as they are conceived as external to the mind. Although Kant's account of space will be interpreted here as a final attempt to reconcile these two approaches, there is still another angle to the historical setting of the problem of space which must be mentioned before turning to Kant.

39. THE DECISIVE SHIFT IN THE PROBLEM OF SPACE. SPACE AN ABSTRACTION

In his *Enquiry into Space, Time, Immensity and Eternity* (1734),[9] Edmund Law took issue with the thought about space found in More, Newton, and Clarke, where space was conceived as absolute in the sense of existing regardless of whether or not things fill it and according to which motion in space is relative not only to other things in space but also to space itself. Law especially objected to More's use of the infinity of space (and time) as an argument for their transcendent reality as well as to the further argument that because absolute space and time are beyond the human capacity of experience they are, therefore, also intelligible and unrestricted by the human condition. Law argued instead that immensity and infinity evince the fact that space and time are ideas of the understanding, and that it is human thought which has the capacity to extend beyond every and any limit. With this an important shift occurs in dealing with the problem of space that will have decisive consequences for modern philosophy.

Equally opposed to the ontological prejudice that the existence of things is inferred from the nature of ideas, images or copies, Law followed Locke in the belief that we can never compare our ideas or images with the things themselves of which they are ideas or images; they can only be compared with other ideas. Accordingly, the existence of space is not contained in the idea or image of space and the concept of space is not a copy of it. Present to the mind is, instead, an abstract idea of a receptacle to which the universe as a whole is referred. Space and time are then rather relational concepts which bestow order on things and events, and the idea of space is but a "convenient measure," as it were, established so as to relate things to one another. Like Hobbes, Law insists that that idea is nothing else but a product of the human imagination, the capacity to make images or copies, ideas. Unlike Hobbes, Law draws the conclusion that with this notion it is possible to explain the distinction between absolute and relative motion, that is, between two systems of reference, one of which is more general than the other.

To be sure, there is still the question of how we aquire the notions of space and time such that they have the connotation of being independent of the mind. Where, indeed, do the "Originals" of extension and duration come from? Law's (and Locke's) answer is that if space and time are the products of the mind, then they have their roots in sensation and reflection, combined through the imagination by means of habit. But such a view would seem to lead to an absurdity pointed out by Berkeley.[10]

Not only in the *Principles of Human Knowledge*, but also in the *Analyst* and *De Motu* Berkeley offered a rather severe criticism of Newtonian and Leibnizian physics and especially of the mathematics that produced them. On the background of Locke and Law, but also Hobbes, the upshot of Berkeley's account is that the idea of space proves to be nothing but an abstract notion hence, for Berkeley, an empty notion. This is because, according to the doctrine of minimum visible and tangible data, the criterion for space (and also for a geometry of space) is rather sense perception; all the rules and laws of mechanics, finally of physics and mathematics, are to be explained on the basis of sense perception confined to finite, sensible quantities. Any other view, such as that of Newton or of Lsibniz, leads to mathematical as well as theological absurdities: there is no more an abstract idea of space than there is of a triangle.[11] Yet, this is still not the whole story of the shift in the problem of space.

Even though Berkeley's criticism of mathematics and of abstract ideas certainly sets him apart from Leibniz, there is nevertheless an important positive relationship. Leibniz distinguished extension from space in the sense of a container and held that space, like time, is nothing but an abstraction. In contrast to Berkeley, however, although they are abstractions space and time are legitimate ones. While Leibniz can agree with Berkeley that space and time do not have any sort of substantial existence, the abstractions are legitimate for Leibniz in the sense of being ideal entities serving as substrata for the necessary truths of reason, hence are substrata for conditions valid for any possible world. It is of course necessary to add that for space and time to be intelligible in this manner the given of sense perception must be disregarded. What, then, is space according to Leibniz?

In his Fifth Letter to Clarke (section 47), Leibniz says that we can consider different things as co-existing in a certain order, and that space denotes the order of simultaneous co-existence. If one of the co-existents changes its relation to the multiplicity of other existents which otherwise retain their relations to each other, and if another body takes on the relation which the one had to all the others, then this change is called "motion." Space comprises all places and is, accordingly, a network of relations of co-existence.[12] What the network looks like is explained later on in the letter where space and spatial relations are compared to generations and genealogical lines of development:

> In like manner, as the mind can fancy to itself an *order* made up of *genealogical lines*, whose bigness would consist only in the number of generations, wherein every person would have his place; and if to this one should add the fiction of a *metempsychosis*, and bring in the *same* human souls again; the persons in those lines might change place; he who was father, or a grand-father, might become a son, or a grand-son, etc. And yet those genealogical places, lines, and spaces, though they should express real truths, would only be ideal things.

With Leibniz, the shift in the meaning of space is accomplished. Space, a network of ideal relations of places valid regardless of content like the metempsychotic genealogical lines, is no longer a container of all things, no longer what defines the real, no longer a transcendent reality or a property of God, no longer a phantasm. What, then, are the consequences for the historical setting of the problem of space in modern science and philosophy?

So far, once we reach Leibniz four possible answers have been given to the dilemma that arises from the Cartesian "freeing of space" by means of idealization and construction:

1. In the first place, there is the view according to which space is construed as a real container, as the medium in which all things are. Thus not only is matter in space, but it is always given with respect to space (Newton, More, Clarke).
2. In the second place, there is the view, basically epistemological in nature, that recognizes that space is a product of the human mind (Hobbes, Locke, Law). Space, on this view, proves to be an all-embracing immensity which is an extension of the segments of perceptual data.
3. In the third place, there is the view according to which the idea of space is an abstraction, and that there is no absolute space (Berkeley).
4. Finally, there is the view that there is but one cosmic space consisting of an ideal relational network (Leibniz).

Aside from their mutual differences, these four views possess in common the fact that the problem of space is developed regardless of the role of the concept of space in physics or mechanics. In other words, the concept of space in each case is established purely on philosophical grounds, and only then does it find its role in physics or mechanics. A broader implication in all four views is that a science such as physics still requires philosophical validation; the Cartesian "freeing of space" is largely a philosophical freedom. For the historical setting of the problem of space precisely this implication in common is of consequence in the sense that Kant's "Copernican Revolution" makes it of no consequence. The historical setting gets reset with Kant and becomes transformed but always with reference to the four views. Equally important, however, is another factor in common of the four views: each of them functions and can be regarded as a "specifying assumption" of one or the other principal lines of thought about space in modern science and philosophy. In the next sections, I shall try to suggest in as brief a way as possible that with Kant (and post-Kantians) that it is not so much a matter of one or the other, or both, principal lines of thought being abandoned as it is of a new "specifying assumption" being introduced to replace those which had been current until Kant.

40. KANT AND THE "INTUITION" OF SPACE

Although it may be somewhat of an exaggeration to say so, the "Copernican Revolution" with respect to the problem of space occurred before the middle of the eighteenth century with Leonhard Euler's "Reflections sur l'espace et le tems" of 1748: "The principles of mechanics," Euler says,

> have already been so solidly established that it would be absurd for anyone to doubt their truth. Even if one is not in a position to demonstrate them by the general principles of metaphysics, the marvelous agreement of all the conclusions ...would be sufficient to set their truth outside doubt. ...For one would be right in rejecting in metaphysics all reasonings and all ideas, no matter how well founded they might otherwise be, if they lead to conclusions contrary to those truths of mechanics. And one would also be right in admitting into metaphysics only those principles which can co-exist with these same truths of mechanics....[13]

The "tribunal of justification" is therefore not philosophy but mechanics ("physics") itself and, indeed, philosophy and the philosophical notion of space must square themselves with mechanics. Nor can space (and time) be treated in isolation, Euler insists, for they are the logical presuppositions of mechanics which decides in the first place if they are to be such presuppositions--the heartbeat of the "Copernican Revolution." Thus if any meaning of reality is to be imputed to space and time it is not because, for instance, they are Divine sensoria but instead because they are the indispensible preconditions for the principles of mechanics. To risk further exaggeration for the sake of extreme brevity, it may be said, more broadly, that space and time are required for the construction of "Newtonian" science and that, therefore, certain concepts of space and time are valid concepts.

It is in the light as well as the spirit of Euler's "reflections" on space and time that Kant, in the "Transcendental Aesthetics" of the *Critique of Pure Reason*, sets forth three characteristics of the idea of space: 1) space is that in which all things are; 2) space is the subject matter of geometry, hence is intelligible; and 3) space is intuitive in the sense that the different parts of space are components of one and the same space rather than, for example, instances of a concept (as in Leibniz). Not only, then, is the way in which space is understood determined by Euclidean geometry, but the results of the philosophical analysis of space are determined beforehand by mechanics. Philosophy must account for the space which the geometrician employs in his construction. What this signifies may best be explained, for our purposes here, in the form of an argument given by Kant in section 12 of the *Prolegomena*.[14] There Kant says, in effect, that before any particular set of things is presented to us we know that they will necessarily be intuited as having spatial *and temporal* properties such that all the propositions of geometry, arithmetic and kinematics will be true of them. Moreover, if space and time were determinations of things as they are in themselves, regardless of whether they affected us and gave rise to sensations, their spatial and temporal properties ought to appear then not as necessary but instead as brute matters of fact. There would, accordingly, be no way of explaining why it is that, before anything is given, we know that our mathematics will necessarily apply to them and necessarily be true of them--just this explanation cannot be supplied by the four views described earlier. However, if it is the case that space and time attach to things not as they are in themselves, but only as they are in relation to our sensibility, then it can be explained how it is that in advance we know that whatever is presented to us must have spatial and temporal determinations such as described by our mathematics.

It is, of course, the second alternative that Kant takes for, if nothing else, it meets the preconditional requirements of mechanics. Thus in the same section of the *Prolegomena* Kant goes on to say that "complete space" has three dimensions, thus making the tacit assumption that any "incomplete space" is the limit of another space. Just as a point is the limit of a line segment, so a line is the limit of a limited surface, a surface the limit of a limited volume, and a limited volume is itself the limit of the volume that lies beyond its surface (the volumes then being called "spaces"). With this second alternative Kant has, in effect, achieved a "reconciliation" of the four views of space, but with a quite different grouping. "Complete space" is certainly the medium in which all things are which, however, are always given, it may be said, "incompletely," i.e., with respect to space. The mind must be such that the second alternative is indeed the case, so that "complete space" is an "immensity" of limited segments of perceptual data, though, to be sure, "complete space" is not an absolute but rather "spaces." Finally, "complete space" is an "ideal network" of relations among spatial elements; the propositions of Euclidiean geometry state those spatial relations that hold among the spatial elements and it is possible to isolate the forms of those propositions so that the concepts of spatial elements and of spatial relations can be substituted by the concepts of "any elements you please" and "any relations you please having such and such formal properties."

41. THE CLUE TO THE PHENOMENOLOGICAL PROBLEM OF SPACE IN ITS HISTORICAL SETTING

The Kantian intuition of space and its "reconciliation" of the four views that comprise the historical setting of the problem of space returns discussion of the problem of space to the two questions that arose in connection with Descartes: 1) How and by what means are spatial characteristics connected with each other; and 2) If "real thing" is defined as that which is in space, what is space itself? Raising the first question presupposes an answer to the second, and the answer to the second would always seem to involve a dilemma, a paradox: the connection of spatial characteristics can only be explained if space defines the real, and if space defines the real, what is space? Kant's Copernican-revolutionary answer, like Euler's, contains an unusual assumption, namely that somehow space itself has already been seized upon and found amenable to the requirements of mechanics (or, more broadly, physics) no matter what they may be--providing that space attaches to things as given to our sensibility. Precisely in this "already seizing upon space" lies a clue to the phenomenological problem of space.

The essential meaning of the historical setting, what makes it a setting and not a mere narrative, are the two Cartesian questions. To raise such questions after Kant (and Euler), whatever other ramifications they may have, is to grant that space is, for instance, three-dimensional and that this proposition is established by intuition.[15] But in addition to tacitly having already seized upon and objectivated a perception of space, from Descartes through Kant there is the further assumption, undercutting even the profoundest differences, that by intuition it is established that the parallel postulate is true, that space can be abstracted from space-time and, above all, that space-time can be abstracted from things in space and time--a tacit assumption entailed by the very questions to which the Cartesian "freeing of space" gives rise. Just this assumption is not obvious.

For it does not follow that if there is a (valid) intuition of space as three-dimensional space is unambiguously given, nor that space can be abstracted from space-time, nor that space-time can be abstracted from things in space and time, nor that, finally, intuited space is either "Euclidean" or "non-Euclidean." To make explicit this tacit assumption immediately raises the question of its ground, identified on several occasions as the "sensation of motion."[16] Together ground and assumption make up what was called, with Husserl, the "naturalistic conception of the world." And like all other cases of natural scientific-philosophic thinking, the "naturalistic conception" assumes as a matter of course that the spatio-temporal form of the real, objective world encountered in daily life exists and that its investigation also involves an investigation of the spatio-temporal form as a "product" of subscientific-philosophic experiencing and also of just that experiencing. The assumption about the intuition of space and its ostensible ground, however, points back to a still deeper assumption peculiar to the "naturalistic conception:" the assumption, namely, that *the spatiotemporal form of the real, objective world presented in subscientific philosophic experiencing is apperceived in the "naturalistic conception" as the self-same spatio-temporal form presented in scientific-philosphic thinking*.[17] Because of this continuity of positing, one might say, the spatio-temporal form of the real, objective world presented in subscientific-philosophic experiencing, which remains just that no matter how interpreted scientifically and philosophically, is taken to be a product of a "truth" established by scientific-philosophic thinking in the first place. The continuity of positing and apperception is the new specifying assumption that replaces the others.

The difference is not difficult to see. For instance, the specifying assumption of positing according to More, Newton, Clarke, Descartes, Spinoza, even Hobbes, despite all other differences, is that the spatial form of the real, objective world given in scientific-philosophic thinking is posited as not only distinct from that posited in subscientific-philosophic experience, but cannot be posited in that experience which is confined to "copies" of unpresented or even non-presentable "originals," or to phantasms, or even to constructions of the imagination and habit; or in the case of Berkeley and Leibniz, the spatial form posited in scientific-philosophic thinking is an abstraction hence distinct from what is posited in subscientific experience--for Berkeley, an impossible abstraction, for Leibniz a valid one providing subscientific positing is disregarded. For the first specifying assumption, space--a container, an immensity--is more real than things in space posited in subscientific-philosophic experience; in the other specifying assumption, things in space are more real than space--a relational system posited in scientific-philosophic thinking granted an "ideal" status of some sort.

There are, of course, many variations on these specifying assumptions which need not be traced for present purposes. What is important here is that in those specifying assumptions prior to Euler and Kant not only are the "motives" for the positings of the spatial form of the real, objective world not obvious--i.e., they are taken for granted--but for all of them space can be treated apart from time as well as from things in space. Although the continuity of positing and apperception of the specifying assumption of Euler and Kant likewise would seem to be taken for granted, the de facto implication of positing the self-same real, objective world in scientific and subscientific thinking and experience is that space cannot be treated apart from time, nor can space-time be treated apart from things in space. It is this de facto development of physics in Euler and Kant, made possible by the new specifying assumption, *that provides a clue* to a correct phenomenological analysis of space and time and of things in space-time. To be sure, how and to what extent it is a clue can only be shown by phenomenological analysis itself (see below, Chapter Eight). But the clue is only a clue in so far as the positing and apperception of this specifying assumption are seized upon in reflection as the positing of which the naturalness of scientific-philosophy and subscientific thinking and experience consist--one motive for refraining from that positing (above, section 9) and therefore one motive, albeit not a sufficient one, for "abandoning" the specifying assumption.[18] And the same is true in the case of the other specifying assumptions. Consideration of the two lines of thought about space and their specifying assumptions results in a necessary motive for not assuming from the start that space is either "Euclidean" or "non-Euclidean," or, to return to the starting point of discussion of the historical setting, that space can be abstracted from space-time, or even that motion is the indispensible condition for space perception; it is not necessary from the start to accept space therefore as a medium in which all things are, or an immensity or an ideal network of relations; and so forth.[19]

But the abandonment is not tantamount to the annihilation or even the "refutation" of the ontic and epistemic claims belonging to the specifying assumptions allowed for by the continuity of positing and apperception of scientific-philosophic and subscientific-philosophic thinking and experiencing. The historical setting remains, but precisely only as historical setting. As such, the historical setting offers a clue to phenomenological investigation: how does the continuity of positing allow for the legitimacy of ontic and epistemic claims in the first place? How is the self-sameness of the real, objective world assumed in the positing of the natural attitude? What are the "motives" for the continuity of the positing? Once the "origin" of the continuity of positing has been made thematic, how are the claims of the Kantian intuition of space, for example, to be assessed critically? Finally, can the original Cartesian questions about space still be raised in any meaningful way in the light of the thematization of the positing belonging to the natural attitude?

In short, the historical setting as a whole is a clue to phenomenological investigation by motivating a return to the very "things themselves" to which the historical setting must answer and in light of which it must demonstratively show its legitimating credentials. Rehearsal of the historical setting is then part of the will to philosophical theory, thus part of the increasing self-conscious exercise of phenomenological epoché.[20] Expressed this way, our historical reflections have a very special meaning. Just as in the first part of this book the actual course of Husserl's philosophic thought was treated as a transcendental empirical example of the self-conscious exercise of transcendental phenomenological epoché (section 9) and, moreover, as an example that revealed itself as a transcendental empirical problem of the necessary motive for systematically developing the procedures of the various unbuilding reductions (Chapters Two, Three), so in the second part of this book the actual course of philosophic thought about space and time, sketched in the preceding sections, likewise is to be treated as a transcendental empirical example of thought about the "origin of the idea of space (and time)" in Modern philosophy and science. And likewise, in this second part, the transcendental empirical example comprises a transcendental empirical problem in the shape of the Kantian formulation of the intuitions of space and time that is a necessary motive (section 41) for unbuilding the real, objective world. It is also a "radical" motive for it occasions unbuilding reductions to the most fundamental substrata of our subconceptual experience of the spatiotemporal forms of the real, objective world. More particularly, the conflicting specifying assumptions of the Kantian formulation of the intuitions of space and time motivate specific unbuilding reductions to the most primitive dimensions of subconceptual, subscientific experiencing of space and time in primary passivity. Implicitly, it would seem, the historical reflections have necessitated the second step of unbuilding reduction (section 23f.) and raised the question of even further such reductions to establish the constituting of the "origin of the idea of space and time" (see below, sections 53, 54).

We therefore begin the next chapter with a precise formulation of the second step of unbuilding reduction with respect to space and time. This formulation will allow, in further chapters, the detailed exploration of still further steps in unbuilding reduction and lay the foundation for building up the constituting of the spatiotemporal forms of the real, objective world. The task is both a difficult and controversial one, but the very history of the problem of space and time in Modern philosophy and science makes it unavoidable.

CHAPTER FIVE

TRANSCENDENTAL PHENOMENOLOGICAL UNBUILDING TO THE TACTUALLY, VISUALLY AND AUDITORILY PRESENTED IN PRESPACE

42. THE POINT OF DEPARTURE FOR TRANSCENDENTAL PHENOMENOLOGICAL UNBUILDING TO THE TACTUALLY PRESENTED IN PRESPACE

Because it critically incorporates most of the modern scientific and philosophic thought about space perception, the Kantian formulation of the intuition of space will serve, in what follows, as the central point of reference to the historical setting of the problem of space. That setting is important not just because its conflicting specifying assumptions motivate exercise of the phenomenological epoché. It is equally important because the setting requires that a phenomenological account of space establish a basis for distinguishing between kinaesthetic and other sensa-processes while at the same time tracing the ways in which the inherently prespatial acquires the "appearance" of the quasi-objective or objective space of the real world (cf. above, sections 27ff., 37). In its Kantian formulation, the historical setting would seem to require that a phenomenology of space begin with the second step of unbuilding reduction. But this requirement is not so much the demand that a certain reflective procedure be carried out to resolve a certain set of problems as it is that the very innermost possibility of the second step of unbuilding reduction be explored and its limits established. The task at hand is not an easy one for its complexity lies as much in the phenomenologically unfamiliar terrain it deals with as it does in the often torturous language needed to render the terrain habitable. To get our bearings in this terrain and render explicit the phenomenological stituation to be clarified, we need a very precise formulation of our theme.

In consequence, then, of the *second step* of unbuilding reduction, my theme, as *transcendental* on-looker, is the noetic-noematic sense, "Primary passive intending to a centerless, quasi-objective phantom world currently enduring."[1] Additionally I can effect a further unbuilding reduction to a world of phantoms and their appearances intended to and meant and posited only from Here.[2] Subthemes arise for me as meditating phenomenologist by objectivating, for instance, just that quasi-objective sense which the here-centered world of full phantoms has as an object of past, present, and future tactual intendings in primary passivity and, at the same time, refraining from the intending to and positing of any sense the world of full phantoms has as object of intendings of any other sort. To be sure, even though the exercise of phenomenological epoché here includes refraining from positing the here-centered spatio-temporal spread as having a place in the centerless spatio-temporal spread, there is still positing of the here-centered spread and tangibles in it as transcendent to the quasi-objective experiencing of them. What therefore remains as theme is the quasi-objective inherently here-centered transcendent spatio-temporal spreadoutness of purely tangible phantoms one of which is perceived as at the center of the spread reaching out from Here as far as any part belonging to it can be extended or might be extended: the quasi-objective phantom tangible organism. Moreover, only those tangible phantoms actually perceived at a particular time are in contact with the phantom organism; the phantom "hand," for instance, can reach out from Here and touch phantoms that were not perceived until it did so. And it was on the occasion of this further step of unbuilding reduction that the question was raised about whether any further unbuilding is possible.[3] Does the own-transcendently-inherent here-centered spatio-temporal spreadoutness allow of further unbuilding?

The clue to further unbuilding, and thus the point of departure for further analysis, may be found in a uniquely anomalous characteristic of the constituting of the organism in primary passivity and which is disclosed in its originaliter presentiveness in consequence of the second step of unbuilding reduction. The basis for clarifying the anomaly of the constituting of the phantom organism has been referred to earlier (section 32) in connection with sketching the phenomenological situation in which systems of quasi-objective perspectival appearances are presented in correlation with free modifications of kinaesthetic flow-sequences. But what is of thematic interest now is not so much the peculiarity of the interchange of presentational modes of Here and There as it is what Oskar Becker called, in this connection, the "produced unity"

[*Leistungseinheit*] of the quasi-objective organism presented as an intricately structured "phenomenon for consciousness:"

> As a produced unity, the organism is given from 'within' as a unique phenomenon, not perchance as one physical thing among others. As a material thing it is constituted only on the basis of so-called 'dual sensations' which arise for the sense of touch when one member of the organism comes into contact with another place on the organism. For instance, one can either 'feel' or 'sense' one's hand 'from within' as a moving, sensing member, or else touch it 'from without' with the other hand. The tactual sensations awakened, simultaneously with the touching, in the touched hand itself provide the underlying basis for identifying the member perceived simultaneously 'from within' and 'from without'.[4]

The anomalous sensory duality in produced unity described by Becker--and in the context of unbuilding reduction to the quasi-objective phantom world--makes it clear that not only is the "field of tactual sensations immediately included with the original produced unity of the organism," but also that just that immediate inclusion and sensory duality differentiates it from visual, auditory and other fields of "sensation."[5] Only the "sense of touch" exhibits the sensory duality. The eye, of course, is a member of the organism, but only in so far as it is an object of touch: "We do not look 'at' or even 'in' the eye but instead we look at the visual Data 'in the visual field' 'with the eye'."[6] The "with" of "with the eye" really refers to the mobility of the "glance" which can be varied at will in correlation with oculor kinaesthesias so that the "anaesthetic" eye (Becker) pertains only indirectly to the organism as an organ. In later sections of this chapter and of the next, we shall clarify this phenomenological situation further. The question at issue here is about unbuilding to the constitutive underpinings of the sensory duality in produced unity peculiar to quasi-objective tactual experience and therefore of the differentiation of tactual, visual, auditory and other fields in quasi-objective here-centered space.

The answer to the question of further unbuilding lies as much in the produced unity as it does in the differentiation of tactual and visual fields: in here-centered space locomotor kinaesthesias and their correlations are "set at zero;" but the produced unity of tactual "within" and without" still entails correlations with other kinaesthesias on which the quasi-objective tactual perceiving of motion of hands, arms, eyes, head, torso, is dependent. If, now, all kinaesthetic flow-sequences and correlations are "set at zero" in the *broad sense* of "setting kinaesthesia at zero" (above, section 28), then I can proceed to objectivate a new phenomenological residuum: a primary passive tactual sensing of a tactual sesum-field of just qualitative tangible change and non-change spread out in "space" and which is not sensed correlatively with changes and non-changes in kinaesthetic flow-sequences, thus not exhibiting the sensory dualtiy in the produced untiy of "within" and "without." Accordingly, as on-looker I discriminate tactual sensing and positing of a prespatial tactual sensum-field presented as having areas of more or less intense touch, separated by non-touch, meant as more or less near one another, perchance changing or non-changing with respect to one another and with respect to the co-presented tactual sensum-field. And although in this further refraining the tactual sensum-field is not meant and posited as located in the quasi-objective here-centered spreadoutness (hence not posited as in and on the quasi-objective phantom organism), nonetheless it is still posited in its own inherent spatio-temporal spreadoutness as transcendent to the primary passive sensing of it. The strangeness of this objectivated residuum is compounded by the fact that, as one of its earliest investigators, Thomas Reid, observed, such tangible area-changes have not been "honored with a name in any language."[7] Perhaps the term, "area," is the best that can be used here because it does not necessarily connote "sharp edges" or "points" even in Hume's sense. But the term, "area," does suggest the peculiar spreadoutness of tangible saliences.

43. PRESPATIAL TACTUAL PRESENTEDNESS

More specifically and most immediately, although not necessarily most originarily, I distinguish primary passive sensing of certain changes in one sequence of touch-sensa presented simultaneously with certain other changes in another sequence of touch-sensa--an "appersensing," to coin a term, of simultaneously presented touch-sensa. Such a primitive identifying synthesis may serve as basis for pairing of the sequences of touch-sensa as, e.g., perfectly alike in intensity so that changes in one sequence would "inhibit" further changes in the other--a limiting case of assimilative and differentiative synthesis.[8] But this is only one, and perhaps the most obvious, presentational mode. Altogether, at least three principal modes of simultaneous presentedness constituted in primary passive assimilative and differentiative synthesis can be objectivated here:

1. Simultaneous presentedness of changes in one sensum-sequence with changes in another such that they are presented as "next to one another," "close-by one another:" "contiguity" of minimally charged intensities (at a much higher level of oriented constitution, running one's hand over a rough surface).
2. Simultaneous presentedness of one sequence of tactual sensa paired as of greater intensity with less intensity and pressure in another sequence of tactual sensa: "contact" (at a much higher level of oriented constitution, the palm of the hand resting on a surface; there is variation in pressure such that contact disappears when the pressure of the hand is decreased).
3. Simultaneous presentedness of changes in the sequences of tactual sensa paired as perfectly alike in pressure-intensity: "resistance" (at a higher level of oriented constitution, a surface rigidity to increasing pressure of the hand).

Between these three cases there falls a whole spectrum of distinguishable change in saliencies within the tactual sensum-field. The difficulties of describing these changes are immense; tactual experience after all reveals little of the object touched or in touch and is therefore notoriously ambiguous. It is very difficult, for instance, to "read" the structure of the object off or across the structure of touch. Moreover, it is customary to consider the first case of presentedness as presentedness of tactual quale, such as rough and smooth, hard and soft; the second case of presentedness is that of a surface, while the third that of impenetrability. Although these distinctions may be kept in mind in the objectivation of the tactual sensum-field and changes in it, the anomalous nature of touch makes it difficult to follow the usual procedure of referring, even metaphorically, findings at one level to another and higher level. Before we go any further, a few words about the difficulties involved are in order.

Relaxing the epoché for a moment, and considering the modes of simultaneous presentedness of sequences of tactual sensa with respect to higher levels of oriented constituting, it can be seen that because no quasi-objective or objective organism is co-intended to and posited at this lower level, the tactual sensa do not display the sense of being extended into a tactual environment, i.e., into that space where, at a higher level, for instance, the phantom organism is located. Thus we speak here of prespace rather than quasi-objective here-centered or centerless space. Nor are tactual sensa, in contrast, presented as located in and on the quasi-objective organism. They are not presented, therefore, as extending into tactual quasi-objective space with that part of the quasi-objective phantom organism in and on which they are located at higher levels.[9] The point of relaxing the epoché and making such observations is that it is then possible to effect the epoché again and objectivate a "closedness" that is peculiar to the prespatial tactual spread: no tactual saliency is sensed or appersensed as entering or leaving the tactual spread,[10] nor is the tactual prespatial spread itself co-presented as bounded or "edged." One, though not the only, reason for this is that, as mentioned before, touch informs us but little of the object touched; to take an example from Katz, if we touch a piece of carved wood we distinguish the grain of the wood, its "natural structure," but not the structure imposed upon it by the wood carver. There is, accordingly, some justice in saying that touch reveals little of the nature of what is touched as an object carved, but rather more of the material characteristics of what is touched--the hardness of softness, roughness or smoothness of the surface, flatness or hollowedness of shape. Thus the procedure of referring to higher levels of constituting is often of limited avail because even at those higher levels touch, "by itself," tells us little of what it is we tactually encounter in the quasi-objective or objective world. As a consequence, we have to proceed as best we can in objectivating this lowest substratum of oriented constituting disclosed by exercising again phenomenological epoché. The language used to express this objectivation may seem odd but, at least in part, the reason is in the nature of touch itself.

Perhaps the most obvious example of touch with which we may begin is that of touching a surface, of tactual intending to a surface. But by touching a surface I can further distinguish, to begin with, a co-intending to intervals in the sensings of touch-areas or patches. Let us take Katz's example of what, at a higher level of oriented constituting, consists of touching the bristles of a brush. That is to say, presented simultaneously a plurality of force-fully prominent areas of touch are intended to and posited as having the sense of being separated by "non-touch" (at a higher level of constituting, between the bristles). "Non-touch," tactual prespace devoid of touchness (but, as Katz notes, not a nothing), is presented as contiguous to touch-areas--the first mode of simultaneous presentedness to be objectivated. Indeed, it is "non-touch" rather than touch that serves as the phenomenological basis for constituting the assimilitative synthesis, "contiguity" for the tactual prespace devoid of touch areas is intended to as the "background" of the touch-areas; the most primitively tactual presentation of touching a surface, any surface, is a "discontinuous" one.

In addition, the prespatial tactual spread is co-intended to as itself intervalled in yet another mode: successively, touch-areas are presented, then touch-areas are not presented. At a higher level of constituting, the bristles of the brush are drawn over the tips of the fingers; but here all that can be objectivated is a succession of touch points; we cannot even refer to what looks to the eye like a continuous, smooth surface. Indeed, successive touch-areas, as Katz noted and it is not difficult to confirm in this example, are not even presented as hard or soft, rough or smooth surfaces. Tactual "qualities" disappear altogether in this case. Succession, in fact, would seem to prove to be a more original mode of pesentation because the intervals are presented in contrast as "non-touch simpliciter," that is, there is no objectivation of intending to tactual prespace as "background" as in the case of simultaneous presentation of intervalled touch-areas. Thus, phenomenologically, we have to say that what is objectivated in this case is a limit of oriented constituting; as it were, each "figure," each touch-area, is its own "ground." In still other words, *there is no objectivated or objectivatable basis for assimilative synthesis of likeness or even perfect unlikeness in contradistinction to the mode of simultaneous presentation of touch-areas with co-intended-to intervals.*

Perhaps another way of expressing the phenomenological situation of tactual prespace is by saying that simultaneous presentation of bristled touch-areas is a case of presentation of the tactual prespatial sensum-spread as "qualitied;" in contrast, successive presentation is presentation of the spread as partially or wholly "unqualitied." Successive "unqualitiedness" of the tactual sensum-spread as a whole is, accordingly, the most original presentedness of the spread. Indeed, it seems ambiguous now to speak of either succession or simultaneity in this case; it is rather more like a null-case of presentation, neither intended to and posited as having or not having contiguous "parts," nor as having or even not having "parts" in contact. This would also count for the limiting case of "resistance" where there can be objectivated no noetic-noematic substratum for assimilative syntheses in mutual transfer of sense;[11] rather than a basis for transfer of sense there is only a mute "impenetrability" in the sense that while contact pressure on ("bristled") surface increases, there is no presented change in the nature of the surface--in the example of the grain of the piece of wood, the greater the pressure on the grain there is still the unalterable "constancy" of the grain-surface. Here we have a clue for further clarification.

Simultaneously presented touch-areas, for instance, are sensed as salient with respect to one another in the sensing of alteration in the degree of intensity of touch. As it were, they are presented with a "coefficient" of relative intensity. To that extent, relative alterations in touch-intensities, presented as contiguous or as contiguously intervalled, are therefore co-presented in "depth." Borrowing terms ordinarily used for describing the visual field, we may say that the limit of such tactual presentation can also be objectivated in the case of "covering" or "un-covering" peculiar to the tactual prespatial sensum-spread. By "covering" here I mean the case of successive presentedness of touch-areas where one tactual sequence of changes is presented, then another is presented; the first is "covered," that is, replaced, by the second which, as it were, "gets in the way" of the first which is now "out of touch."[12] Perhaps the situation may be made plainer by relaxing the epoché again to objectivate, at the next higher level of oriented constituting, quasi-objective tactual perceivings of "smooth" wood grain on the basis of sensing the first sequence of touch-areas, then of "rough" wood grain on the basis of sensing the second sequence by touching the area with greater pressure; moreover, this would be the case where, on the basis of sensing the first and then the second sequence, "smooth" and "rough" are perceived as "at the same place" on the basis of appersensing prominent grain differences of the wood in sequential changes (touching an area first with greater, then with lesser, pressure). But at the lower level of constituting under consideration, I find no basis for the constituting of the "same place" at which "smooth" and then "rough" coincide; there is only one after the other, only replacement of one sequence by another, only "covering" and "uncovering" in succession, thus not even a basis for objectivating the one as "smooth," the other as "rough."

Successive "covering" and "uncovering" provides, in turn, a clue to the most originary constituting of the prespatial tactual sensum-spread undergoing qualitative change.

44. PRESPATIAL TACTUAL SENSUM-SPREADOUTNESS

In the simultaneous presentational modes of "contiguity," "contact," and "resistance" of touch-areas the sensum-spread itself is co-intended to and posited as partially or wholly filled--"wholly filled" in the sense of being presented as "background" for intervalled touch-areas paired in primary passivity that, in turn, serves as the basis for assimilative syntheses of likeness or unlikeness. In contrast, in successive "covering" and "uncovering," successive "replacement," nothing stands out tactually to be paired for no interval is presented as "background" in successive "replacement. "Simultaneously presented intervalled or non-intervalled touch-areas (as in the cases of "contiguity," "contact," "resistance") are syntheses based upon primary passive pairing of salient touch-areas differing in their co-efficients of qualitied intensity (e.g., now rough or smooth, soft or hard, sticky or damp). The substratum co-intended to that serves as basis for pairing is best referred to as an "empty" tactual sensum-spread, a spread presented as devoid of "touch areas" to use Katz's expression. Because, however, the tactual spread is not a nothing, it is perhaps best to refer to it as an "empty" closedness of spread in contrast to a "filled" closedness in its role as constitutive basis for assimilative syntheses. To express the situation with less precision, we may say that the interval is transformed into a "background" of synthetically paired touch-areas inseparable from the interval (thus sensed as standing out to be paired in the first place).

What I mean is that a constitutive substratum is now objectivated which would seem to allow of no further phenomenological discrimination. The sensed and intended to sensum-spread, the sensed and appersensed touch-areas of the spread, do not "point beyond themselves;" the palm of the hand in touch with the bristles of the brush is "not yet" a palm, the bristles "not yet" bristles let alone "bristles of a brush." Rather all that is presented in succession is a slightly humid surface salient with pointilistic touch changes of differing intensities and pressures to the extent that there are not even any "qualitied" characteristics of the changes. As it were, the touch-areas are presented as self-confining. As such, no noematic sense of "paired" or "to be paired" is presented so that, strictly speaking, the prespatial tactual sensum-spread, qualitied or unqualitied by touch-areas, offers no horizons or perspectives. Yet even in the case of those touch-areas constituted in assimilative syntheses on the basis of pairing, they are still not presented as entering or as leaving the co-presented sensum-spread; they are discontinuously present or not present, replaced or not replaced at all. Thus with respect to such noematic senses as "horizonless," "unpaired," "not leaving or entering the spread,"[13] it is necessary to speak of the sensum-spread as closed.[14] But "closed" is an entirely peculiar sense in that it does not connote a boundary; the spread is closed but boundaryless. For, as will be seen later, presentation of the spread as boundaried presupposes co-presentation of the mobility of touch-areas in which case the spread itself acquires the appearance of a "ground" for moving sensa[15] (i.e., intending to sensa as identical at different places in the spread, as being located elsewhere than here). But at the level of constituting where no basis for pairing is objectivated or objectivatable, a "background" of even the crudest sort cannot be discriminated.

To be sure, in a rather different but still relevant connection, Katz suggests that there is some sense in speaking of a "ground" where the tactual sensum-spread is presented as intervalled;[16] but even here it does not seem appropriate to speak of a background for changes in any strict sense, at least of an "immobile background" for actual and possible identical moving or resting touch-areas. Later it will be seen that these touch-areas must be correlated with certain kinaesthetic flow-sequences for there to be a basis for constituting of the spread as background[17] (in one of our examples, of the palm of the hand at rest while the bristles of the brush move over it). One reason for this, as Katz points out, is that constituting of the spread as background presupposes the *reversibility* of "figure-ground" relationships.[18] But even granting for the moment the legitimacy of speaking at all of "figure-ground" in connection with the domain of tactual constitution, all that I objectivate at this lowest level is successive replacement *and not reversibility*. Something analogous to "figure-ground" relationships will be found in the quasi-objective tactual field in which the phantom organism is located. Only in this field will a basis be found, correlatively, for objectivating a "background" for identical tactual saliencies: the "closedness" of the prespatial tactual sensum-spread, however, will be proven to found the noematic sense, "periphery of the phantom organism," but not the sense, "background" of the quasi-objective phantom tactual field.[19]

Successively enduring touch-areas of the tactual sensum-spread presented as one replacing, "covering," the other, now comprise my phenomenological theme. As it were, difference and similarity are "lost;" indeed, it would almost seem that the very successiveness of presentation has disappeared so that nothing is presented as standing out together in the tactual sensing. Sheer succession of non-touch, "momentary" non-touch is presented just as such in the sensing of unqualified "spreadoutness all around."[20] Sheer succession of the tactual prespatial sensum-spreadoutness is most originally presented in replacing of non-touch by out-of-touch. The "space-form" is, we might say, a "spread-in-ness" in contrast to "spreadoutness." The archontic objectivated characterization of the tactual prespatial sensum-spread is the spread presented as spread in upon itself, as "closed." At the lowest objectivatable level of constituting, tactual prespace is therefore spread out in upon itself ("depth").

By way of summary three constitutive substrata of tactual prespace may be singled out for mention:

1. The substratum of non-touch, out-of-touch. So far as I can see, this substratum comprises the most original presentedness of tactual prespace. (At the next higher level of oriented constitution, intending to and positing tactual prespace characterized in this fashion "motivates" the "reintroduction" of corresponding kinaesthetic flow-sequences.)

2. The substratum of successive presentation of touch, out-of-touch, i.e., a founded presentational mode consisting of "covering" and "uncovering," of replacement. (At the next higher level of oriented constitution this substratum acquires the "appearance" of the near-sphere peculiar to the constituting of the quasi-objective phantom organism.)
3. The substratum of simultaneous and successive presentedness of a qualitied sensum-spread characterized by "contiguity," "contact," and "resistance." In the case of this further and founded substratum assimilative syntheses are discriminated based on the mutual transfers of sense among touch-areas. (At the next higher level of oriented constitution this substratum acquires the "appearance" of tangible phantoms undergoing change and motion in the quasi-objective phantom spatial field, thus establishing, for example, a basis for the distinction between the grainy surface of the piece of wood and the piece of wood as carved.)

All three objectivational substrata are presented as "three-dimensional," depth being intended to in each case as a facet of the "closedness" of the prespatial tactual sensum-spread. To be sure, in making this statement I find myself at odds with Becker for whom the tactual--as well as other--prespatial spread is not only "inhomogeneous," i.e., here-centered and hence located in quasi-objective phantom space, but also "two-dimensional." Later in Chapter Seven, we shall return to this issue in critically revising Becker's "transcendental deduction" of the spatial form of the real, objective world. At the moment, the next task is to objectivate the visual prespatial sensum-spread.

45. TRANSCENDENTAL PHENOMENOLOGICAL UNBUILDING TO THE VISUALLY PRESENTED IN PRESPACE

Examination of the oriented constitution of the prespatial visual sensum-field requires steps of unbuilding that are counterparts to those carried out in the case of the examination of the prespatial tactual sensum-field. Thus I begin with that oriented constitution where the tangible is presented as visible as well as audible, then discriminate the possibility that the phantom organism changes its quasi-objective spatial locus and uncovering thereby the here-centered quasi-objective world of phantoms. I now proceed to discriminate only the noetic-noematic sense which the here-centered quasi-objective world of full phantoms is intended to and posited as having as quasi-objective things of past, present, or future visual perceivings, while at the same time refraining from intending to and positing those noetic-noematic senses belonging to it as being quasi-objective things of perceivings of all other sorts. My phenomenological theme is now the here-centered quasi-objective world of solely visible phantoms presented in their own inherent visible spreadoutness. To be sure, this segment of a quasi-objective phantom world is something that, at a higher level of oriented constitution, is apprehended as the visible appearance of a centerless quasi-objective world from somewhere, namely from wherever the phantom sight-organs happen to be located at a certain time.

This observation is worth emphasis here because, within the scope of the epoché now effected, it may be asked if anything that is the object of just and only quasi-objective visual perceiving is intended to and posited as the percipient organism, let alone as organs of sight.[21] For what in fact I find are visual intendings to something meant as visible and only as visible at the "center" of the visible here-centered spread, yet which is not presented with the sense of being a percipient (visual) organism--a visible something rooted here, the rest and motion of the visible parts of which are correlated immediately with (oculor-motoric) kinaesthetic processes among others.[22] The visible something is anamolous in its own peculiar way so that to speak of it as "organism" would now be misleading. This need not imply, however, that oriented constitution of the organism is therefore fundamentally tactual, its visibleness but an analogue of the tactual.[23] In Chapter Seven I shall consider reasons for developing an idea of the organism which is quite different and which, I believe, does greater justice to its constitution as an anomaly. For the moment, however, it is necessary to turn to several examples the analysis of which will provide clues to the nature of visual quasi-objective space and, finally, visual prespace.

The first example is one mentioned by Ernst Mach in discussing the peculiarities of the constituting of visual space: once, when walking at night in an unfamiliar district, he was continuously in fear of walking up against a large black object. The object turned out to be a large hill several miles distant.[24] The other related example is one suggested by H.H. Price in a similar context:[25] In speaking of the perspectival visual appearances of a distant and irregular hillside sloping upward at a gentle angle, Price observes that "stereoscopic" vision is possible only within a relatively small and proximate range. Outside this proximity there is what he calls a "collapse" of planes, and the visual appearances of the hillside undergo a variety of "distortions." For instance, the hillside appears as vertically flat "like a scene painted on cardboard." As I shall try to show, however, the word, "flat," is not quite the appropriate term to describe this phenomenal situation. Rather what occurs is that the "convexity" of what is seen in its proximity becomes increasingly bordered and tends toward "concavity." As it were, the visual appearances become more and more hollow.

In terms of here-centered quasi-objective space, the examples may be reformulated in the following way. With locomotor kinaesthesias "set at zero," the visible quasi-objective phantom organism is presented as "fixed" at one spot; quasi-objective spatial visual things are not presented as far although seeming to appear as near under certain circumstances (such as the unfamiliarity of the terrain, the darkness of the night, and so forth); nor at this level of oriented constitution is the modification which the quasi-objective visual appearances of the hillside undergo, a change from a quasi-objective phantom visual appearance of something far into a quasi-objective visual appearance of something near. So far as Mach's example is concerned, all that can now be objectivated is the qualitative change of something visual as "looming larger," i.e., it becomes increasingly convex to changes in the direction of the glance; the quasi-objective visual appearances "no longer" undergo what, at a next higher level of constitution, would be perceived as change from visual appearances of something far into something near as, e.g., when in walking the hillside is approached. Instead, the quasi-objective phantom hillside is presented as a concatenation of collapsing planes and distorted visual appearances, of qualitatively "flattening" and "hollowing" visual fields.[26]

As a result, at best all that can be said is that the quasi-objective visual field in here-centered, quasi-objective space tends to become ellipsoid rather than distorted; one part of the field qualitatively "grows," becomes "elongated" correlative to changes in posture and glance; the visual field seems always to spread beyond the area within which a change (such as the "distorted vertical flatness" of the hillside) is salient; and the spread beyond that area is presented as ellipsoidal,[27] i.e., where visual concavity and convexity coincide in simultaneous givenness (the limiting case). Just this phenomenological situation is the point of departure for further unbuilding to uncover a visual sensum-field in which identical units of visual saliencies at rest or in motion are sensed visually correlative to changes and non-changes, for instance, in oculor kinaesthetic processes.

Moreover, I can go a step further and not only "set at zero" oculor but all kinaesthetic processes such as was done in the case of unbuilding to tactual prespace (the broad sense of "setting kinaesthesias at zero"). With this further step I objectivate, as phenomenological on-looker, a prespatial visual sensum-field of purely qualitative change and non-change spread out in prespace.[28]

46. PRESPATIAL VISUAL PRESENTEDNESS

The phenomenological theme is now visual prespatial spreadoutness, the de facto edges of which are "no longer"[29] presented and posited as having the noematic sense acquired in and through actual and possible visual sensings of something beyond them that could be brought into the visual sensum-field. In other words, areas "at the edge" of the spread do not admit of being made central; the possibility, furthermore, is disregarded that, when such areas are made central new ones become visible at the one edge while others at the opposite edge are no longer visible. Rather all that I now find is a visual sensing, going on in primary passivity, of the "closed" uncentral, unedged visual spreadoutness undergoing mere qualitative change. Moreover, I find, as on-looker, no basis for identifying visual saliencies in the visual spread as remaining at rest or moving across the spread; there is only continuous sensing of qualitative change such as from very dark to very light, from less dark to lighter here and then there. Thus visual sensing of a heap of color-patches salient in the prespatial visual spread is not ipso facto a visual meaning of them in isolation from one another; they are sensed together as similar or dissimilar. Yet, although areas of color are not presented in isolation from one another, they are still presented as "outside" one another so that the noematic sense, e.g., dark red, actually seen, is transferred to the noematic sense, light red, visually seen elsewhere in the spread. (Strictly speaking, one ought, perhaps to say "visually sensed" and "visually appersensed" in the spread.)

Additionally it is also possible to find cases of partial cancellation of the transferred sense as well as, simultaneously, a partial confirmation that serves as basis for assimilative and differentiative synthesis: the two areas or patches acquire the noematic sense, "qualitatively similar but also different." (See above, section 11f.) The mutually transferred sense, light and dark, cancel each other out in the overlaying of noematic sense; but the noematic senses, red, found an assimilative synthesis of similarity. There is also to be found the transfer of noematic sense, "part of the visual spread," which likewise founds an assimilative synthesis of similarity. In other words, the colored areas or patches are not only constituted in syntheses of identity and numerical distinctness ("individuation" in the prespatial visual sensum-field), but they can also be constituted in assimilative syntheses as similar and different. Presented as "outside" one another, they are sensed visually as thick and porous, as dense and solid.

Yet it is important to note that to say that assimilative and differentiative syntheses of similarity and dissimilarity are based on identifying and distinguishing syntheses only signifies here that the areas or patches of color salient in the prespatial visual spread are presented originaliter as "outside" one another and individuated. It does *not* signify that they are identified as the "same" throughout qualitative change (scl. constituted as identicals undergoing qualitative change). Simultaneously co-presented as "outside" one another and appersensed visually as "parts" of the visual spreadoutness, the color patches cannot, in any strict sense, "disappear from the spread."[30] This feature of visual prespace suggests a clue for more precise characterization of changes of the visually presented in prespace.

An area of color is presented as salient in the prespatial visual field. On the one hand, it is easy to imagine that the green area, let us say, gradually loses its saturation until it merges with the rest of the field; or a change in the brightness of the area may be such that it becomes perfectly like the brightness of the field and merges with the field in that respect. On the other hand, neither the saturation nor the brightness changes, but the area of the color diminishes until it becomes indistinguishable from the field. The examples of the hillside, in fact, illustrate two sorts of disappearing, although not in any strict sense: 1) the case where, e.g., a segment of the field fades to a brightness similar to that of the rest of the field--the area or patch is presented as solid, dense, a bulge on the field, then gradually "loses" its solidity and density, its bulginess (as the planes collapse); 2) the case where the periphery of an area of color "shrinks inward," such that the spread is presented as becoming more and more "superficial," unsolid and insubstantial in tending gradually (in the example of the hillside seen in increasingly great proximity) toward becoming "hollowed out." The term "fading" will be used to name the first case, the term "diminishing" to name the second case. What is signified in either case, each in its own way, is that the colored areas and segments lose the difference constituted in the assimilative syntheses of similarity and difference, yet retain their identity and distinctness with respect to the co-intended-to spreadoutness surrounding them. Indeed, even the phrase, "with respect to," is somewhat misleading for the colored areas are presented as mutually incomparable--the one is convex, the other as concave--without respect to one another. Even more than in the investigation of the tactually presented in prespace, language expressing the objectivation of prespatial visual phenomena gets stretched to its limits.

Yet having taken this step in unbuilding discrimination, it is worth while making the attempt to further describe prespatial visual presentedness. For example, various modes of fading-presentation can be distinguished, such as mutual "contamination" of color patches--for example, where, at another level of oriented constitution, one would say that the sky "allows itself" to be "contaminated" by color-structures close to it but without imposing on those structures any configuration of its own. Thus in Merleau-Ponty's words: "I do not propose to bestow the term 'visual thing' upon a phenomenon which, like areas of colour, presents no maximum visibility through the various experiences which I have of it, or which, like the sky, remote and thin on the horizon, unlocalized and diffuse at the zenith, allows itself to be contaminated by the structures closest to it without setting over against them any configuration of its own"--in other words, we deal with "lived perspectives" rather than geometrical ones, thus the term "contamination" seems appropriate[31] as well for Mach's example of the hill at night looming into the sky at a distance.

Nor is it of necessity the case that when extensions of visually sensed color patches "shrink inward," quality also fades; its self-identity is still meant through the change, and it is still meant and posited as spread out in the field, thus as localizable and qualifiable along the line of vision.[32] It is rather that the color patch acquires a density and solidity which, as it were, in its salientness "splinters" the visual sensing. Still more particularly, visually sensing a plurality of qualitatively distinct saliencies as "outside one another" is, at the same time, sensing (or appersensing) intervals between them, but without any motivation for overlaying of noematic sense; they are simply presented as standing out in depth.[33] An illustration of this is the phenomenon of "collapsed" planes mentioned earlier.

Gradual changes are sensed in the field presented as more and more "hollow" and, correlatively, as less and less solid, as less dense and massive: the "flatness" referred to by Price is the continual change toward "hollowness" as though the convex were turning inside out, becoming concave. To be sure, visual saliencies are continuously meant as self-identical in the sensing of them, even though there is no basis for the overlaying of noematic sense and thus no basis for the saliencies being constituted in assimilative syntheses. Rather on the continuum of changes approaching complete "chromatic hollowness" of the field, lesser and lesser "contamination" is correlatively presented: at the same time, there is fading of, e.g., brightness and saturation and diminishing of the colored patch. The "verticality" of the salient change objectivated in visual quasi-objective space in conjunction with the "collapse" of planes here is presented as bulgy, prespatial looming correlative to less "contamination" of colors. The less "contamination" signifies that fading is reaching a limit: concavity. As the area of change fades, there is a correlative decrease in saliency of surrounding field presented as increasingly "hollowed out." Thus in the example suggested by Mach, the color patch still is presented as standing out in depth despite the fading owing to co-presented "hollowing" of the field which, beyond the area of change, tends toward "flatness." The hill looms up as though in his way. Another way of expressing the situation is to say that the field spread out beyond the change-area "shrinks inward." Visual nightmares though these phenomena may seem, they nonetheless point to the most primitive of visual prespatial presentedness: "covering" and "uncovering."

47. "COVERING" AND "UNCOVERING" AS THE MOST PRIMITIVE PRESENTEDNESS OF VISUAL PRESPACE

In terms of the oriented constitution of visual prespace in the narrow sense of the term, "prespace," Mach's and Price's examples consist of the transformation of visual convexity into concavity. In turn, that transformation may be further clarified under the heading of "covering" and "uncovering." And it is in this connection that Oskar Becker first raised the question of the "dimensionality" of the prespatial visual field.[34] The reason is not difficult to find: the phenomena of fading and diminishing, of the change of convexity into concavity, include their presentation "in depth." And according to Becker, if the prespatial visual field is ostensibly objectivated as "three-dimensional," the phenomena of "covering" is also (at least implicitly) objectivated at the same time. Yet, as Becker himself notes, the phenomenon of "covering" is possible only in connection with a correlative "uncovering" that, in turn, occurs only when specifically oculor kinaesthetic processes stand in functional correlation with visual sensa that eventually found quasi-objective (or objective) perceivings of visual appearances of things.[35] For example, kinaesthetic processes and sensings of visual sensa are so correlated that two visually sensed patches of color, one presented as covering the other, are now visually sensed as moving away from each other. How, then, can an account be given for such a situation when all moving about in "space" has fallen to the epoché in unbuilding, hence when all correlations between kinaesthetic processes and sensing of visual sensa have been "set at zero"? Are the phenomena of "covering" and "uncovering" to be found and objectivated at all in visual prespace in the narrow sense of the term, "prespace"?

At issue for Becker in his discussion of visual prespace is the question of the constituting of prespatial visual data standing out at a distance, that is, in depth. Essential for the constituting of depth, he says, is at least potential visual intending beyond the de facto edges of the spread meant as indefinitely spread out, and just this sort of intending and meaning constitutively entails co-intending to the percipient organism in motion functionally dependent on actualization of kinaesthesias and the "hierarchy of appearances" of which is correlated with the concatenation of appearances of something visual in the field. This would be the case where, for instance, a visual appearance of something far is transformed into a visual appearance of it as near so that by the organism moving around the "covered" something it can be seen as distinct from other things and thus "uncovered" by actualizing still other series of visual appearances. At still higher levels of constitution a somewhat different sort of uncovering is achieved when, for example, a seeable place is co-presented as a tangible place, where, by reaching around one thing that covers another, that other thing is both touched and seen. This makes possible the originary presentation of the covered thing as uncovered tactually but with the sense, "the same as what is covered and then uncovered visually." Yet what Becker seeks to describe hardly pertains to visual prespace where there is "not yet" constituting of an organism which moves around and reaches.[36]

To respond to Becker a counter example is perhaps in order. Consider the seeing of a round, earthenware jar;[37] if I look at it close up, my line of vision begins to break up and, as it were, glide around the sides of the object as though it were taking possession of the rotundity of the jar. Reducing the theme now to visual prespace, what I objectivate as visually sensed is a rotating patch or colored segment of the visual spread successively appearing and diminishing;[38] a "curved" segment of color-patch is replaced by another--for instance, there is sensing of the qualitative change of "dark" becoming "light," of "light" becoming "dark." In other words, what I objectivate are sensings of bulky, filled voluminosity of a segment of the visual spread replaced by sensings of lesser bulkiness, lesser voluminosity and, correlatively, of more surface, greater superficiality: a spectrum of qualitative change from "bodiedness" to disembodiedness." Correlatively, there is appersensing of less compactness and becoming merely chromatic without mass.[39]

Of special interest is the objectivated correlation of voluminosity and superficiality for at the same time that the former is succeeded by the salient color-change of the latter in the spread, the line of vision peculiar to visual sensing of bulky, filled, compact voluminosity is "lost;" in the further course of successive visual sensings and appersensings, however, it is regained. Thus the line of vision proper to prespatial sensing is that line of vision precisely only in the sensing of *optimal* voluminosity: the visually sensed, it may be said, is the "center" of the line of vision. But when, in further course of sensing, voluminosity becomes superficiality the "center" is "lost" so that it may be said that the purely visual "here" is the "center" rather than the visual "there." But that signifies that the visually sensed and appersensed object, as it were, is presented as "retracted into here" and, paradoxically, yet remains "there;" it remains "at a distance," yet is not "at a distance" in its superficial hollowness, its unsolidity. In other words, the convexity of the compact, volume-filled segment of the spread has been replaced by concavity; correlatively, the density of spatial "presence" has been replaced by spatial "distance" as the line of vision retracts upon itself.[40] There is retracting-expanding visual "centricity" but certainly no basis for distinguishing between "near" and "far" appearances in and through which a visual something is exhibited.

At the constitutional level of quasi-objective or even objective space, the first case, convexity, proves to be seeing of the jar nearby as it approaches closer and closer. In the other case, where the line of vision retracts, shrinks back upon itself, concavity, there is seeing of the jar as at a distance, receding from Here. But at the substratum of visual prespace, the hollowed superficiality of the retracting line of vision "centering" back upon itself is already the sensing and appersensing of a "curved" prespatial segment of the spead (convex or concave) in which the segments of the spread are presented originaliter as "at a distance." The retraction or expansion and shift of the center of the line of vision correlative to the "shrinking inward" or the "hollowing outward" of the visual spread at the "edges" of change are precisely visual intendings to depth of the prespatial spread so that in this respect it can be said that the prespatial visual sensum-field is characterized by primitive "three-dimensionality" peculiar to its spreadoutness.[41]

Indeed, more generally it would seem that the various qualitative changes in the prespatial visual spread, on a continuum of change between convexity and concavity, such as "chromatic contamination," fading," "flattening," "collapse of planes," are all presented originaliter in depth. Perhaps the phenomenological situation would be more accurately expressed by saying that they are all presentational modes of changes in depth of the prespatial visual spread at a level of constituting more fundamental than that at which assimilating syntheses are constituted. On the other hand, as will be seen in Chapter Seven, at this "pre-assimilative" substratum of oriented constitution, precisely the transformation of convexity into concavity "motivates" the constituting of assimilative syntheses based on the overlaying of noematic senses so that oculor, cephalic, locomotor and other kinaesthetic flow-sequences are actualized in such a manner that, e.g., visual things at a distance are presented as "solid" in and through quasi-objective and objective far-appearances.

48. THE VISUAL PHENOMENON OF "RESISTANCE." CONCLUDING REMARKS

The examples of qualitative change in visual prespace mentioned in the previous sections involve still another phenomenon: "resistance." By "resistance" in visual prespace I mean one or another mode of opaqueness co-intended to in the visual sensing of compact, volume-filled visual changes in the visual spread. Correlative to sensing of opaqueness there is appersensing of less and less "contamination." In contrast, there is "fading" as tending toward "perfect contamination" of colors, hence of penetrability and lack of "resistance"--the change toward concavity, of fusion, translucence and mere chromaticism. In these terms, "contamination" of colors is then a primitive mode of "covering," and translucence an equally primitive mode of "uncovering." But they are presented in conjunction with one another so that, at the next higher level of oriented constitution (visual prespace in the broad sense), "diminishing" is an "uncovering" with respect to "contamination" constituted in assimilative syntheses of similarity and difference; whereas in "fading" the loss of difference is a mode of presentation of "resistance" (scl. a mode of "covering"). Such a case would consist of changes in those sets of visual sensa on the basis of which (in the example adapted from Mach), the distant hill is perceived as looming up ahead. When the changes in the sensum-field "shrink inward," there is sensing of "decontamination" such that, as it were, color patches are "separated out." "Fading," in contrast, consists in part of a "contamination" correlative to an increase in the area of change, hence of greater covering of the field.

In other words, "uncovering" of the visual prespatial or sensum-field ("diminishing") is "covering" of qualitatively changing saliencies "resisting" one another, becoming opaque, dense, compact instead of translucent, hollow, superficial. "Looming up ahead" is, so to speak, a case where one visual saliency is presented as increasingly opaque with respect to other co-presented saliencies while translucence decreases along an area of change that "shrinks inward." For example, areas of dark blue and green are simultaneously presented surrounded by a light field; together they are constituted in assimilative syntheses, based on identifying syntheses, as alike in brightness and saturation; and so forth. Thus the area of dark blue "diminishes," translucent chromatic "contamination" with the green area decreases until a moment is reached where there is "no longer" a basis, again, for constituting color-areas in assimilative syntheses.[42] The dark blue "looms" both over against the green patch and the sensum-field itself intended to beyond the "diminishing" area of change. Because of the "looming" in the "diminishing," or "resistance," the area of change "shrinks inward," the line of vision retracts back upon itself: "penetrability."

Finally, it must be added that "empty visual prespace" signifies optimal concavity, chromatic "decontamination" of color and "hollowed" superficiality, but always on the spectrum of qualitative change from optimal "filledness," convexity and compactness. Similarly, "prespace between," intervaledness of the visual prespatial spread, is co-presented in the sensing of optimal "loss" of difference in the spread.

In later chapters we shall return to these prespatial visual phenomena for the principal thematic interest is how the objectivational substrata of qualitative change in the visual prespatial spread acquire the "appearance" of quasi-objective and eventually objective space. More specifically, the prespatial visual phenomena of "increase/decrease," "rotation/rest," and "covering/uncovering," will be found to have a central role in the building up of the spatial (and temporal) form of the real, objective world. Before drawing a number of conclusions from the objectivation of prespaces, it is necessary to examine one more prespace for the sake of at least partial completeness: auditory prespace.

49. SPACE AND SOUND

In his 1907 lectures on "Physical Thing and Space," Husserl points out that auditory objectivations are only determinate in a secondary and dependent way so far as the oriented constituting of something physical in space is concerned.[43] I can, for instance, see or touch the same physical thing again and on different occasions; the same physical thing is presented through these and other tactual and visual appearances, and it is irrelevant for presentation of the same whether the series of tactual and visual appearances are from the right or the left, or from left to the right. Just this situation does not prevail in the case of hearing a melody, a succession of musical sounds, for instance: I cannot reverse the ordered series of auditory appearances without altering the presentation of the same melody. Indeed, to speak of the "same melody" is quite distinct from speaking of the "same" visual-tactual thing.[44] The role of auditory sense perception is thus secondary when it comes to the constituting of the spatiality of physical thingness. To be sure, Husserl adds, this does not signify that sound has nothing to do with the constituting of space and of things in space; but, for instance, auditory direction and localization always presuppose the "primal" determinations of sight and touch (Husserl's example is that of the whistle on the steam engine which only indirectly presents the steam engine). And the same may be said for other sensuous fields such as that of temperature.[45] Nevertheless, unbuilding reduction of the oriented constitution of sound has the special advantage of revealing the perhaps otherwise overlooked nature of the constituting of temporal objectivities in contradistinction to spatial ones (thus frequent use of examples of listening to sounds and sound-sequences by Husserl in his account of the consciousness of "internal time").

Yet even granting the peculiarities of auditory perception, and aside from problems posed for Husserl's own theory of space constitution,[46] the question still remains as to how the perception of sound figures into the oriented constitution of the spatial and temporal form of the real, objective world. A brief examination of the domain of audition is therefore worth while to see if there are features basic to space constitution. To this end it is necessary to return again to the fully constituted phantom quasi-objective world in which there is a fully constituted quasi-objective phantom organism presented as having front and back, right and left, above and below. The quasi-objective tangible organism is likewise presented as visible and audible, warm and cold, and the like, as is the quasi-objective phantom surrounding world. To be sure, with respect to de facto courses of seeing, there is no originary visual presentation of change in depth "as in back;" visual depth is presented most originally as only "in front," i.e., along the line of vision. Indeed, it would seem that depth "as in back" *and* "as in front" are spatial determinations basically rooted in touch--a problem to be considered later (Chapter Seven). For the moment it suffices to note that depth of the quasi-objective (or objective) auditory field is as ambiguous and ubiquitous as is auditory localization in general. In turn, this suggests that definite auditory localization and directionality is established only in the case of quasi-objective "inter-field" spatial spreadness (for example, where a touched or touchable and seen place is co-intended to and posited as the "source" of certain sounds at one or the other side, in front or in back, near or far[47]).

50. UNBUILDING TO THE QUASI-OBJECTIVE AUDIBLE PHANTOM WORLD

However this may be, and whatever may be the manner in which the hearing of sounds contributes to the oriented constitution of the real, objective world, the task at the moment is to objectivate those determinations peculiar tothe spread of the auditory sensum-field. The phenomenological theme, again, is a quasi-objective spatially "centerless" phantom-world. As before, I reduce my theme to uncover as phenomenological residuum a here-centered quasi-objective world of phantoms, then a here-centered quasi-objective world of *audible* phantoms by objectivating that noematic sense which the here-centered quasi-objective here-centered world of full phantoms has as an object of actual and possible past, present, or future auditory perceivings and, at the same time, refraining from positing the noematic sense belonging to it as an object of quasi-objective perceivings of any other kind.

Using Husserl's example of the whistle on the steam locomotive, and listening to a sequence of blasts from the whistle, the first thing to note is that quasi-objective phantom sound-appearances are presented as lying between an auditory Here and a sounding There (at a higher level of oriented constitution, between the auditorily percipient organism and the object making sounds). In a sequence of such sounds, there is to be discriminated in addition a crude and perhaps ambiguous localization. In his pioneering study, Young has noted, apropos a similar situation, that "Auditory localization depends upon some physical difference between the stimulation of right and left ears. *Visual*-auditory localization depends upon an adjustment of the organism with respect to a visible or known position in phenomenal space. When a deliberate search is made for the source of a sound without the aid of vision and touch subjective changes in the loudness and timbre of the auditory experience in relation to head and gross bodily movements assume the leading role."[48]

In this respect and even at the constitutive level or substratum of quasi-objective perceivings of auditory phantoms, the difference between right and left sides along with the correlation of auriculomotor kinaesthesias and auditory phantom appearances allow for the auditory perceiving of different directions and change in location. Cephalic kinaesthesias also play a role here and, indeed, it is difficult to distinguish between auriculomotor and cephalic kinaesthesias in this situation--turning the head not only to "hear better," but to effect a more precise localization. Similarly, the presentation of loudness and timbre are constituted in syntheses of similarity and likeness based upon identifying syntheses of sounds in their sequence of appearances, with a co-perceiving of a "vibrating" rhythm of phantom sound-appearances correlative to perceiving "movement" of sound-sequences. And whereas hearing the difference in loudness and timbre is also an intending to the intervals between sounds in the sequences, hearing sameness or difference of pitch is an intending to sounds as "above" or "below" the binaural auditory plane tied down, as it were, Here.

I can therefore objectivate a primitive directionality, more specifically, a directionality of opposites ("right" and "left") and orientation in the here-centered auditory phantom spatial spread. There is likewise auditory perceiving of "movement" based on the perceiving of rhythm and vibration--although that "movement" would always seem to be repesented as rectilinear rather than, for instance, rotating. And this would seem to hold whether or not sounds are presented sequentially or simultaneously. Finally, the auditory quasi-objective phantom here-centered spread is presented as in depth, i.e., as filled or fillable. Although quasi-objective phantom sounds may be said to be presented as "atmospheric," "filling the distance" between Here and There, the quasi-objective auditory spread continuously exhibits articulation into directions and orientations.

Still other facets of the quasi-objective spread of phantom sounds will be considered later. It is necessary now to turn to a brief account of changes in the prespatial auditory spread.

51. UNBUILDING TO THE AUDITORILY PRESENTED IN PRESPACE

"Setting at zero" all kinaesthetic flow-sequences and correlations in our broad sense of "setting kinaesthesias at zero," I proceed to objectivate a prespatial auditory sensum-field of "moving" and "resting" auditory saliencies, of qualitative auditory change and non-change. I now discriminate an auditory sensing of qualitatively changing and non-changing sound-sensa presented as filling the auditory spread with their diffuse voluminousness, "inherently" there. Moreover--remaining with the example of the locomotive whistle--I can further discriminate intendings to areas into which sound-sensa recede and fade away, or from which they emerge and increase in intensity. Or, considering still other sorts of change and variation, sequences of sound-sensa are distinguished which, at a higher level of oriented constitution, are perceived as "cords" of one kind or another: a continuous blending and merging of sound-sensa, "contaminating" one another in much the same way as the color-patches in the visual sensum-field. Sound-sensa can also be auditorily sensed as "outside one another," as "resistant" to blending and merging, fusing.

To be sure, this case of simultaneous presentation of sound-sensa is difficult to distinguish from another, namely the case where sound-sensa are presented as "inside one another."[49] Or one sound-sensum is given here, another there, between which a silence or a sound of lesser intensity is co-intended to; for instance, a case where the pitch is more or less equal, but the intensity undergoes variation. In contrast, in the case of "inside one another," the sound-sensa are presented as at the same "place" but of different pitch. The cases of "fusion" or "contamination," of "outside one another," and of "inside one another," are all presentational modes constituted in syntheses of similarity and difference, based on identifying and numerically distinguishing syntheses. Interesting as these cases may be, it is perhaps more worth while for the purpose at hand to examine two other presentational modes which would seem to establish the limit of the various auditory assimilative and differentiating syntheses: the case where, instead of, as it were, a little sound here and there and all the rest of the spread silent, for instance, there is presented a whole auditory sensumfield with no silence anywhere in it; or, the opposite case where sound-sensa go over into stillness: "But even stillness is a filling out of the auditory field; it is now empty, but *empty of sounds*...it is empty, but it has not disappeared."[50] These two cases are phenomenological clues to the constituting of the inherently spreadoutness of auditory prespace in the narrow sense of the term, "prespace."

That is to say, in both those presentational modes I find more or less sharp discrimination of sound-sensa which, at the level of, e.g., quasi-objective space (either here-centered or centerless), found the quasi-objective hearing of auditory phantoms presented as lying in a horizontal binaural plane. Only on the basis of correlations of sensings of certain sound-sensa with cephalic kinaesthetic flow-sequences is there a basis for perceiving and sharply discriminating of auditory phantoms above or below this plane.[51] But at the substratum of auditory prespace in the narrow sense of "prespace," the only sound-sensa discriminated at all distinctly as to direction and localization are those presented as lying in the horizontal plane. This is not to suggest, however, that the auditory prespatial sensum-field is "two-dimensional" in contrast to the tactual and visual prespatial sensum-fields. The two different cases of "filling out" the auditory field lying in the horizontal binaural plane exhibit as well a "depth" peculiar to the auditory:

1. Auditorily salient areas lying in the horizontal plane may be presented as intervalled by sounds of lesser intensity or by stillness, the areas being "surrounded" in a way akin to darkness in the visual prespatial field or to non-touch in the tactual field. Increase in intensity of certain sound-sensa "cover" other sensa, or "loss" of timbre of certain sensa "uncover" still other sensa.[52] In either case, there is correlatively an appersensing of "depth" in the "covering" and "uncovering."

2. "Stillness," the other sort of "filling out" of the auditory field, comprises a change from the auditory to non-salience. In this case, auditory sensing of sound-sensa which found, at a higher constitutive stratum the quasi-objective hearing of phantoms lying above or below the horizontal plane, are presented in auditory prespace as an area in which nothing stands out auditorily in an auditory omnipresence, as it were.

In the first case of variation in the sensum-field, the more sharply sound-sensa are presented in the sensing of them, correlatively the more sharply the appersensing of intervals between them, the more the co-intended to field is presented as "diminishing" is size, as "shrinking inward" from different if not opposite directions in "filling out." In contrast, in the second case, in the sensing of decrease in auditory saliency in the horizontal plane the spread of the sensum-field is co-intended to as "extending outward" toward complete "voluminosity." It may be said, then, that in the one case an "approaching" is objectivated while in the other case a "receding" of sound is objectivated. Both cases exhibit primitive determinations of "depth" in the auditory prespatial spread.

52. CONCLUDING REMARK

The foregoing account of auditory prespace is, admittedly, even less complete than those of tactual and visual prespaces. Yet in each account I tried to objectivate a "world"--not much of a "world," to be sure, but still a "world"--in a manner sufficient to carry forward the pupose of the inquiry into the nature of the archontic constitutive substrata of the spatial form of the real, objective world by examining some of the ways in which that form "shapes up" in the first place. Before it is possible, however, to consider the "building up" ways in which the various prespaces acquire the "appearances" of and enter into the secondarily constituted quasi-objective and objective spatial forms of the real, objective world it is worth while making explicit the results of the inquiry so far for transcendental phenomenology. This may best be accomplished by reflecting on a number of problems that have arisen in connection with objectivating the prespatial spreads.

53. PROBLEMS ATTENDANT UPON OBJECTIVATION OF THE ORIENTED CONSTITUTION OF TACTUAL, VISUAL AND AUDITORY PRESPACES

In Part One I sketched the various steps of the transcendental phenomenological epoché to be exercised in examining the spatio-temporal form of the real, objective world.[53] In addition, some of the unbuilding procedures were developed that are to be employed upon effecting phenomenological epoché at each step. Finally, the respective transcendental reductions were made thematic with respect to the oriented constitution of the spatio-temporal form of the real, objective world as the noematic correlate of my transcendental mental living apperceptively experiencing itself as in and part of the real, objective world. In other words, the theme of inquiry has been not just my transcendental mental living as such in its transcendental "purity," but also my "mundanized" transcendental mental living. "The epoché," Husserl says, "can also be said to be the radical and universal method by which I apprehend myself purely: as Ego, and with my own pure conscious life, in and by which the entire Objective world exists for me and is precisely as it is for me. Anything belonging to the world, any spatio-temporal being, exists for me--that is to say, is accepted by me--in that I experience it, perceive it, remember it, think of it somehow, judge about it, value it, desire it, or the like."[54]

In this connection it was noted that, living straightforwardly in the transcendental natural attitude, I ascribe in habitual ways "fields of sensation" to my animate organism.[55] It is worth quoting Husserl's words again:

> Among the bodies belonging to this 'Nature' and included in my own peculiar ownness, I find my *animate organism* as *uniquely* singled out--namely as the only one of them that is not just a body but precisely an animate organism: the sole object within my abstract world-stratum to which, in accordance with experience, I ascribe *fields of sensation* (belonging to it, however, in different manners--a field of tactual sensations, a field of warmth and coldness, and so forth)....[56]

This ascription of "fields of sensation" disclosed by phenomenological epoché is a fundamental component feature of the mundanization of my transcendental mental living; that is, the ascription itself is a mode of self-apperceiving of my transcendental mental living as in and part of the real, objective world. Within the general framework of transcendental phenomenology as defined earlier (section 5), this signifies that the ascriptive, self-apperceptual component of the mundanization of my transcendental mental living must be treated in terms of what Husserl called the "contingent apriori" and the "formal apriori."[57] And because the unbuilding analyses of the prespatial spreads co-intended to and posited in the tactual, visual, and auditory sensings of their respective sensa and sensum-fields make up part of the phenomenological analysis of the ascription of "fields of sensation," they too would seem to belong within this same framework. This is not immediately obvious, however, and one of the first problems that arises attendant upon the unbuilding analysis of prespatial spreads is that of its place in the transcendental phenomenological scheme of things.

To be sure, Husserl's writings provide some indications. In the *Formal and Transcendental Logic*, for instance, Husserl raises basic questions concerning the meaning of transcendental as well as formal logic in general, and of the relationship of transcendental phenomenology and transcendental logic in particular. As is known, he arrives at the conclusion that transcendental logic is not merely a logic of propositions, but also one of the consciousness of possible propositions themselves.[58] I refer to this conclusion of Husserl especially because one of the examples he provides to illustrate it is drawn from the analysis of the field of hyletic data. If we consider this field, Husserl suggests, we can discover certain apriori truths about all such possible data. Propositions about hyletic data accordingly comprise a body of truths about the "contingent Apriori:" that is, if vision, for example (or touch, or hearing) has certain properties, such as described above (in section 25, or in quite a different way in sections 46ff.), it then may be said that the propositions stated are true. But it cannot also be said that it is a necessity that all transcendental mental livings in the transcendental natural attitude have vision.

In contradistinction, however, the "formal Apriori" expresses the necessity that such mental livings must have visual hyletic data (or visual sensa salient in a visual sensum-field): by "formal Apriori" in the broadest sense of the term Husserl understands all the circumstances basic to transcendental mental living as transcendental mental living. If, now, I include the restriction in *Formal and Transcendental Logic*, section 6, that this mental living be a rational one, i.e., able to make syntactical judgments that are verifiable, then it can be said that transcendental logic, transcendental theory of the rational, is identical with transcendental phenomenology of any knowing-verifying transcendental mental living whatever (above, section 4), which is to say, the critical-epistemological understanding produced by the theoretic will and striving to return to the things themselves. Now, transcendental logic must also include an account of the limiting cases of transcendental mental living where, for instance, the hyletic data are so chaotic that judgements cannot be verified--such as the case where all primary passive apperceptions result in syntheses of disagreement rather than agreement.

While I find no reason to take issue with Husserl's general scheme just sketched, it is clear from the unbuilding analyses of prespatial spreads that not only must the "content" of the contingent Apriori be altered, but also the formulation of the limiting cases to be included in the transcendental logic (see below, Chapter Nine). For if, now, I render explicit the phenomenological epoché I have sought to employ in the preceding sections, this signifies that I have refrained from positing my "fields of sensation" as having loci in the "real space" of the real, objective world as "spatially on the surface of, or inside, this real physical thing: my animate organism." If the unbuilding reductions to the tactually, visually, and auditorily presented in their respective prespaces are correct and consistent, then the transcendental phenomenological epoché and reductions effected signify two things:

First, they signify that just as refraining from positing my mental living as "having a locus in the real time of the real, objective world" does not entail a refraining from positing my transcendental mental living as "having its own inherent temporal form," so the refraining from positing my "fields of sensation" as "having loci in the real space of the real, objective world" does not entail refraining from positing those fields as "having their own inherent spreadoutness."[59] The phenomenological result is not without consequence, especially when contrasted with Husserl's view expressed, for instance, in the lectures on the consciousness of "internal time." To give just one example of what happens when the two refrainings are identified, Husserl says, in effect, that upon "exclusion of 'objective space'" one discovers the "primary contents" of the visual field and that those contents yield the continuum of the visual field. This would imply, constitutively, that the "primary contents" are fundamental to the continuum of the visual field, or at least the continuum is constituted by unification of the given "primary contents" of vision. *However, if the one refraining does not entail the other*, then, as I tried to show, the continuum of the visual field is constitutively fundamental to the "given primary contents," that is, that "shrinkage" and "expansion" of the visual field are fundamental to "covering" and "uncovering" of visual saliencies in the field, that indeed "shrinkage" and "expansion" of the visual field can proceed regardless of whether there is sensing of visual "primary contents" in the field: a phenomenological result diametrically opposed to that of Husserl.

Second, the epoché and reductions signify, when applied across the board, that just as I reflectively experience the inherent quasi-objective temporality of my transcendental mental living, so I reflectively experience the inherent quasi-objective spatiality and prespatiality of the sensum-fields, and their inherent temporality, apart from any ascriptive, apperceptually experienced mundane spatio-temporal locatedness of those "fields" relative to my animate organism.

The fact that the one refraining does not entail the other, but rather makes thematic the positing of my "fields of sensation" as inherently "spatial" (and "temporal") allows for making explicit the multistoried oriented constitution of that inherent spatiality (and temporality) in transcendental mental living with respect to its mundanization, its being in the world. More particularly, the concern was to explicate and clarify the prespatial sensum-fields in isolation from one another, that is, apart from their different ways of belonging to the animate organism in addition to refraining from positing any sense they have acquired or might have acquired from standing in functional correlations with kinaesthetic flow-sequences. This latter refraining follows of necessity because the several prespatial sensum-fields are considered apart from any apperceptually experienced locatedness relative to the animate organism. As a consequence, the procedure of "setting kinaesthesia at zero" necessarily follows from refraining from positing the "field of sensation" in real, objective space. Perhaps the most significant phenomenological finding, as a consequence, is the uncovering of the "three-dimensional" spreadoutness characterizing the continua belonging to each of the prespatial sensum-fields analyzed. At the same time, this finding gives rise to still another problem, for am I, then, as transcendental on-looker, left with the sort of "worlds" to which the famous Berkeleyan apple is consigned?--"an" apple which, "in reality," is a tactual apple, a visual apple, an olfactory apple, only called "one" apple by convention?[60] Is it, indeed, only "by convention" that I intend to and posit the real, objective world with one space-time as the same world intended to and posited by others? Is it not therefore the case that even though, for the sake of analysis, I "isolate" and explicate the various constitutive substrata of the real, objective world that that world is nonetheless an organic, living whole? Or is it only by convention that one would say with Cézanne that the painting of the landscape contains within itself the smell of the landscape, the sound of the landscape, in short, the response of all of the other senses?[61] All that may be as much by convention as by necessity; yet, even so, by virtue of what does one prespace "express" and reveal another prespace?

In one way or another, these and similar problems all lead back to the question at issue in the examination of prespaces in "isolation" from one another: How do the various prespaces acquire together and at the same "time" the "appearance" of *one* quasi-objective space, ultimately of *one* space (time)?

Earlier in discussing the problem of space perception understood in a Kantian fashion, it was noted that for a phenomenological account to respond to the problem it is necessary, first, to establish a foundation for distinguishing between kinaesthetic and other sensa-processes and, second, to establish how the inherently prespatial acquires the "appearance" of, and enters into, the quasi-objectively spatial as something secondarily constituted.[62] The rudiments of the first task have been indicated in this chapter (and will be further developed in the next). Moreover, it remains to make explicit the distinction between kinaesthetic and other sensa-processes so far as they figure into other aspects of oriented constitution; thus the next chapters will make thematic the ways in which prespatial saliencies and their own-inherent spreads play a role in the oriented cosntitution of quasi-objective spatial things and their quasi-objective spaces (and times). Needed here, by way of conclusion to this chapter, is a further critical reflection on what is meant by "setting kinaesthesia at zero" (see above, section 27).

54. THE PHENOMENOLOGICAL CONCEPT OF "KINAESTHESIA"

In Part One specific changes were introduced into Husserl's ideas of "hyletic data" and kinaesthetic data."[63] And in the course of discussion in this chapter still other changes were implicitly introduced. Indeed, it has become increasingly clear that, phenomenologically, Husserl's idea of "kinaesthesia" is severely limited. One reason for the limitation is that Husserl subsumes two inconsistent concepts under the heading of "kinaesthesia." An early, yet characteristic, expression of the inconsistency is the following: "Presupposition: I have a freely controllable [*verfügbares*] total system of kinaesthesias which, at any moment, may be freely actualized by holding still or being set going. The kinaesthesias function in a double way: as constituting phantoms (things pertaining to the senses) and as constituting motion, self-motion (of the organism) and motion of other bodies, thus as constituting place and possible change of place as universal determinations of phantoms...."[64] To make kinaesthesia constitutive of quasi-objective phantoms is to make them a special case of hyletic data and thus to include them--following Husserl--as non-intentive components really inherent in mental living.

There is therefore no basis for distinguishing between kinaesthetic and hyletic data so far as Husserl's own account is concerned. More precisely expressed, there seems to be the assumption that some data can be construed as quasi-objective or objective appearances in and through which quasi-objective physical things are presented both as extended in space and as occupying a place in space. Those data that cannot be so construed are kinaesthetic data which stand, to be sure, in functional correlation with the other data that are indeed so construed and "animated" as the basis for perceiving place and change of place.

The only conclusion to be drawn, then, is that kinaesthetic data are both construed and not construed at the same time as adumbrative of quasi-objective or objective appearances of physical things extended in space. As it were, to make the distinction one must dare not admit it. The double meaning of "kinaesthesia" is then not only inconsistent, but it is also self-contradictory. Claesges notes the difficulty in his several discussions of kinaesthesia[65] and, as mentioned before,[66] he seems to uncritically follow Husserl in this respect in so far, however, as it applies to the constituting of one's own organism. "Consciousness of my own organism," he says, and "kinaesthetic consciousness" (an apt term) are the "'same', mediated by itself in the case of the consciousness of the organism, but immediate in the case of the system of kinaesthetic capacities [*Vermöglichkeiten*]...."[67] It then follows, Claesges says, that because *immediate* data are related to the "center of all intentive activity," the difficulty is resolved; in other words, kinaesthetic data turn out to be non-intentive components really inherent in mental living.

But, again, to resolve the difficulty the distinction between hyletic and kinaesthetic data is vitiated. The only way out of this dilemma, I believe, is the procedure followed in this chapter of "setting kinaesthesias at zero" in both the narrow and the broad senses of "setting kinaesthesias at zero." There is then no reason for imputing to kinaesthetic data the kind of constitutive function that Husserl and, following him, Claesges do. Nor do I then find any confirmation of the theory of hyletic data in general. All that I find, instead, are primary passive sensings of sensuous saliencies and appersensings of their correlatively presented spreadoutnesses, the continua of which are, in each case, substrata of oriented constitution more fundamental than those substrata at which sensuous saliencies are constituted.

As a consequence, what Husserl develops in this regard, and what Claesges continues, is clearly the idea that "space perception" depends solely upon the "sensations of movement."[68] In transcendental phenomenological terms, that signifies that to refrain from positing my "fields of sensation" as having loci in real space entails--*with one exception*, *scl.* the animate organism--of necessity refraining from positing those "fields of sensation" as having their own-inherent extension. *To make that "exception" begs the question and, in turn, makes the effecting of the transcendental phenomenological epoché inconsistent.* To be sure, by rendering the epoché inconsistent, the inconsistency (and self-contradiction) of the notion of "kinaesthesia" can be maintained, although at the price of having to overlook the anomalous constituting of the organism (and, correlatively, "Earth"). The double meaning of kinaesthesia must be rejected, therefore, and the distinction between kinaesthetic data and other sensuous data adherred to. Adherring to the distinction rules out the assertion that "space perception" depends entirely on the "sensation of movement." It will be one of the tasks of the next chapters to clarify further the distinction by examining the oriented constitution of place and motion in quasi-objective space.

CHAPTER SIX

TRANSCENDENTAL PHENOMENOLOGICAL BUILDING UP OF QUASI-OBJECTIVE SPACE IN PRIMARY PASSIVITY

55. THE POINT OF DEPARTURE FOR BUILDING-UP ANALYSES OF QUASI-OBJECTIVE SPACE

The last chapter made thematic some of the ways in which several prespatial sensum-fields are intended to and posited as continuously spread out wholes with differentiated parts. Objectivation of those spreads provides a basis for distinguishing between a broad and a narrow meaning of the term, "prespace." The former consists of positing the spreadoutness of a sensum-field as undergoing changes in sensa meant and intended to as identical in correlation with co-intendings to flows of kinaesthetic sensa-sequences; it is the phenomenological residuum attendant upon discriminative refraining from positings of the quasi-objective unities that function as quasi-objective appearances of the phantom organism, phantom tactual, visual and auditory appearances. The latter consists of positing the spreadoutness of a sensum-field as undergoing changes that do not include co-intendings to correlative flows of kinaesthetic sensa-sequences; it is the phenomenological residuum attendant upon still further discriminative refraining designated by the phrase, "setting kinaesthesia at zero." And just this last refraining is a clue to the building up analysis of the oriented constitution of the real, objective world and its spatio-temporal form.

The clue may be expressed in the form of an analogy: The various steps of unbuilding reductions of the constituting of the spatial dimension of the real, objective world disclose that just as every perceiving of something spatial is ipso facto an apperceiving of the space in which that something appears, so every sensing of something prespatial is ipso facto an appersensing of prespatial spreadoutness of which that something is a part. But the converse of the analogy does not hold. For while every apperceiving of space is ipso facto perceiving of something spatial, it is not necessarily the case that every appersensing of prespatial spreadoutness is ipso facto sensing of something prespatial. As a consequence, "movement" and "change of place" are not constitutively fundamental to space perception. In other words, not only are prespaces in the narrow meaning of "prespace" intended to and posited as "three-dimensional," but also as, each separately, continua "founding" identifying and assimilative syntheses of sensa in correlation with flows of kinaesthetic sensa-sequences. "Movement" and "change of place" are "founded" phenomena.

Thus although the point of departure for building-up analyses will be prespaces in the broad meaning of the term, "prespace," the first task is to ascertain the *motivation* for accomplishing the constitutive transition from prespaces in the narrow meaning to prespaces in the broad meaning of the term. In Husserl's words, this is the task of establishing the motivation for "attaching kinaesthesias" to particular sensum-spreads in such a manner that changes in the spread acquire the appearance of, and enter into, "movement in a field of loci" and the sensum-spread itself acquires the appearance of a "bodyless [*körperlosem*] background for bodies."[1] This is equivalent to objectivating the difference between immediately and mediately controllable sensaasthe necessary condition for constituting identical sensa in sensum-fields as moving or at rest over against a background. Moreover, such a background to which Husserl refers is not meant as itself in motion; it is meant neither as a background object nor as made up of background objects.

Husserl's example of such a sensum-field in the broad meaning of "prespace" is the oculomotor sensum-field. This visual field, he says, is presented in the visual sensing of it as a whole in which something salient may be moving; then, owing to strivings to control flows of kinaesthetic sensa-sequences, it is constituted as that "in which the change of becoming prominent, of appearing as a 'body' at rest and then moving, or the like, takes place. This, however, concerns part-phenomena pertaining to the whole; but these bodies are moving indeed 'in' the 'whole'--which, itself, is not a whole made up of bodies. (Such a whole, after all, would be a body.) It is the 'absolutely' resting basis for bodies (their 'ground')."[2] It is not obvious from the context of his statement that Husserl has properly reduced his theme, the real, objective world, to visual prespace in the broad sense of the term, "prespace."[3] Nevertheless, the "'absolutely' resting basis for bodies" is just that immobile background for visually sensed and meant identical sensa in motion presented with changes such as, for example, covering and correlative uncovering. Shortly a closer look will be taken at this and similar phenomenological situations exhibited by other prespaces as well.

Here the second task arises which touches upon the "feeling" for the problem of the "genesis" of space perception for it is necessary to discover the motivation for building up from prespaces to the constituting of quasi-objective spaces, finally to a single, unified "intersensorial" quasi-objective space (and time) of the fully constituted phantom world. In other words, the task is to show how something "primordial," namely prespaces, take on the appearance of, and enter into, something "secondarily" constituted, namely quasi-objective spaces, in each case with a new noetic-noematic stratum of sense.

Yet the moment the tasks are stated a whole host of other problems arise, not the least of which is the need for some sort of clarification of what often goes under the heading of "data of sensation."

56. A PHENOMENOLOGICAL CLASSIFICATION OF THE "DATA OF SENSATION"

At least since the time when Descartes first melted a piece of wax, if not before, clarification of "data of sensation" has been a problem for philosophy in the modern period. Husserl's response to the problem with his theory of hyletic data has been previously discussed with respect to the specific context in which the theory was formulated.[4] For various reasons, both psychological and transcendental, Husserl's idea that "data of sensation" are really inherent components of mental life-processes such as perceivings of physical things was rejected. What, instead, seems to me to be the case is that, as Dorion Cairns expressed it,

> *sensing of* sensation data are really immanent moments in the unities of all those perceptual intuitings of physical things that I experience reflectively. ...In the mental process, perceiving this white sheet of paper...I myself find, by a suitable turn of mental regard, *not the sensation datum white*, *but an intentive sensing of* the sensation datum, white. And it is not

> the sensation datum, white, but *rather the intentive sensing of it*, which is something belonging inseparably to the proper essence of the concrete perceiving of this white sheet of paper, and belonging thereto as a really immanent component part.[5]

Accordingly, Cairns says, the correct question to ask is "whether *sensings*, as processes in the reflectively experienced stream of mental processes are, throughout, substrata within more complex intentive processes, and the question whether every intentive process which is not exclusively a sensing has a 'foundation' that *is* a sensing."[6] Sensings of sensa which are "foundations" for mental life-processes (such as quasi-objective and objective perceivings of physical things) I shall call "sensing of perception-founding sensa."

Yet, not all sensings of sensa so understood are substrata within the more complex mental life-processes comprising quasi-objective and objective perceivings are "foundational" in the same way. It is necessary to distinguish sensings of those sensa, for instance, which exhibit the anomalous sensory duality and which are substrata within the concrete quasi-objective or objective perceiving of the organism. These sensings of primarily tactual sensa I shall call "sensings of organism-founding sensa" (see above, section 42). Still another sort of sensings of sensa that may be distinguished in some complex mental life-processes are sensings of sensa that serve to make known or manifest the quasi-objective or objective organism as "one's own organism," as one's own "lived body." Terminologically, to distinguish these sensings from the others I shall call them sensings of "feeling-sensa," which roughly translates Husserl's term, "*Empfindnisse*." The terms, admittedly, require greater precision and include a wide variety of sensings of sensa as "feelings" of pleasure, pain, comfort, discomfort, sentiments, even tension and relaxation.[7]

The way in which sensings of these sensa is foundational I find in their being substrata within the complex of mental life-processes that are likings and dislikings, even in primary passivity, which motivate willings and strivings (also in primary passivity) to bring about changes in certain sensings of sensa. A sensing of pleasant sensa, for instance, motivates strivings to control the presentational modes of such sensa, to increase their actualization, and the like. Shortly this phenomenological situation will be examined in detail.

Here, with respect to this third sort of sensing of sensa, two things must be noted. The first is that making explicit the sensing of feeling-sensa introduces specifically that intentionality which Husserl called "non-doxic intentionality." And in *Ideas*, First Book, Husserl develops the idea that hyletic data are subject to being animated by both doxic and non-doxic mental life-processes. To be sure, he does not suggest that there is something like a hyletic pleasantness; instead, what he says that he finds, is a liking, for example, based on a seeing of something which is liked, and the hyletic data serve as the basis for the seeing and not for the liking. Clearly I disagree with Husserl in this respect. What, in contrast, I find is that being-pleased, taking-pleasure-in, etc., even in primary passivity, are mental life-processes which fall into the class of processes designated valuings. What is intended-to is value, and value is presented and posited in the same fashion as being in the sphere of doxic intentionality; in other words, value, like being, is a posited or acceptive characteristic. Thus I find that the first changes effected with respect to the sensing of pleasant sensa in primary passivity are changes in presentational modes indirectly controlled for the sake of actualizing pleasant sensa. Perhaps in the further course of experience the ego is "stimulated" to advert volitionally as well as doxically to what is sensed as pleasant. Or, to take a rather different example, there is the distinction between a sensing of a feeling-sensum such as pain and a feeling the pain, that is, a sensing of the unpleasantness of the pain. The latter, we shall say, is motivated by the former.[8] In any case, non-doxic intentionality and positionality extends, even in primary passivity, to sensings of certain sorts of sensa.

Finally, we must also note here that there is yet a fourth sort of sensing of sensa, namely the sensing of those sensa which have been called "kinaesthesias" (or sensings of kinaesthetic sensa-sequences).[9] Sensings of these sensa likewise belong to the domain of non-doxic intentionality in contradistinction to the first two sorts plainly belonging to the domain of doxic intentionality.

The four-fold distinction of sensings of sensa falls, then, into the cross-classification of doxic and non-doxic intentionality. The first and fourth sorts of sensings have been examined in some detail with phenomenological results in many respects distinct from those of Husserl. But inquiry into the first sort of sensings resulted in the fourth sort of sensings being "set at zero." The transcendental phenomenological significance of these inquiries is, as noted before, that they mutually entail each other.[10] Thus some mental life-processes that are not exclusively sensings but have foundations that are sensings, also have foundations that are sensings more basic than others. In so far as the "context" of the contingent Apriori is concerned, this signifies that it would seem to be true that the substratum, sensing of perception-founding sensa, is more basic than sensing of kinaesthetic sensa or sensings of organism-founding sensa or sensings of feeling-sensa. In short, the archontic substratum is not only doxic, but one specific sort of doxic intentionality. Although it is too early to develop the transcendental consequences of this finding, the foregoing distinctions and classifications are sufficient to allow for taking the first step of building-up analysis of the oriented constitution of the real, objective world. In taking that first step in the sections to follow, there will be ample opportunity to clarify further our distinctions and improve their expression.

57. THE FIRST STEP OF BUILDING-UP ANALYSIS OF THE ORIENTED CONSTITUTION OF THE REAL, OBJECTIVE WORLD. THE CONSTITUTION OF "HERE"

In making the distinctions between the various sorts of sensings of sensa and in making thematic their classification, I have implicitly relaxed that refraining which I called "setting kinaesthesia at zero" in the broader meaning of the phrase, "setting kinaesthesia at zero" (section 28). Thus I now institute either actually or in phantasy a course of sensing of kinaesthetic sensa of some part or parts of my quasi-objective organism as moving relatively to the rest of the organism and as moving relatively to one another. Clearly it now would no longer seem to make sense to speak with Husserl of kinaesthesias as "attached" to prespatial sensum-fields.[11] "Kinaesthesia" signifies sensing of kin-aesthetic sensa--a presentive awareness quite distinct from the presentive awareness of being in or out of touch as well as from that of sensing, e.g., pain or pleasure here or there. Sensing of processes of kinaesthetic sensa is then equally distinct from strivings to control that (or, indeed, any) sensa. What, now, do I reflectively find first and foremost when I take this first step of building-up of the oriented constitution of the real, objective world?

First and foremost I find and objectivate the noetic-noematic substrata of quasi-objective perceivings of some part or parts of my quasi-objective organism as moving relatively to the rest of the organism or as moving relatively to one another: the sensing of tactual sensa together with sensing of kinaesthetic sensa regularly accompanying the sensing of tactual sensa. The most *immediate* case that I find is the sensing of kinaesthetic sensa which is motivated by sensing of contiguous touch-areas most originally presented as not extending into each other ("resistance"[12]), and these sensings of kinaesthetic sensa regularly accompany the sensing of touchareas no matter if any change or non-change in one or the other sequence of sensa is controllable mediately or immediately, no matter if they are "available to volition" (Becker). To be sure, this case of constitution in the prespatial tactual sensum-field is an assimilative and distinguishing synthesis just as are the other cases that were considered before (such as sensings of identical touchareas that comprised the phenomena of contiguity and contact).[13] Thus in a concrete case of quasi-objective perceiving such as extending my arm and touching a surface with my hand, I find, as substrata, the immediate and limiting case of assimilative and distinguishing synthesis founded on sensing of one touch-area not only as touching another area, but also as being touched.

Most immediately I objectivate, as proceeding in primary passivity, an identifying and numerical differentiating of the sensed touch-areas: each sensed as meant as identical to itself and distinct from the other. On that basis, the touched areas are meant as alike in quale, for instance, but distinct in place so that they are "paired" as similars in one respect, as dissimilars in the other, as numerically distinct. That much seems obvious when the hand touches the surface, e.g., moves over the grain of a piece of wood, the threads of a piece of linen or the bristles of a brush. But what is of even more importance in this situation is that it entails objectivation of dissimilarity in yet another respect for sensing of one of the touch-areas is sensing of it not just as similar in quale and numerically distinct from other areas, but the hand sensed as being touched in the touching, as coinciding with feeling-sensa is sensed at the same time; in the pressure of the hand on the grainy surface of the wood the hand "feels" itself being touched in the touching as we would have to say at a much higher level of constitution; the touching hand "feeling" itself touching a surface is both "background" and "foreground" of the tactual phenomena (a "bipolarity" Katz calls it). But here, at this level of constitution, we can only say that there is a synthesis of perfect dissimilarity of the touch-areas (the grains of wood, the threads of linen, the bristles of the brush--discontinuous surfaces), but a synthesis of perfect coincidence of the one touch-area with a feeling-sensum and of non-coincidence with the other touch-areas in the mutual transfer of sense. In short, presented are the sensing-substrata of quasi-objective perceiving of tactual surface and of quasi-objective co-perceiving at the same time of the hand touching the surface.

In other words, *it does not seem possible to objectivate the sensing of tactual-sensa and feeling-sensa as coinciding without objectivating at the same time the sensing of kinaesthetic sensa in correlation with and as regularly accompanying them*; "motion" is essentially included in the tactual phenomenon constituting the quasi-objective surface as rough or smooth, for example. *But the contrary need not be the case*, for I can objectivate sensings of tactual sensa without objectivating sensings of feeling-sensa if the sensings of kinaesthetic sensa have been "set at zero" (if there is no "motion" of the hand over the surface). Thus the "produced unity," the bipolarity, peculiar to the constituting of the organism is always a "*founded phenomenon*," never a foundation of quasi-objective co-perceiving of the organism. Part of the relevance of this phenomenological result for space perception is that it signifies that the prespatial tactual sensum-field is itself intended to, then, not as a field spread in upon itself but instead as spread out *from where* the co-intended-to feeling-field *as a whole* coincides with a sensed *part* of the tactual sensum-field. Co-intending to the noematic sense, "from where," is likewise the foundation of the quasi-objective perceiving of something nearby. On that occasion, "from where" acquires the "appearance" of the uniquely anomalous "here" of the quasi-objective phantom organism. More generally stated, sensings of feeling-sensa may be assigned the function of making known or manifest something as the organism.[14] Sensing of such sensa, such as a "pain," is quite distinct from sensing of other sorts of sensa. Just as a tactual sensum is not presented in the sensing of it as, e.g., "rough," but rather as the foundation on which there is a quasi-objective perceiving of rough, so the sensing of the sensum which is the foundation of a quasi-objective perceiving of a bump on the head is not a sensing of a pain; that sensum is rather presented as coincident with the feeling-sensum, in the co-sensing of which the bump is presented as painful. Such a conclusion, however, suggests that feeling-sensa have a primacy of localization over other sorts of sensa.[15] If, now, I objectivate the sensing of feeling-sensa in their field coincident as a whole with the sensing of tactual sensa as parts in their field, the uncentered spread of prespatial tactual sensa figures as substratum in the quasi-objective perceiving of *centered* localization, the sensing of which, in turn, is a foundational substratum in the quasi-objective perceiving of the phantom organism as here. To be sure, the locus, here, is not presented then as one among other loci; to borrow an expression of Merleau Ponty, it is presented as the "here of situation" rather than the "*here of position.*"

58. FURTHER REFLECTIONS ON THE FIRST STEP OF BUILDING UP OF THE REAL, OBJECTIVE WORLD

Basically, there are two reasons for speaking of the "here of situation." In the first place, to objectivate sensing of the coincidence of feeling-sensa and perception-founding tactual sensa is to objectivate a substratum of the quasi-objective perceiving of a multiplicity of organismic fields, each of which, roughly speaking, corresponds to an organ of the body--an organismic field of the arms, or of the legs, or of the hands, and the like. In the second place, to objectivate the sensing of coincident sensa-fields is, at the same time, to objectivate the sensing of them as regularly and harmoniously accompanied by sensings of fields of kinaesthetic sensa. Thus still another reason arises for speaking of a primacy of localization that belongs to the sensing of feeling-sensa: objectivation of sensing of kinaesthetic sensa is not ipso facto objectivtion of *strivings* to control kinaesthetic sensa immediately and changes correlative to perception-founding sensa mediately. For instance, at higher levels of oriented constitution I can objectivate strivings to move the quasi-objective phantom arm; I can distinguish not only substrata of sensing of tactual sensa as foundations of quasi-objective perceivings of the arm as it rises, but I can at the same time distinguish sensing of feeling-sensa such as "muscle sensations," "tension sensations," and the like, as the arm reaches out and the fist clenches. Changes in both perception-founding sensa and feeling-sensa are then objectivated in correlation with various changes in the sequences of kinaesthetic sensa.

There is also the rather different case to be considered of reaching out and touching a hot surface with the hand, an intentive process that has as substratum the sensing of parts of the tactual sensum-field as becoming warmer and warmer and, correspondingly, more painful. Here, however, I can objectivate yet other substrata such as strivings to actualize certain sequences of kinaesthetic sensa so that there are changes from warm to less warm and less painful sensa. The point of the example is that, here, the sensing of kinaesthetic sensa is not just sensing of them in correlation with sensing of tactual sensa and feeling-sensa, but the sensing of kinaesthetic sensa is motivated by preferring, in primary passivity, of one rather than another course of sensa expected which, in turn, motivates strivings to control changes in kinaesthetic sensa (on which withdrawal of the hand from the hot surface is functionally dependent) and in tactual sensa (the sensing of which is a substratum of the expectant perceivings of the surface as less warm).

Although not formulated in all detail, the examples are sufficient to allow for drawing a few important conclusions about the constituting of change in kinaesthetic sensa correlative to change in tactual and feeling-sensa. So far as localization is concerned, correlation of kinaesthetic sensa and feeling-sensa is fundamental; but so far as constituting appearances of quasi-objective things, and those things themselves, is concerned, the correlation between kinaesthetic sensa and tactual sensa is fundamental. In other words, I cannot objectivate the sensing of feeling-sensa without, at the same time, objectivating them as intended to coincident with co-intended to tactual sensa; and I cannot objectivate those syntheses of sensa-coincidence without seizing upon them as regularly accompanied by kinaesthetic sensa. But this is not ipso facto an objectivating of preferrings, let alone strivings, to control any of the sequences of sensa immediately or mediately in the prespatial sensum-fields.

That this is so receives confirmation from a very different angle, namely limiting cases found in the fully constituted real, objective world, such as jumping on and off moving vehicles or being carried by something moving. To be sure, the phenomenological foundation for the constituting of somatic as well as other sorts of motion has not yet been established; the various sorts of kinaesthetic flows of sensa directly involved have yet to be considered and described--e.g., locomotor kinaesthesias. But even without carrying through the complex building up analyses of these phenomena, it is clear that these are cases where, without preferring one or another course of kinaesthesias (such as locomotion), let alone striving to actualize one or another preferred course of kinaesthesias, there is a correlation between changes in the ways in which things appear and in aspects or perspectives under which they appear relative to my somatic states. Similarly, in the case of constituting of prespatial sensum-fields in the broad meaning of the term, "prespace," there is not only correlation of changes in the tactual field coincident with changes in the feeling-field along with other changes in the tactual field regularly accompanied by changes in the kinaesthetic field without motivation for preferring one set of kinaesthetic changes over another and without striving to actualize any other sort of kinaesthetic change.[16]

In short, what is "normal" at one level of oriented constitution proves to be an "anomaly" at another. In later chapters these and other "anomalies" will be considered.[17] Here several questions arise in the light of these reflections. The first concerns the manner in which prespaces acquire the appearance of quasi-objective spaces and space, for clearly the answer does not lie simply in the correlation of kinaesthetic sensa with other sorts of sensa. "Space perception" is not therefore most primordially rooted in "sensation of movement." The second question concerns, then, the manner in which there is motivation of preferring one sequence of sensa to another and, in turn, motivation of strivings to control changes in sequences of sensa. Of course, nothing has yet been said about visual and auditory sensa and their spreads; thematic interest has been confined to certain substrata of the "produced unity" belonging to the prespatial tactual and feeling-fields. Objectivation of their co-intended to prespatial spread disclosed a primitive centeredness, the phenomenological significance of which has yet to be developed. But already some of the significance is beginning to emerge in the form of the phenomenological problem of how the prespatial tactual sensum-field is constituted, in part, as located in and on the phantom organism. Or, to express the matter in a different way, it would cerainly seem from what has been said so far that the same sensing of tactual sensa is as much the foundation of quasi-objective perceiving of the phantom organism as it is of the quasi-objective perceiving of tangible phantoms. This suggests that there is a relation, perhaps better, correlation, between perceiving of changes and non-changes in qualities of tangible phantom things and perceiving of changes and non-changes in different parts of the tactual phantom field located in and on the phantom organism as a whole. Stated with respect to what has been objectivated of prespatial sensings, sensings of the tactual prespatial sensum-field as a whole are sensings of that field as having parts which coincide with the feeling sensum-field sensed as a whole. Whatever else this objectivation may signify, it is a clear indication that the manner in which the phantom organism is constituted is quite distinct from the manner in which other tangible phantoms are constituted.[18]

It is also clear that investigation of prespatial fields from other angles is necessary to discover their meaning for constitution of the spatial form of the real, objective world. So far some clues have been offered, but further development of them is required to get at the root of "space perception."

59. PRESPACES AND IDENTICALS IN PRESPACES

According to Husserl, the moment I, as transcendental on-looker, proceed to objectivate the various sensum-fields, I objectivate a heterogeneous diversity that is nothing else but a chaos.[19] To explain this Husserl says, in effect, that the diversity of fields is a diversity because they lack any sort of "hyletic unity" which they can only acquire when "animatingly" construed as adumbrations of quasi-objective appearances of phantoms in quasi-objective space. These "animating" construings, he adds, are themselves syntheses of identity and assimilation constituted, in turn, in "temporal syntheses" (to use his rather misleading term). Their principle constitutive role is to unify the various sensum-fields into one quasi-objective field of experience. To be sure, because Husserl's theory of hyletic data has been set aside in favor of a rather different description of mental life-processes, building up analysis of the real, objective world takes on a quite different shape. This is best seen by returning to an example used before and then introducing a more elaborate one.

The previous example is that of "resistance" in the sensing of and meaning the prespatial tactual sensum-field spread in and upon itself. In this connection, a number of intentive substrata were noted; thus in the case of the hand touching the hot surface, there is an intending to a coincidence of changes in the feeling-field and in the tactual sensum-field along with preferring, in primary passivity, of one regularly accompanying course of sensing of sequences of kinaesthetic sensa to others; in turn, that preferring motivates strivings to control this or that or another sequence of kinaesthetic sensa. By fulfilling such strivings to immediately control sequences of kinaesthetic sensa regularly accompanying changes in the coincidental sensum-fields, there is presented actual or possible mediate control of those changes in the sensum-fields. If I examine this objectivation further, I find that implicitly I have objectivated such sensum-field changes *intended to and posited*, at the same time, *as "identicals,"*[20] as the same tactual and feeling-sensa previously salient in the fields, now controlled (mediately) here and there, now at this time, then at that time. In other words, I cannot objectivate intendings to and positings of tactual and feeling-sensa correlative to intendings to and positings of regularly accompanying, immediately controllable sequences of kinaesthetic sensa without there being, at the same time, objectivating of intendings to and positing of those tactual and feeling-sensa as identical. And it is precisely and only this intending to and positing of sensa as identicals that is the foundation of quasi-objective perceivings of quasi-objective appearances and of quasi-objective physical things presented in and through those appearances. To make this phenomenological situation thematic it is most useful for our purposes to employ a rather different sort of example; it is drawn from Sartre's *L'Imaginaire*, and is particularly intructive when expressed in terms of the building up analysis of the oriented constitution of the real, objective world.

I begin, as before, by exercising phenomenological epoché, reducing my theme and refraining from positing the visual sensum-field and the actual or possible functional correlation of changes in that field. The "anesthetic" eyes are closed (above, section 42). As Sartre observes, not only, then, have changes in the visual sensum-field "disappeared," but, in the narrow meaning of the term, kinaesthesias "are set at zero." To be sure, some further classification is required because Sartre says that what is still presented in an original mode is the "trajectory of movement as a form in the process of being made:" "If I trace a figure *8* with the end of the index finger, what appears to me is an *8* in the process of being constituted, somewhat like the letters of titles in the movies which form themselves on the screen. To be sure, this form is given *at the end of my finger*. But it does not appear as a kinaesthetic form. It appears as a visual figure."[21] In contrast to the tactual *8*, the visual *8* is not presented on the basis of a serious quasi-objective or objective perceiving of visual qualities; rather it is meant as what would be seen were the "eyes not closed:" "it is a visual form in image," a fictive visual intending to *8*. From this example Sartre concludes that the phenomenological situation consists of the correlation of sequences of kinaesthetic sensa with changes in the tactual field, on the basis of which the figure *8* is felt at the finger tip in such a way that those kinaesthetic sensa-sequences function, he says, as an "analogue" or "analogical substitute" of the *visual figure proper*. In other words, kinaesthetic sensa-sequences intended to correlative to something "identical" in the tactual field are exhibited at the same time as a kinaesthetic analogue of an identical in the visual perceptual field. Thus Sartre's example. Expressed now in terms of my reduced phenomenological theme, the example describes those perceivings that include as substrata the sensing of kinaesthetic sensa-sequences on which movement of my index finger tracing out the figure *8* is dependent and which motivate expectation of an "evidence in whose systhesis everything that is still unfulfilled expectant intention, in the particular evidences founding a synthesis, would attain fulfillment."[22] Motivated are expectant intendings to changes in the quasi-objective visual field which, in correlation with certain oculomotor kinaesthesias, would be basic to the quasi-objective visual perceiving of *8*. Still adherring for the moment to Sartre's account, that which would fulfill the expectant intending is founded on a fictive or feigned sensing of visual sensa in the form of the figure *8* in the visual sensum-field regularly accompanied by a fictive or feigned sensing of kinaesthetic oculomotor sensa-sequences. In still other words, intending to changes in the tactual sensum-field regularly accompanied by sensing of kinaesthetic sensa-sequences on which movements of the finger, tracing out the figure *8*, are dependent, motivates expectant intending to changes in the visual sensum-field regularly accompanied by sensing of oculomotor kinaesthetic sensa-sequences--expectant intendings fulfilled, i.e., actually included either in fictive or feigned quasi-objective perceivings of something identical in the visual field or in serious quasi-objective perceivings of something identical.

Sartre's example consists of moving a finger over a surface tracing out the figure *8*. Our restatement of his example within the frame of a reduced theme reveals the following complexity:

1. The distinction between sensing of tactual sensa coincident with co-sensing of feeling-sensa--"resistance," and in Sartre's example pressure at the tip of the finger--simultaneous with sensing of kinaesthetic sensa-sequences regularly accompanying sensing of changes in the tactual sensum-field that are substrata of quasi-objective perceivings of an identical moving finger; moreover, sensings of regularly accompanying kinaesthetic sensa-sequences found strivings to control those tactual changes by virtue of controlling immediately one preferred sensa-sequence rather than another kinaesthetic sensa-sequence.

2. In tracing out the figure *8* there are, rather than one, two surfaces in contact, two touch-areas, sensed at the same time. In the one case, the tip of the index finger, objectivation of the sensing of tactual sensa presented as coincident with feeling-sensa, correlated mediately with attendant sensings of kinaesthetic sensa-sequences of specific sorts, discloses that it is *also* sensing of tactual sensa, the *8*-being-traced, not presented as coincident with feeling-sensa, nor which are presented as regularly accompanied by sensing of kinaesthetic sensa-sequences and thus are not correlated with preferrings and strivings to control sensa immediately and mediately. In other words, these latter are changes in the sensum-field included as substrata in the quasi-objective perceivings of the surface on which the index finger traces out the *8*; the quasi-objective surface-*8* appears as "foreground," we may say, over against the "background" of the quasi-objective surface of the finger-surface.

3. Thus I distinguish: (a) a sensing-identifying synthesis of tactual sensa intended to and posited with the noematic sense, "coincident with co-sensing of feeling-sensa, regularly accompanied by sensing of kinaesthetic sensa-sequences," *and* numerically differentiated from other tactual changes simultaneously intended to

as salient in the tactual sensum-field; (b) the mutual transfer or "pairing" of specific noetic-noematic senses founded on those sensing-identifying and differentiating syntheses: first of all, there is full cancellation of the sense, "tactual changes coincident with feeling-sensa in correlation with sensing of kinaesthetic sensa-sequences." But there is only *partial* fulfillment of the sense, "tactual changes;" these are simultaneously presented as similar to other tactual changes such as are included as substrata in perceiving the surface of the tip of the index finger as like the surface of the *8* being traced out; for example both are perceived as smooth.

Two assimilative syntheses are constituted, then, of central importance for their role in further building up of oriented constitution. Moreover, they presuppose foundations in which tactual saliences are meant and posited as of quite different sorts. The "fate" of such mutually paired noematic senses and its significance for space constitution may be expressed by saying that, owing to full cancellation of the sense, "tactual changes coincident with feeling-sensa in correlation with sensed kinaesthetic sensa-sequences," and *partial fulfillment* of the sense, "tactual changes," the tactual sensum-field is intended to and posited as spread out "background" for changes in tactual and feeling-sensa as well as for changes presented at the same time as components of the tactual sensum-field. This is, of course, not the whole story of the examination of the complexity concealed in Sartre's example, and important though it is, the conclusion still is not obvious. Further clarification is required to bring to light the inner workings of these basic substrata included in "space perception." In what follows, the analysis of Sartre's example will serve both as a point of reference as well as a springboard for analysis of other and limiting cases of prespatial constitution.

60. LIMITING CASES OF PRESPATIAL CONSTITUTION

Relaxing for the moment the various steps of the unbuilding epoché that disclose prespace in the broad meaning of the term, "prespace," consider the tactual-visual perceiving of one's own fully constituted percipient organism as in motion or as at rest; it is not only tactually but also visually presented. Nevertheless, some parts of the organism only can be tactually presented in an original way; for example, the back and sides of the head and almost all of the face are not presented visually in an original way. It is of course conceivable that more can be seen of the organism than is actually seen at a given moment. By bringing about changes in somatic states on which motion of the organism is dependent parts of the organism can be made visible. For example, if kinaesthetic sensa-sequences on which turning the head is dependent are actualized in strivings, some of my shoulder is made visible. Yet, this making the "invisible" visible cannot be accomplished in the case of the back of the head. Certainly this does not signify a sort of visual non-being of the back of the head for me; rather it signifies that there is lacking a course of sensing of kinaesthetic sensa-sequences that would otherwise be correlated with the sensing of visual sensa on which would be founded the seeing of the back of the head. To express the matter another way, it is a case of an intentive process not exclusively a sensing which in this case does not have a foundation that is a (visual) sensing.[23]

The lack of a course of kinaesthetic sensa-sequences correlated with sensing of visual sensa is similar to the case where sensing of the tactual *8* in correlation with sensing of appropriate kinaesthetic sensa-sequences is an analogue of the sensing of the visual figure *8*. The back of the head is meant as visible though unseeable. In other words, there is a progressive harmonization of systems of kinaesthetic sensa-sequences correlated with actual and possible changes in different sensum-fields; and in some cases (the visual *8* or the back of the head) there is sensing of kinaesthetic sensa-sequences that serve as analogues for other sensa-sequences that are lacking in the specific sense that *immediate control* of kinaesthetic sensa-sequences regularly accompanying sensings of other sorts of sensa is not or cannot be correlated with *mediate control* of those other sorts of sensa (e.g., tactual changes "out of reach," visible but unseeable sensa, and the like). At one and the same time there is both the *limiting case* of correlations of regularly accompanying sensing of kinaesthetic sensa as well as the *further harmonization* of kinaesthetic sensa-sequences that, as analogues, and only as analogues, refer beyond themselves and thereby motivate expectant intendings. However, in the case of the visual *8*, expectant fictive or feigned or even possible serious intendings are motivated; in contrast, in the case of the back of the head it would seem that only and just expectant fictive intendings are motivated. In short, the phenomenological situation is anomalous, and a brief consideration of its contrast with another example will make it possible to formulate the phenomenological significance of these limiting cases.

Consider the case of sounds heard more or less equal in intensity with the head in a fixed position, auditory diffuseness being centered more or less in the horizontal plane of the binaural axis. When the head turns, some intensities increase and others decrease; there is more and more distinct hearing on one side simultaneous with less and less distinct hearing of the sounds on the other side. With respect to prespatial oriented constitution, auditory perceivings include as substrata sensing of changing auditory sensa as localized in the auditory sensum-field correlative to sensing of kinaesthetic sensa-sequences on which, at a higher level, perceiving of the head as turning is dependent. (To be sure, here there is, strictly speaking, no perceiving of the head; presented are only sensings of muscular, tactual and feeling-sensa.)[24] When the head turns, direction in the auditory sensum-field changes; sound "moves with the body much as one's shadow when walking around a light."[25] When the head turns, correlative differences of auditory intensity are presented "to the right" and "to the left."[26]

Without developing them further, but expressing them much more precisely, it is clear from this and the other examples that just those kinaesthetic sensa-sequences, the sensing of which is a substratum of those intentive strivings to turn the head and which are correlated with sensings of changes in the auditory sensum-field (from, e.g., equal or lesser intensity to greater intensity), are sensed and posited as the *same* kinaesthetic sensa-sequences correlated with sensings of changes in the visual sensum-field (perhaps in conjunction with sensing of oculomotor kinaesthesias). Or, it may be said that those kinaesthetic sensa-sequences, the sensing of which is a substratum in the intentive striving to turn the head so that auditory identicals are localized, is sensing and positing of the self-same kinaesthetic sensa, the sensing of which is a substratum in striving to turn the head to bring those auditory sensa before the eyes[27]--just as those kinaesthetic sensa-sequences, the sensing of which is a substratum in striving to move the eyes to see the visual form, *8*. In brief, the *same* kinaesthetic sensa-sequence is correlated with changes in both visual and auditory sensum-fields;[28]

or, if not intended to as the same, then they are intended to as analogues (in Sartre's sense of the word, "analogue").

An important methodological consequence follows, *for it therefore makes no difference that the initial step of building-up analysis is objectivation of the sensing of the tactual sensum-field regularly accompanied by sensing of kinaesthetic sensa-sequences; it can just as well be any of the prespatial sensum-fields.*[29] Stated universally, any intending to and positing a sensum-field and its changes is ipso facto co-intending to and positing any other prespatial sensum-field and its changes presented in an original way with the sense, "regularly accompanied by sensing of kinaesthetic sensa-sequences." And in so far as there is sensing of regularly accompanying kinaesthetic sensa-sequences, the utterly heterogeneous diversity of prespatial sensum-fields is constituted as a unity that is the basis for the oriented constitution of quasi-objective space. Before turning to the motivation and clarification of this next stage of building-up analysis it is worth while restating four conclusions that may be drawn from the examples:

1. Identicals in the tactual, auditory and visual sensum-fields as well as the feeling sensum-field are constituted and can only be constituted correlative to constituting of kinaesthetic sensa-sequences as self-same.

2. Correlation of the self-same kinaesthetic sensa-sequences, or their analogues, with any one sensum-field founds the mutual appresentation of other sensum-fields, e.g., tactual and visual sensum-fields and their respective changing or non-changing identicals in them. In this respect, it may also be said that diversity of prespatial sensum-fields is presented in so far as they are intended to and posited with respect to sensing of the self-same kinaesthetic sensa-sequences.

3. Objectivation of each sensum-field has also made thematic those assimilative syntheses constituted on the basis of mutual transfer of sense so that, in each sensum-field, some parts are sensed as coinciding with co-sensed feeling-sensa while, at the same time, other parts are sensed as not so coinciding. The coincidence and non-coincidence are mutually dependent; the latter are presented in an original way from here, but are there; the former are always presented in an original way from here, but are never there. As a consequence, not only is each sensum-field co-presented as spreadout and extended, hence as "background" for changes or

non-changes in the field, but it is presented as already constituted with the "directionality" of here and there.

4. Those sensum-fields that are intentive substrata in the quasi-objective or objective, perceived surface of my organism are constituted, first of all, in those assimilative syntheses where, for example, in the case of the tactual sensum-field, there is mutual cancellation of sense between those tactual sensa coincident with feeling-sensa (and presented as regularly accompanied by kinaesthetic sensa-sequences subject to immediate control by intentive strivings) on the one hand, and, on the other hand, those tactual sensa not coincident with feeling-sensa (and presented as only mediately controllable by intentive strivings). There is, then, partial fulfillment, similarity, between some tactual sensa, but cancellation of the senses "coincident with feeling-sensa" and "immediately controllable sensa." Secondly, those prespatial sensum-fields that are intentive substrata of quasi-objective or objective, perceived surface of my organism and the sensing of which is regularly accompanied by the sensing of kinaesthetic sensa-sequences, are characterized by an "incompleteness," by a "lack" of the sort noted in the examples of the visual sensum-field. Owing to the anomaly, the constituting of the organism is "completed" by intending to things other than my organism.[30] The phenomenological significance of this conclusion can only be made thematic later (Chapter Seven).

Taken together, these four conclusions express the phenomenological foundation for constituting those noetic-noematic strata or stories peculiar to quasi-objective space (phantom space) and the quasi-objective organisms in quasi-objective space as well as quasi-objective things and events of other sorts. It is this constitution which will be developed in the next chapters. To formulate these four conclusions I have tried to restrict the language expressing them just to the constitutive levels in question, but with minimal reference to a higher level of constituting. While this ensures a certain degree of precision and accuracy, it also introduces an aura of "abstractness" into the whole discussion. This is unfortunate but, it seems to me, necessary at this point. In order to dispel some of that aura, and by way of conclusion to this chapter, some critical reflections are in order to establish, at least in part, the phenomenological significance of these conclusions for the meaning of "space perception." At the same time, the phenomena under discussion will become more "concretely" recognizable.

61. CONCLUSION: CRITICAL REFLECTIONS

With respect ot the first and second conclusions of the preceding section, I shall seek to show in the next chapters that intending to and positing the correlation of prespatial sensum field-changes with the self-same kinaesthetic sensa-sequences motivates quasi-objective perceivings of various appearances of quasi-objective spatial things in what, first and foremost, is presented as here-centered quasi-objective space. Strictly speaking, the only *unity* constituted in consequence of the founding intendings to prespaces, in the broad meaning of "prespace," is the unity of quasi-objective space and of quasi-objective spatial things. The many diverse prespaces of the many and diverse sansum-fields acquire the *appearance* of *one* quasi-objective space; *in and of themselves*, the prespaces *do not form a unity of one prespace*; strictly speaking, the term, "inter-sensorial space" is a misnomer. They are, rather, constituted as the "appearances" of one quasi-objective space in so far as their diversity is intended to and posited with respect to the sensing of the self-same kinaesthetic sensa-sequences. The diversity and multiplicity of prespatial sensum-fields are, therefore, not chaos.[31] Later we shall see this is not quite the whole story when we examine the constituting of spatio-*temporality* and the meaning of "self-sameness" of kinaesthetic sensa-sequences--a meaning so far left implicit in our discussion (see below, sections 62, 82ff.).

This view shares some similarities with Husserl's account, but differs sharply in other respects. It is worth while, for the sake of contrast and perhaps partial confirmation, to sketch one version of Husserl's account that would seem to be generally representative of his thought.

In his 1925 lectures on phenomenological psychology Husserl states that the data of the various hyletic fields, by themselves, have no hyletic or "sensuous unity with one another." It is only when "apprehension-characteristics" are included and thus concrete appearances of things are examined, that a unity is constituted. More specifically, the unity has its basis in the "intentional characteristics" of hyletic data by virtue of which the data become adumbrations:

> What is produced here is that unity of visual appearances with one another, of tactual appearances with one another, and

> indeed of visual appearances with all others, whereby the one Object continuously appears in them....To each homogeneous group of hyletic Data there corresponds a synthesis peculiar to it in which, as belonging peculiarly to this synthesis, a stratum of the Object of perception appears as an aspect of the physical thing itself....[32]

With respect to the field of "tactual sensation," Husserl notes that not only is it totally distinct constitutively from other fields, but it

> can never mix and unite with the field of visual sensation to make up *one* field. All unification takes place here by virtue of apprehension-functions which make appearances-*of* out of Data of sensation. There, indeed, a visual Datum and a totally different Datum, a touch-Datum, can have unity in that the one is given with the character of an object-adumbration, more particularly, a visual adumbration, and the other, similarly, as a tactual adumbration of the same Object.[33]

On the one hand, in contrast to the view developed here, Husserl holds that what I have called "prespatial sensa" not only adumbrate appearances but also the quasi-objective things presented in and through those appearances. On the other hand, as these and other similar passages in Husserl suggest, hyletic data are independent of their "animating," apprehension-functions. In earlier chapters I have tried to indicate reasons why it is not necessary to accept either the doctrine of hyletic data or the apprehension-functions in order to adequately account for prespatial sensa and their fields *founding*--not adumbrating--quasi-objective perceivings of phantom-things in quasi-objective space. Nonetheless, I agree with Husserl that there is no one prespatial spreadoutness in either the broad or narrow meaning of the term, "prespace." To be sure, there is no agreement that what is accordingly disclosed is a unity of quasi-objective space connected with adumbrational functions of certain sensum-fields and their saliencies. As a consequence, for Husserl, but obviously not here, the task of building-up analysis consists of progressive reintroduction of "animating" apprehension-functions as operations performed on hyletic data said to be really immanent in certain mental life-processes.

With respect to the third and fourth conclusions of the preceding section, it may be noted that in manuscripts written around 1927 Husserl raises the question about the empathic origin of intersubjectivity.[34] In so doing, he makes several observations significant for the first step of building-up analysis undertaken here. For instance, he describes a situation similar to the Sartrean example of the tactual and visual form *8*. Husserl's example concerns the "eyes" of the phantom organism: in effect, he notes that all of the organism cannot be seen, although most parts of it can be touched; as in Sartre's example, the visual "image" is transferred to the tactual "image" of an unseeable part of the organism; yet it is, so to speak, an "unrealizable" "image." The eyes, after all, cannot be seen.[35] It is of course possible to imagine one eye seeing the other in much the same way that one hand touches the other; but although imaginable, there is no "realization" of one eye seeing the other. A "contradiction" arises in the constituting of the produced unity that is the quasi-objective organism. In other words, there are cases where there is not even an analogue of the visual "image."

This points to a difference arising the moment when these examples are compared. The Sartrean example is intructive because it helps make thematic that in a unified quasi-objective spatial field the tactual is appresentatively extended beyond itself by quasi-objective co-perceiving of the visual. But in the example of the eyes, or of the back of the head, there is no visual extension (no realization of the analogue). The constituting is therefore "incomplete" or, as Husserl says, a "contradiction." As will be seen, this is but one facet of the anomalous constituting of one's own organism. And yet another facet, already noted, is that the quasi-objective place of the quasi-objective organism is one that always remains Here. Thus building-up analysis of the spatial form of the real, objective world really begins, now, with objectivation of here-centered quasi-objective space. And while both these facets of constituting the quasi-objective organism will be considered in some detail in the next chapters, it may be noted now that quasi-objective and objective perceiving of one's own organism is "completed" by perceiving of quasi-objective and objective things other than it. The "incompleteness" is de jure tactual and, as it were, "room is made" for things other than one's own organism. To be sure, this is hardly the whole story of the origin of intersubjectivity with respect to the oriented constitution of the quasi-objective phantom substratum of the real, objective world. But it is an aspect of central importance. Certainly because of tactual "incompleteness" from Here, tactual experience--adherring not only to the surface of the organism but also incapable of being directly unfolded before it because, for instance, there is lack of "visual extension"--of the phantom organism is radically distinct from experience of all other tangible phantoms.[36]

CHAPTER SEVEN

THE TRANSCENDENTAL PHENOMENOLOGICAL BUILDING-UP OF PHANTOM QUASI-OBJECTIVE SPACE. THE TRANSCENDENTAL PHENOMENOLOGICAL "DEDUCTION" OF SPACE

62. THE TEMPORAL FORM OF PRESPACES AS PHENOMENOLOGICAL FOUNDATION OF THE ORIENTED CONSTITUTION OF QUASI-OBJECTIVE TIME AND SPACE

So far the focus of phenomenological objectivation has been on those mental life-processes which include as substrata "my" *actual* sensings of "Data of sensation." As yet, the focus has not been on those actual sensings that are foundations of mental life-processes belonging to others, although, as has been emphasized, the actual sensings belonging to others are implicit in "mine." How this is so will be the topic of Chapter Eight. Here it is necessary to turn the objectivating regard to something else, the thematizing of which has been postponed long enough: To begin, as I have done, with "my" actual sensings signifies that the *constitutive clue* for building up the spatial dimension of the phantom world is the present or current phase of "my" sensing-mental living, the present phase of "my" consciousness of sensum-fields and their salient changes. Drawing on Husserl's lectures on the "consciousness of internal time," and in anticipation of the "deduction of time" in the next chapter, the phenomenological situation that I have in mind may be represented graphically as follows:

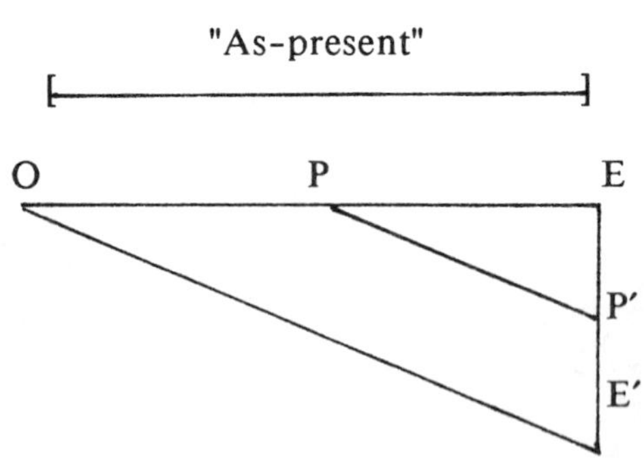

In the diagram, [——————] represents the constitutive clue pointing back to the "impressional" intending to (*scl.* sensing of) the "object;" in addition, the diagram represents a set of distinctions to be made in the temporal stream of the intending: O to E represents a series of "now points;" O to E′ represents the receding of such a "point" into the past, P to P′ the receding of yet another "point" into the past; and E to E′ represents what we may call with Husserl a continuum of phases.[1] With respect to the manner of presentation, e.g., the prespatial tactual (or visual, auditory) unity is presented "as-present;" this Husserl calls the "mode of temporal orientation" of the unity in question. It is of course possible to distinguish a continuity of other modes of temporal orientation (represented, for example, on O to E′), such as O as just-past, or just-just-past; and so forth. As it were, the "date" in time of the enduring prespatial "object" is kept, but, in analogy with spatial orientation, Husserl says, the "distance" in time increases (represented in the diagram by the contimuum of phases E to E′). Husserl also refers to it as the "difference in the mode of orientation to the present." As will be seen in the next chapter, the analogy can be carried further so that it is possible to distinguish a "non-oriented" enduring temporal flow of the present phase which points to *earlier and later*, *but not to past and future*. In other words I can, in contrast, objectivate and seize upon the mode of temporal orientation as the "mode of recession" ["*Ablaufmodus*"],[2] a "now point" becoming more and more past. Indeed, to examine something "as present" it is necessary to consider it both in its temporal orientation (how it is present) and its mode of recession (how it becomes more and more past).

But the intending, such as the tactual sensing, likewise has *its* temporal horizon back to which the "as-present" points: for example, the future sensing of something enduring "as present"--a protentive sensing that "recedes," "sinks back," into the continuum of phases or extents of sensing as it proceeds. Moreover, the *future phase* is itself intended to by the present "impressional consciousness" as *having been protended to* in its "manner to come;" while the past protentive phase is "still there" as subsequently "retained." In other words, the primary passive sensings in the various examples used before all are presented as having horizons of possible future phases of the same sensing; the future phase is meant as having been expected in its manner to come; at the same time, there is going on a continuous "retention" of (or: "retrotending to") the "impressional" extent or phase of sensing as it recedes into the past (O to E′ in the diagram). Accordingly, I may say that the "impressional" phase of a given intending to and positing of something enduring as the same throughout its own temporal extent in primary passivity is continuously a retrotending to itself all the while protending to itself in its manners or modes to come: the "impressional" phase *continuously identifies itself as the one phase of the sensing previously protended to and subsequently retrotended to.*

The "impressional" phase of the sensing includes, therefore, both identifying as well as differentiating syntheses--the "present of the present" becomes identified with the "future of the past" and the "past of the future;" the "past of the present" is identified with the "present of the past," while the "future of the present" is identified with the "present of the future." And in each of these cases there is a continuity of modes to be objectivated and seized upon: for example, the "future of the future," the "past of the past;" and the like. Similarly, there are also differentiating syntheses such that no matter how much alike, the phases are temporally individuated; if successive phases are alike, then the phase is presented as enduring; in other words, the unifying in syntheses of identity of the enduring phases is constituted *on the basis of temporal individuatings.* This signifies that in both sorts of intentive synthesizings there is a universal principle of synthetical constituting operative: if there is intending to (*scl.* sensing of) two (or more) "objects"--be they immanental or transcendent to the temporal flow of primary passive mental life-processes--they are *not intended to in isolation of one another* but the sense of the one is "automatically" *carried over to the other*, be it a phase of mental living, be it a mode of temporal orientation, be it a plurality of sensa. In short, nothing whatever is intended to in isolation but always already as similar, as perfectly alike, or as utterly unalike, just to mention only the limiting cases.

Now, in the situation and affairs objectivated and seized upon in the last chapters, the *temporal dimension* of oriented constitution just sketched signifies, in addition, that the *constitutive clue for building-up analyses* consists of precisely those assimilative and differentiative syntheses of utter difference founded upon sensings and co-sensings of various sensa and kinaesthetic sensa-sequences that make up the prespatial and temporal sensum-fields, i.e., the "spreadoutness" of sensum-fields simultaneously presented by virtue of being founded on the simultaneous presentation of the particular founding "contents" (at the lowest constitutive level objectivated, prespatial "spreadoutness;" at the next higher level, prespatial identicals over against prespatial field-backgrounds). Appearing in primary passive intendings to them, the fields are constituted in assimilative and differentative syntheses not only as alike in being simultaneously presented and as utterly heterogeneous, but also as alike in being sensuous. That is: alike in being simultaneously presented; "generically" different; "specifically" alike as sensuous. On the basis of *assimilative and differentiating* syntheses, the prespatial sensum-fields are constituted as *generically different and specifically* alike.[3]

As a consequence, at the level of prespace in the broad meaning of the term, "prespace," both generic and specific non-formal or "material universals" (above, section 33) are already constituted in primary passivity. And *this is the founding motivation, the "as-present" in simultaneity, for building up one concrete real-world space and time*. Most immediately, it is the basis for oriented constitution of *quasi-objective space and time*: it is a *quasi*-objective "as-present" because it is related *only to "my"* actual sensings and perceivings and "not yet" to those of other mental lives--another essentially necessary facet of the "incompleteness" of oriented constitution noted before.[4] It is important, too, that emphasis is given again to the fact that there is no presented substratum of so-called "intersensorial space" lying between the intentive substrata of constituting characterized by sensing of sensa and sensum-fields, on the one hand, and the substrata of constituting characterized by quasi-objective perceiving and apperceiving of phantoms and phantom-fields on the other hand. At best, "intersensorial space" would seem to be an adumbrational fiction derived from the theory of hyletic data.

In conjunction with the constituting of generic and specific "universals," the other factor of importance to be considered now is the motivation for striving to fulfill more than "my" actual experience of space. That is, on the one hand, I discriminate the foundation for constituting one and the same space and time of which "my" actual present is but one possible example; on the other hand, I discriminate, in primary passivity, striving to "complete" actual sensings correlative to the self-identifying and differentiating of the temporal stream of my mental life-processes. In the remainder of this chapter, and in part of the next, special attention will be paid to the constitutive "completion" of prespatial "incompleteness." Two basic forms of "completion" will be considered in the building-up analysis of the real, objective world: here-centered phantom space and centerless phantom space.

63. PRELIMINARY EXPLORATIONS OF QUASI-OBJECTIVE SPACE

It is worth while to begin with the "obvious:" as related to "my" actual sensings, changes and non-changes in the assimilatively synthesized fields are functionally dependent on "my" somatic states; they stand in correlation with primary passive strivings to make such changes *near*, to "complete," e.g., the sensing of one and the same change. For example, we discriminate strivings to control what is presented immediately, hence what is preferred in sensing something relative to an always actually sensed and posited here so that a color patch remains constant throughout different lightings, or so that the surface of a touch-area remains smooth and "in touch," or so that sounds remain constant in pitch. In other words, striving to control what is presented immediately is, *in primary passivity*, *striving to achieve optimal presentation--clarity*, "*knowing touch*" (Katz), *acute hearing*. "*Dependency*" therefore signifies in this context optimal presentation relative to the background of an invariable and actually unvarying Here--the most primitive meaning of *near*. One and the same sensum presented in the sensing of it as undergoing qualitative change relative to "my" actual Here is, at the same time, most originally presented as an aspect or perspectival appearance, but also as one among many possible actualizable aspects correlative to actualizing qualitative changes or non-changes actually Here, controllable immediately. It is possible to discriminate systems, "sets," of correlated appearances--the same under varying aspects correlative to still other changing aspects of the same head or eyes as turning, where the one set of aspects is immediately controllable as actually present, while the other set is mediately controllable or presented as an analogue of the mediately controllable.[5]

The most immediate phenomenological datum is therefore a *centeredness-here* with respect to which all sensed changes and non-changes in any given sensum-field whatever are presented as relative to Here and only to Here. To be sure, what comprises our *phenomenological theme* now are not mere sensed changes and non-changes but instead aspects of the same now, e.g., touched as here, now as over there; now as seen here, now as seen over there but from here. And it is this "perspectival relativity" from Here that serves as motive for actualizing certain kinaesthetic sequences in preference to others likewise regularly accompanying sensing of various sorts and on which, say, turning the head or eyes or moving arms or hands depend in their turn. For instance, oculomotoric and cephalomotoric kinaesthetic flow-sequences are actualized such that eyes and head are sensed correlative to the sensing of something remaining at the center of the prespatial visual field. And those "somatic" states sensed or co-sensed are, in turn, dependent on the actualizations of such kinaesthetic flow-sequences. Indeed, the actualization of these and other kinaesthetic sequences, motivated by primary passive preferrings of clarity over obscurity, of smooth over rough, low over high in pitch, based upon mutual transfer of sense, is presupposed throughout the assimilative-synthetical constituting of sensum-fields. In turn, these mutual transfers of sense and assimilative syntheses found quasi-objective perceivings, co-perceivings and apperceivings of something as one and the same throughout changing aspects, appearances. Let us consider another example: At the level of the fully constituted real, objective world, standing next to the table.

At the level of oriented constitution made thematic here we can objectivate the following: A plurality of prespatial shapes and sizes are presented in the visual sensing of them at different places; there is mutual transfer of sense: shape and size There and shape and size Here; mutual likeness and also difference are constituted with respect to place: at There shapes and sizes are presented with respect to Here where "zero" shape and size (Husserl) are presented, a shape and size never otherwise than Here, thus presented as having a constancy always relative to Here; and the "zero" shape and size Here are transferred There.[6] Moreover, such sensory constitution I find included as foundation in *quasi-objective* seeing of the shape and size of the animate organism Here; more particularly, I objectivate a partial seeing of "my" quasi-objective organism that also co-intends to unseen portions of shape and size Here and, likewise, the seen shape and size There are co-intended to as having unseen portions (the underside of the table, for instance). Thus two shapes and sizes: the "zero" shape and size of "my" organism Here, and the shape and size of the table There. In mutual transfer of sense the difference arises which is more than mere difference of Here and There because, in the one case, seeing of the quasi-objective animate organism, some of the unseen portions co-intended to cannot be seen no matter what kinaesthesias are actualized; thus "my" eyes can turn, but cannot be seen. Kinaesthetic sequences can be actualized on which turning of the eyes depends, but they can only be touched and not seen; in contrast, the touched underside of the table can be seen and is seeable by virtue of actualization of certain kinaesthesias on which bending the head to look under the table are dependent.

The difference that arises in this transfer of sense, then, is not just a matter of dissimilarity but rather of conflicting noematic senses that found cancellation of partial, seen similarity of shape and size. It is just this conflict in the midst of sense-cancellation, so to speak, that motivates still further carry over of noematic sense relative to the "zero image" of the animate organism Here and what we may call the "zero + n" image of the table There. More specifically, as we shall see in the next sections, what is motivated is the preference for actualization of locomotor kinaesthetic sequences on which movement from place to place depends (movement under "my own power").

64. SOME PHENOMENOLOGICAL CHARACTERISTICS OF THE CONSTITUTING OF QUASI-OBJECTIVE SPACE

Moreover, were the phantom organism never moved from place to place "under its own power," were it never to transform a There into a Here, were it never to effect a "mobility" in its hereness, here-centered quasi-objective space would never take on the appearance of a centerless quasi-objective space. But were the phantom organism only carried from place to place, the centered quasi-objective space would be intended to as a "stretching out" of here-centered quasi-objective space rather than as a genuine appearance of centerless quasi-objective space; in other words, it would not be otherwise intended to and posited as one possible Here among a multiplicity of possible Here's. Otherwise here-centered quasi-objective space would remain an anomaly. Only by virtue of the actualization of sensed sequences of locomotor kinaesthesias is it possible to transform, e.g., quasi-objective visual appearances of a distant phantom into the quasi-objective visual and tactual appearances of the same phantom as near, as of the same shape and size whether or not it is actually presented as near or far from Here. This is the *second* phenomenological characteristic of a quasi-objective centerless space: not only is the same intended to and posited at differing places, the *first* characteristic, but as the same no matter if near or far. A *third* characteristic is that the same quasi-objective phantom unity can appear as one and the same in a reverse order of appearances; e.g., it is seen and posited as the same shape, size, and the like, even if it returns to its initial place. This third characteristic requires further clarification in terms of what will be called, in the next section, the "phantom core-world." For the moment several other characteristics must be mentioned.

Although it makes no difference, phenomenologically, which of the prespatial sensum-fields in the broad sense of the term, "pre-space," is to be "first" objectivated, there are nonetheless important differences that accrue to them as appearances of quasi-objective space. For instance, the presentation of the centerless quasi-objective phantom space is peculiar in so far as visual constituting is concerned. It is peculiar because even with the sensing or co-sensing of locomotor kinaesthetic sensa-sequences that are actualized or actualizable, the visual phantom animate organism is not presented, as a unity, in and through a quasi-objective perceiving of it as "moving." This phenomenological situation is accurately described by Claesges when he says that

> The oculormotoric constitution of the organism is delimited in a distinctive manner; this delimination is a multiple one. In the first place, there is with the organism no movement flowing from <the organism> 'in and of itself' ["*von Selbst*"] as coming closer to and moving away <from itself>; nor is there any <movement> of the sort that I can correct by a suitable kinaesthesia such as is possible for all other sorts of phantoms. In the case of the visual organism, the cyclical

> kinaesthesia necessary for the constitution of a closed surface figuration [*Oberflachengestalt*] is impossible. (I cannot look at my organism from all sides.)[7]

Claesges bases his account on the following statement of Husserl:

> <In its visual constitution, the phantom organism has> its members which are experienced as moving or as not moving; but as a whole the organism cannot be experienced optically by perspectival constitution (in connection with constitution by relief) now as resting, now as moving <--it cannot be so experienced> by that constitution by which we experience external physical things as original and genuine realities in their real world.[8]

Making some allowance for a certain degree of ambiguity in the last part of Husserl's statement (is he contrasting the phantom organism with other phantom bodies, or with physical things in the fully constituted real, objective world?), it would seem that his point is that, *as a whole*, the quasi-objective *visual* phantom organism is presented as neither "at rest" nor "in motion;" the visual phantom organism can appear neither as receding nor as approaching itself--regardless of actualization of sensed locomotor kinaesthetic sensa-sequences. As it were, its "incompleteness" is confined to an "absolute" center of nearness. Similarly, the parts or members of the phantom organism, for instance, a phantom arm moving up or down, "moves" only within certain optimum limits; strictly speaking, they cannot disappear from the visual field as can other visual phantoms and their parts.[9] The nearness-determination is one that persists with the phantom organism throughout all change of place. Moreover, the *tactual* quasi-objective phantom appearances of the animate organism similarly refer to Here--to the center of nearness not only from "within" but from "without" as well (the hand touched is always localized here). Directly and immediately, all localization occurs within this "absolute" nearness. In short, it may be said that at this level of oriented constitution all systems of perspectival appearances and aspects refer to those of the striving, percipient animate organism at the center of the phantom field of experience.

For example, let us suppose that a phantom appearance is seen at a distance from here--a quasi-objective seeing, let us say, of a qualitied red patch at a distance from here. At the same time that I objectivate the seeing, I objectivate a primary passive striving to actualize a course of locomotor kinaesthetic sequences in consequence of which mediate changes are effected in the presentational mode of the red patch so that it appears as something near both visually and tactually. At a much higher level of constitution, I see a red tomato on the vine in the garden and go over to touch it in order to see whether it is ripe. With respect to the level of constitution under discussion, however, the most that I can say is that when actualized, sensed locomotor kinaesthetic sequences, as it were, amplify the continuum of quasi-objective seeing, of actual seeings that extend beyond just any one "point of view."[10] Specifically, the seen approach to the phantom is presented both visually and tactually as one among many other segments of the whole co-perceived quasi-objective phantom perceptual field. In consequence of strivings to bring about these rather than other phantom appearances, the seen limit of distance, for instance, is, as it were, pushed farther and farther away (with corresponding increase in "angular dimensions" of vision as the phantom appearances of the patch become nearer and nearer). Although the quasi-objective appearances of the phantom organism (e.g., visual and tactual) remain unvaried as the phantom organism approaches the patch, those of the patch undergo continuous alteration of far-appearances into near-appearances.

Accordingly, changes and alteration in qualitied, perspectival appearances are directly correlated with changes in quasi-objective spatial orientations relative to the seen and touched line of approach; in turn, these orientations vary with the changes in co-perceived somatic states of the phantom organism which are themselves functionally dependent on continued actualization of co-sensed locomotor kinaesthetic sequences. This interlocking and correlative series of quasi-objective appearances holds not only between the self-same phantom patch, presented now as far and now as near, but also between the patch and the particular members of the phantom organism (e.g., the hand reaching out to touch the red patch): one and the same phantom organism (or: one and the same member of the phantom organism) presented in and through quasi-objective *non-perspectival* appearances (*scl.* the "zero image") relative to one and the same phantom patch itself presented in and through quasi-objective perspectival appearances. As a result, "room" is made, so to speak, for the presence of something other than "my" organism: the combining of quasi-objective, say, visual appearances that are, to be sure, incomplete, but which allow of no perspectivalization into near and far, into approaching and receding, hence in and through which the phantom organism is presented and posited as permanently near in the visual field on the one hand, and, on the other hand, the perspectival quasi-objective visual appearances that are presented as far and then as near-appearances.

It is this combining which is objectivated phenomenologically as the essentially necessary foundation for constituting the visually one-sided "confrontation," which characterizes quasi-objective visual perceiving. Moreover, in this case the visual "image" is not an analogue of tactual perceiving but rather an "image" in its own right. Indeed, combined of necessity with the system of tactual quasi-objective appearances by virtue of the functional dependency on kinaesthetically-based somatic states, the visual-tactual system of quasi-objective appearances of the phantom organism as a whole, there is in these unifications a *differentiation* of the phantom organism from other actual phantom things such that "through my body I go to the world, and tactual experience occurs 'ahead' of me and is not centered in me..." (Merleau-Ponty[11]). The interlocking non-perspectival appearances funtionally correlated with phantom somatic states found the constituting of things and qualities other than the phantom organism. In short, the "self" is constituted mutually correlative to "other things" generally, some of which turn out to be "other selves." But it is too soon to develop this phenomenological finding. Nevertheless, it is worthwhile noting that included in this level of oriented constituting are the rudimentary elements of the "self" and "other selves" as senses of certain intendings proceeding in primary passivity.

65. ORIENTED CONSTITUTING OF THE QUASI-OBJECTIVE "CORE-WORLD"

The *non-perspectival* visual (but also tactual, auditory) constituting of the animate phantom organism, and its unification with tactual and other quasi-objective appearances of the same may be said to comprise a *quasi-objective "core-world,"* a specious proximity of the phantom organism to itself, a hereness and uncentricity that establish a "zero point" of orientation because the organism is presented most originally as a non-oriented phantom unity in its quasi-objective appearances. The "zero point" is either here-centered or centerless depending on actualization or non-actualization of locomotor kinaesthetic sensa-sequences: the "zero point," for instance, will acquire different appearances at higher levels of oriented constitution. In contrast to a here-centered quasi-objective spatial field, a centerless quasi-objective spatial field is then one intended to and posited as a continuously varying segment of the field. Equivalently stated: only part of the centerless quasi-objective spatial field is presented at any given moment, and different parts are presented when the phantom organism moves from place to place, but always only from the *actual* standpoint of the phantom organism. This presentational mode must be sharply distinguished from that of parts of the quasi-objective spatial field relative to the actual standpoint of the phantom organism that does not move from the spot, as it were, or only turns around at its place. In such terms we may restate what has just been objectivated by saying that that "part" of the quasi-objective spatial field presented from the actual standpoint of a mobile phantom organism here, "absolutely" near, is a "core-world" in the pregnant sense of the term, "core-world."[12] It is a core-world in the sense of a here-nearness that is intended to and posited as transferred and transferrable from place to place in quasi-objective space.

Fundamentally, the quasi-objective phantom core-world is presented as having the noematic sense of Near in a double meaning of the term, "near:" in the first place, visual appearances of the phantom organism are not spatially oriented as near and far; they do not recede or approach and, e.g., visual size remains constant (is both intended to and posited as constant); and so forth. In the second place, the core-world is intended to as tactually perceivable regardless of whether or not regularly accompanying locomotor kinaesthetic sequences are actualized. In the ongoing course of mutual carry over of sense, this sense, Near, is transferred "automatically" to other parts of the co-perceived quasi-objective spatial field as a whole. But there is also discrimination of a partial cancellation, a negative fulfillment, of the transferred sense, Near, in the assimilative syntheses of similarity: the core-world is presented throughout as having optimal boundaries beyond which phantoms disappear or at which they enter one or another part of the field.[13]

Thus it can be observed that when phantom unities are presented, e.g., in the quasi-objective seeing of them as receding from Here, they continuously appear in a manner similar to those phantom unities enduring nearby; far-phantoms "behave," as it were, like phantoms nearby with respect to their presentedness with determinations of spatial orientation, except that they appear as having a "smaller format."[14] They are meant, in other words, in primary passivity as similar to phantoms in the near-sphere, e.g., with respect to size and shape, but with a smaller format *so that only the format, and not the size and shape, are meant as changing* (Husserl: "like pictures in a picture book").

As will be seen later, what is objectivated here is the intentive foundation included in all perceivings of so-called "constancy" of size and shape of things in the real, objective world. To be sure, the synthesizings here are assimilative *and differentiative*: the far visual phantoms, for instance, are also constituted as dissimilar in so far as their modes of quasi-objective spatial orientation are not immediately variable, unlike visual phantoms nearby. As a consequence, as phenomenological on-looker I have to say that the *most originary visual mode of presentation of the phantom visual core-world is the sense*: *"here-centered quasi-objective spatial appearance of a centerless quasi-objective space."*

So far the emphasis has been on the visual dimension of the core-world. It is necessary to balance the phenomenological ledger with a few comments about the tactual dimension, especially with respect to the oriented constitution of the animate phantom organism. This is because there are radical differences in the oriented constitution of the animate phantom organism despite the many similarities with the oriented constitution of any other quasi-objective phantom thing. Two such radical differences may be singled out for mention here.

The first difference to note concerns the fact that the phantom organism is the only quasi-objective physical thing on which any of "my" fields of "sensation" are spread out. The phantom hand, for instance, touched by the other phantom hand, is itself something both perceived and on which a part of "my" tactual field is spread out. The one hand, when touched by the other, is simultaneously presented as something tactually perceived and as something with which the other hand is perceived. And this holds *mutatis mutandis* for all other members and parts of the phantom organism regardless of what other function they have or might have. In short, the quasi-objective animate phantom organism is constituted as a quasi-objective physical thing and as an "organ of perception." To be sure, it is necessary to emphasize that the *philosophical significance* of this constitution is that the phantom organism is already the embodiment (or: mundanization) of transcendental mental living. Indeed, to speak of its oriented *constitution* is to speak of nothing else but the "consciousness of" the phantom organism. Shortly we shall return to this significance.

The second radical difference to note concerns the fact that there is no analogy between tactual sensa on the one hand, and, on the other hand, tactual qualia. In other words, for example, tactual sensa founding quasi-objectively perceived tactual qualities are not themselves presented as rough or smooth. Only the surface of a quasi-objective physical phantom thing, perceived on the basis of sensing of tactual sensa, is presented as rough or smooth.[15] In the example of one phantom hand touching the other, two such surfaces are sumultaneously presented, the quasi-objective tactual perceiving of which is founded on the sensing of the same tactual sensum-sequence. Moreover, at the level of quasi-objective space, this sensum-field and its salient sensa, the sensing and co-sensing of which found the quasi-objective perceiving not only of the phantom organism but also of other physical phantom things, are located in and on the phantom organism. As a consequence, what I objectivate are changes and non-changes in the quasi-objective appearances of other tangible phantoms besides the phantom organism that are correlative to changes and non-changes in differing areas of the tactual sensum-field located in and on the phantom organism as whole.

To be sure, this sensing of the whole sensum-field founds but part of the quasi-objective spatial field of tangible phantoms. Thus, in so far as sensing of the whole tactual sensum-field located in and on the phantom organism founds quasi-objective perceiving of the appearances and concatenations of appearances in and through which the phantom organism is presented, as well as those quasi-objective appearances and their concatenations in and through which other tangible phantom unities are presented, "tactual experience" always occurs ahead of the absolute nearness of the phantom organism. The reasom should now be obvious: the *founding whole* is *but always only a part* of the quasi-objective spatial tactual field.

And, as noted in the case of the quasi-objective visual spatial field, the whole co-perceived tangible field, presented most originally as part of the quasi-objective spatial field as absolutely near, appears "non-perspectivally." The "non-perspectival" tactual, and the perspectival visual quasi-objective appearances in and through which the same phantom organism appears, characterized by the "discrepancy" between tactual sensa and tactual qualia on the one hand, and by the "incompleteness" of visual constituting on the other hand, comprise the motivation for oriented constitution of *primordial space*. At the same time, the phantom organism clearly is not constituted as but one phantom body among others; at the very root of the "genesis" of its oriented constitution, the phantom organism is always objectivated as already "privileged" as foundation for constituting the "lived body," the "*corps propre*."

66. THE PHENOMENOLOGICAL SIGNIFICANCE OF THE FOUNDEDNESS OF THE PHANTOM CORE-WORLD

It is worth recalling that the basic sort of qualitative change objectivated in the tactual prespatial spread in the narrow sense of the term, "prespace," is change on a continuum of sensa-sequences replacing one another (sections 43, 44). Two sorts of presentational modes were distinguished in this connection: presentation of touch surrounding non-touch, and presentation of "closed surroundingness" (e.g., tactual sensing of contiguous touch-areas). Also objectivated was the limiting case of the most original presentation of the prespatial tactual spread: "resistance." Intending to these changes that do not "extend" into each other founds intending to those areas of tactual change as "coinciding" with co-intending to fields of "feeling sensa." Constituting of the unified feeling-field and prespatial tactual sensum-field motivates strivings to actualize such co-sensed kinaesthetic sensa-sequences as regularly tactual sensa, e.g., of "lesser intensity" or even of "zero-intensity." Indeed, whole systems of actual kinaesthetic sequences are presented at this level or substratum of oriented constitution as correlated with changes in tactual sensa.

An example of the constituting of weight will serve to illustrate the "systematization" of kinaesthesias in this context. The sensing and co-sensing of a plurality of touch-areas pertaining to the tactual sensum-spread found quasi-objective perceivings of "part-appearances" (Husserl) of the phantom organism presented as in contact with segments of the external phantom quasi-objective world. Certain parts of the phantom organism are presented, more particularly, as in contact with, e.g., the phantom "floor" and another tangible phantom being lifted, e.g., the "table." The phantom hand, let us say, lifts the phantom table. Simultaneously, qualitative changes are perceived and co-perceived over other parts of that segment of the tactual sensum-spread located in and on the phantom organism and which are not in contact with that part of the sensum-spread founding the quasi-objective perceiving of the table and hand undergoing changes in degree of "pressure;" on the basis of sensing changes in these other parts, "weight" is "felt" in the unseen but tactually co-sensed phantom back of the organism, in the phantom forearms and forehead. Thus at the same time that, so to speak, "fore and aft" are determined in the tactual sensum-field coinciding with the feeling-field located in and on the organism, "an impression furnished by <e.g.> the muscles of the fingers is equivalent to an impression furnished by the whole hand."[16] Precisely by virtue of this "equivalence" the localizations of back and front are determined. Eventually the "equivalence" includes the phantom organism as a whole, "systematically embracing all its parts."[17]

More accurately expressed, what may be called a synthesis of "equivalence" founds the presentation of the "synergic unity" peculiar to the constituting of the quasi-objective phantom organism. Thus intendings to patterns of kinaesthetic sensa- sequences sensed correlative to sensing of those sensa that found quasi-objective perceivings of "non-perspectival" tactual appearances of the phantom organism include intentive "feelings" in which the phantom organism is presented as a percipient, sentient whole: the most original presentational mode of the quasi-objective phantom organism as phantom percipient organism. And preferring, in primary passivity, one course or pattern of kinaesthetic sensa-sequences to another regularly accompanying tactual sensing of change (or non-change) is the substratum included in strivings to control immediately certain tactual changes and certain other tactual changes mediately. Tactual changes presented in and through non-perspectival appearances are precisely and only those immediately controllable; tactual changes presented in and through perspectival tactual appearances are precisely and only controllable mediately.

As a consequence, the phenomenologically objectivated phantom core-world in all its complexity is throughout a founded phenomenon; the very constituting of the phantom organism, of that system of non-perspectival and "incomplete" visual and tactual appearances designated as "organism" in the fully constituted, real objective world is a founded constituting. So far as the philosophical significance of this fact is concerned, it signifies that the oriented constitution of the phantom organism in surrounding quasi-objective space results in an original possession of a "world" standing in functional correlation with the phantom organism. But owing to prespace (and, as will be seen, pretime), access to the fully constituted real, objective world is already, as it were, made available; the possibility of "my" belonging to the world "through" my animate organism is already established in primary passive intentionalities. The problem of access to the real, objective world is the shape of the problem of space that results in the Cartesian setting of space "free."[18] In the remainder of this, and in the next, chapter, I shall seek to lay the phenomenological groundwork for developing the problem of access as the problem of the oriented constitution of the real, objective world. That groundwork will serve, in turn, as the foundation for the "perception of other minds" in Chapter Nine. To make way for this groundwork, some supplementations are required for our building-up analysis.

67. THE ORIENTED CONSTITUTION OF HERE-CENTERED AND CENTERLESS QUASI-OBJECTIVE SPACE. SUPPLEMENTATIONS

The last sections examined some details of the phenomenological foundations or substrata of the more complex mental life processes in which the quasi-objective spatial dimension of the quasi-objective phantom world is constituted. Some supplementations are required now, one of which concerns the constituting of the visual quasi-objective space peculiar to the here-centered spatial spread. Recall that when discussing the unbuilding to the visual sensum spread in the narrower meaning of the term, "prespace," it was noted that that spread is presented most orignally as having the noematic sense of "closed" and as being presented with the determinations of either convexity or concavity (section 46f.). With respect to the present stage of building-up analysis I find that, on the basis of visually sensing certain changes in the prespatial field a roughly convex or concave spreadoutness is presented and posited as substratum of quasi-objective perceivings of a roughly "spherical" quasi-objective perceptual field of visible phantom things. Such perceivings also have a foundation that is, in part, sensing of kinaesthetic flow patterns accompanying intentive sensings of changes that are, in turn, substrata of quasi-objective co-perceivings of the phantom organism as "turning on the spot" until visual sensa that have "disappeared" again "reappear" at the opposite side of the perceptual field. In other words, there is a correlation between sensing those kinaesthetic sensa-sequences actualized in "turning" or "rotating" and sensing of the appearance and disappearance of visual sight things from the field regardless of whether those kinaesthetic sensa-sequences also accompany sensings of other sorts of sensa that are substrata or foundations of quasi-objective co-perceivings of any other motion of the phantom organism (e.g., locomotion).

By way of illustration of this phenomenological situation we may borrow an example used by Heinrich Hofmann: the seeing of a locomotive pull into the station.[19] With respect to experience in the real, objective world this would be a case of seeing a locomotive approaching us as we stand on the platform of the railroad station to watch the train go by. But with respect to the substratum of *here-centered*, quasi-objective space there is a seeing and positing of a phantom unity in and through a multiplicity of quasi-objective phantom visual appearances; the seen phantom locomotive-thing is perceived as continually "growing" (Becker)[20] as it "approaches" on the line of sight. Slowly at first, then more and more rapidly as it "comes closer," the apparent quasi-objective seen sizes pertaining to the sight thing are presented as "larger" and "larger." Strictly speaking, the phantom locomotive-thing is seen not as "moving" along a "straight path" but as "approaching" along a "curved path." With the phantom organism rooted to the spot in here-centered quasi-objective space, a sequence of changing seen sizes along a curved path are meant and intended to as a multiplicity of visual determinations of one and the same sight thing waxing in size until a limit point is reached. Equivalently stated: the phantom locomotive-thing is presented through seen apparent sizes "rotating" around a continually changing "axis of sight" until the visible surface lies in a plane roughly perpendicular to the line of sight. When the visual changes cease, the quasi-objective visual appearances are no longer presented as "rotating." Graphically it is possible to represent the phenomenological situation in this way:

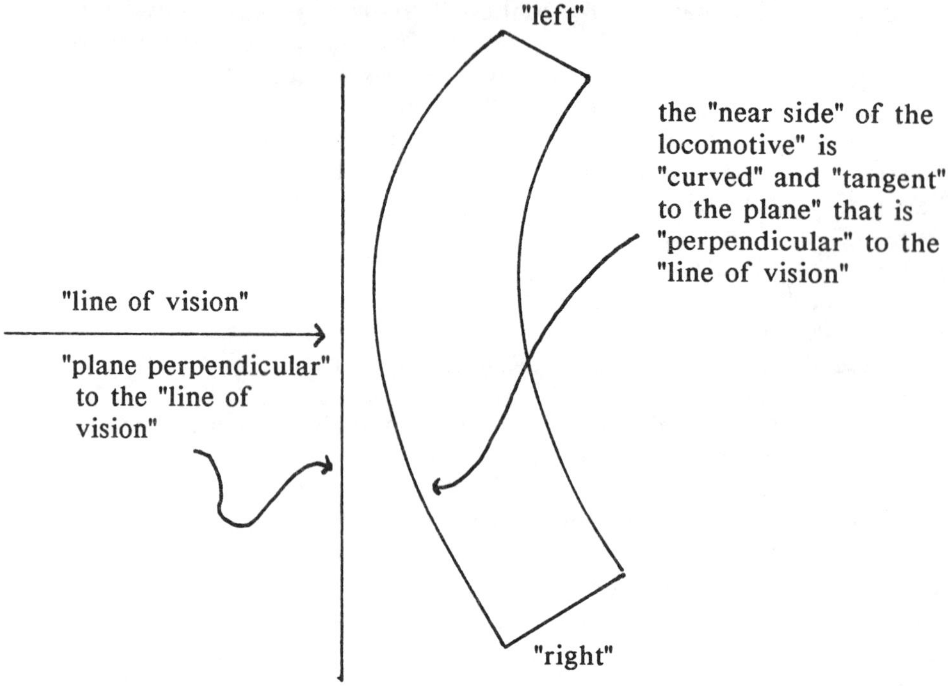

Leaving aside for the moment certain questions concerning the constituting of "size constancy," the graphic representation may be elaborated as follows: The sensing of sensa in visual prespace that are substrata of the quasi-objective seeing of the phantom unity, locomotive, undergoing variation in seen size in here-centered quasi-objective space, are similar to the sensing of sensa that are substrata of the quasi-objective seeing of the looming phantom hillside (section 46); it may also be said that the sequence of changes in seen size is "not yet" constituted as a unified and harmonious sequence of *far*-appearances being transformed into *near*-appearances. Seen changes of size--or seen color changes, heard auditory changes (such as high and low changes in pitch, changes in pitch intensity)--are, instead, all presented within an optimal "zone" (Hofmann). Accordingly, it is necessary to speak of the affairs objectivated as "growing" such that what, perceived at a higher level where, e.g., patterns of locomotor kinaesthesia have been "reintroduced" in correlation with changes in appearances, at the level of here-centered quasi-objective space only looms larger, becomes more and more convex as part of the visual field sensed becomes ellipsoidal and curved along the line of vision. The visual sensa sensed as moving or at rest relative to each other and also to the central segment of the prespatial visual spread found what is objectivated as apparent turning or rotating of the sequences of changing surface sight-things in and through which the phantom unity, locomotive, is seen. In particular, that phantom unity is an example of the seen phantom moving along a curved path and seen less and less obliquely until the visible side is tangent to a plane perpendicular to the line of vision. When the phantom unity no longer undergoes the sequence of changes, e.g., in apparent size, no longer "rotates," it is presented as at rest.

More particularly, the phenomenological foundation in prespace, in the *broad* meaning of the term, "prespace," is the following: salient segments of the visual sensum-field are sensed as "disappearing" and "looming" in functional correlation with certain sensa-sequences of oculomotor and cephalomotor kinaesthesias on the basis of which sensa-identicals are constituted and which, in turn, found quasi-objective seeings of less and less oblique appearances of the surface of the phantom unity which, in the fully constituted real, objective world would be objectivated as the side of the locomotive moving from there to here and turning on an axis at a changing distance of vision. Correlative to sensing of the segment of the sensum-field as "disappearing" and, *simultaneously*, of another segment as "looming," there is visual sensing of the former as "shrinking inward," as becoming concave. Such a change is present most originally as progressing toward the center of the visual spread. Sensing the "progression" then founds the quasi-objective seeing in here-centered quasi-objective space of the apparent rotating motion of the phantom unity, Locomotive, along a curved path.

To be sure, not all phantom unities in here-centered quasi-objective space are presented as approaching Here; some are presented in ways that found a seeing of them as moving away; still others are presented as not undergoing any changes at all, for instance, in seen size (as at rest). Our example here is but one of many possible examples, and the qualitative changes sketched are but a few actual ones on a continuum of possible changes. What specific purpose, then, does our example serve for clarifying quasi-objective space at this stage of building-up analysis?

68. QUASI-OBJECTIVE HERE, THERE, BETWEEN, AND DEPTH

In the first place, the example of the phantom unity, Locomotive, serves to indicate the way in which the changing visual phantom unity is presented and posited relative to a fixed here-centeredness. As noted before, here-centeredness is an "absolute nearness" not simply because no locomotor kinaesthesias have not yet been introduced into building-up analysis, but because of the peculiarities of the *tactual* constituting of the phantom animate organism, and, indeed, of all phantom unities--including such things as the localization from "within" and "without," the non-disappearance of certain phantom sight things (parts of the phantom animate organism), and the lack of transformation of far-appearances into near-appearances, all of which point to the fact that here-centered quasi-objective space can always only be objectivated as visual-tactual quasi-objective space. As a consequence, qualitative changes in the visually presented are by necessity such as they are relative to a largely unseen (and, in part, unseeable) but tangible locus. In short, the example of the locomotive is one of original presentation of the quasi-objective phantom world seen, touched, heard from here. In Becker's words, "orientated <*scl.* here-centered> space becomes, after all, a space--whereas everything earlier in the hierarchical series of constitutional strata is still prespatial--

precisely by virtue of the fact that it assigns the subject <*scl.* the quasi-objective phantom animate organism> a definite place, namely, 'Here,' the zero-point of orientation."[21] To Becker's statement it must be added that at all higher strata of oriented constitution the animate organism is presented originaliter as the zero-point of orientation, although, to be sure, as a *mobile* zero-point. This constitution will be made thematic in the next chapter.

In the second place, while, on the one hand, the prespatial fields of touch and vision are constituted as non-selfsufficient moments of one visual-tactual quasi-objective space, on the other hand the sensed motion and rest of prespatial identicals found the constituting of patterns of qualitative change that are *constant within optimal limits*, in other words, always and only from *here*. Equivalently stated: the same phantom unity can only be meant and, for example, seen as moving or at rest in so far as its far-appearances can be transformed into near-appearances. Precisely this experience of an optimal limit *motivates strivings to actualize other and different sorts of kinaesthetic sensa-sequences* (such as those of ambulatory kinaesthesias). But the situation is not quite so simple in here-centered, quasi-objective space.

Strictly speaking, "motion" and "rest," in any ordinary sense of the words, are not presented as such at this stage of building-up analysis because not all qualitative changes in quasi-objective phantom appearances can be "compensated" kinaesthetically, for instance, either by those kinaesthesias accompanying eye-movements or by those accompanying leg-movements whereby the animate phantom organism would leave the spot. The example of the phantom locomotive-thing reveals that certain changes in the visual field, such as changes in seen size, are presented instead as rotations orientated with respect to a central point in the visual field. The phantom unity, meant as the same throughout these changes, Hofmann appropriately calls a "skiagraph"[22] because, presented as rotating, the skiagraphic locomotive is not co-presented in the seeing of it as essentially changing its distance from Here. To be sure, other and quite different motions are possible: rotation toward Here, rotation away from Here--although these are changes that cannot be mediately controlled by actualization of one or another set of kinaesthetic sensa-sequences. Even though founded on the sensing of identical sensa-changes in, e.g., the prespatial visual spread, changes of the sort indicated in quasi-objective here-centered space lack any phenomenological determinations with respect to which they are or would be apprehended as changes in and at a distance essential to the constituting of something as moving. At best these objectivations discover only what maybe qualified as *skiagraphic motion and rest.*

In the third place, the example of the locomotive illustrates the constituting of the quasi-objective core-world presented most originally with respect to the actual standpoint of the quasi-objective organism in here-centered space. It is, of course, not yet objectivated as a mobile actual standpoint, and this suggests a further peculiarity of quasi-objective here-centered space. Phantom quasi-objective space is presented as one visual-tactual space, and tactually the here-centered perceiving of the skiagraphic locomotive signifies for this constitution that only one, as it were, of the series of appearances is accessible to touch--when the phantom unity is in reach--while all the other appearances in that series are accessible to touch only if ambulatory kinaesthesias are actualized. But this is not the case in here-centered space. Consequently, the visual presentation of what ought, so to speak, "to be in touch" but is out of touch, out of reach, motivates a striving in primary passivity to effect a change in the actual standpoint of the phantom organism to "compensate" for what should be, yet is not, tactually accessible in the core-world. The motivation, in other words, includes as foundation a pairing of Here and There, a mutual transferring of Here and There in consequence of which the core-world acquires the "appearance" of being spread out in centerless quasi-objective space to accomodate, as it were, visually and tactually quasi-objective appearances of, in our example, the phantom locomotive. We may also speak here of a disharmony featured in the various correlations between appearances of phantom unities so that the visually meant unity is apperceived tactually, and the tactually apperceived, meant as horizonally "within reach;" thus the rotating aspects of the phantom unity, the skiagraphic locomotive, are seen and apperceived as far-appearances undergoing alteration into near-appearances in changing "distance" from Here.[23] Or, to express the same situation differently, the presented *actual* standpoint of the phantom organism is meant as but *one possible standpoint*, as but one possible here-centeredness among a multiplicity of possible standpoints. The core-world is visually-tactually presented as but one possible "appearance," then, in and through which the quasi-objective spatial spread is presented relative to Here determinable, now, as possibly any Here you please.

The example of the phantom unity, Locomotive, illustrates, accordingly, the features that are essential to the oriented constituting of here-centered quasi-objective space as well as the motivation for not remaining with this constituting and acquiring instead the "appearance" of a foundational substratum proper to a more complex intentionality. The "world" of skiagraphic phantoms, taken in and of itself, contains no basis for permanency; as it were, it must resolve itself into another and essentially different "world," a "secondarily constituted world" co-intended to as horizon of the here-centered, quasi-objective "world."

In the fourth, and final, place, the example of the phantom unity, Locomotive, involves the tactual and visual-tactual co-perceiving of the "ground" to which the phantom organism is rooted. Specifically, what is involved is the tactual co-perceiving of resistance. Leaving aside for the moment the task of describing the constituting of the members of the phantom organism both separately and as a unified, harmoniously correlated whole in here-centered quasi-objective space, we may objectivate that feature of quasi-objective perceiving of here-centered space where the "feet" and "ground" are presented most originally as in contact with each other, exerting "pressure" on each other.[24] With this feature, perhaps more than with any other feature of our example, it is possible to see the necessity for building up of a new stage of the constitution of the real, objective world. For the skiagraphic locomotive and the percipient phantom organism are co-presented as on the same phantom ground which, it itself, is presented not as simply somewhere or other, but There with respect to Here and from Here meant with a primitive direction (for instance, the skiagraph rotates toward Here, in this rather than another direction). And, with the reintroducton of ambulatory kinaesthetic sensa-sequences, changes in the quasi-objective appearances in and through which the phantom ground is presented are specifically correlated with periodic changes in the quasi-objective appearances of the phantom organism ("step by step" walking over the ground) in one or another direction. That segment of the phantom ground immediately perceived tactually is continually amplified in one or another direction.[25] In other words, quasi-objective perceiving of the ground under one's feet in moving from place to place is equally co-perceiving of a directional approach which brings the phantom organism into contact or at least within reach of other phantom unities presently out of reach but reachable by travelling over ground meant as common ground on which all phantoms are situated. It is this situation which must be briefly examined.

That is to say, phantom feet of the quasi-objective animate organism do not merely move over the ground but step on it and the place stepped on is continually meant as a segment of the whole ground extending beyond the place where the feet actually step.[26] Thus presently unreachable seen phantoms, for instance, are appresented not simply as tangible, but at the same time as in contact with the common phantom ground on and over which they move or are at rest. Moreover, the quasi-objective visual appearances through which actually unreacheable phantom unities are apperceived as tangible are correlated with the corresponding quasi-objective appearances of the ground on which the phantom organism must travel to reach those phantoms so as to transform the far visual appearances into near visual-*tactual* appearances in the core-world. But this also signifies that there is co-intending to the *space between* phantom unities and the phantom organism; more particularly, space between them is co-intended to as *changing or changeable distance* from Here to There. In other words, "space between" in a certain *direction of here-centered*, quasi-objective space, determined by skiagraphic rotation toward or away from Here, is meant as changing or changeable distance in *centerless* quasi-objective space. In the case of here-centered quasi-objective space, where the phantom organism does not leave the spot, the quasi-objective "space between" from Here to There is not intended to as traversed or traversable, thus strictly speaking phantom unities are not presented as "at a distance." But in the case of centerless quasi-objective space just the contrary is the case.[27] In the former, the "space between" is "filled in," that is to say, it "increases" or "diminishes" correlative to changes in seen size, for instance, of a rotating phantom unity such as the approaching skiagraphic locomotive; in the latter, in contrast, "space between" is presented as changing distance.

To objectivate "space between" presented in centerless quasi-objective space is at the same time to objectivate *depth* as an individuating dimension. But "depth" has here a rather different meaning from that in here-centered quasi-objective space. To be sure, in our example it is possible to discriminate "space between" Here and There gradually "filled in" as the skiagraphic locomotive appears in the mode of rotating toward the fixed Here along a curved path over the phantom ground, gradually changing in seen size, "growing" larger and larger. Prior unbuilding already disclosed the founding basis for this sort of "depth perception" in prespace (see sections 43, 47): the presentation of a primitive sensum-field the salient segments of which are "covered over" and "uncovered" with respect to one another, hence presented with a prespatial "interval" between the salient areas. It is the intending to this "interval" in prespace that is objectivated as a necessary foundation of the quasi-objective perceiving in here-centered quasi-objective space of the "space between" Here and There. What, then, differentiates "depth perception" of here-centered quasi-objective space from that of centerless quasi-objective space?

In here-centered quasi-objective space the phantom organism is fixed to the spot relative to which all other quasi-objective places in the spread are presented, relative to which all other changes in phantoms occur. Suppose that a phantom unity approaches, that is, is seen as undergoing a rotational series of changes in size along a curved path on the common phantom ground. At the same time it is seen, the phantom is meant as a quasi-objective tangible surface, apperceived as tangible until it comes into direct contact with the phantom organism. Suppose, now, that instead of being co-perceived as rooted to the spot, the phantom organism is co-perceived as *at rest* in the centerless quasi-objective spatial spread. The visual size-appearances of the approaching phantom, meant and apperceived as tangible, are similar to those seen and meant in here-centered quasi-objective space except that the distance traversed by the other phantom from There toward Here is co-intended to visually and tactually as *distance likewise traversable by the phantom organism.*

Thus the *first difference* to be noted in the two cases of "depth perception" is that, in the centerless spread, distance is meant as traversable *both* from There to Here and from Here to There. This suggests the *second difference*: unlike changes presented in here-centered quasi-objective space, changes in quasi-objective size-appearances of the approaching phantom in centerless quasi-objective space are *controllable*; the phantom organism can move to another place by successfully actualizing sensa-sequences of ambulatory kinaesthesia in order to maintain optimal presentation of the phantom locomotive, for instance, by approaching the phantom locomotive. In other words, the series of changing appearances of seen size are kinaesthetically compensated. Change of place now signifies exchange of place, but also correlation between actual and possible ambulatory kinaesthetic sensa-sequences and, say, visual and visual-tactual appearances whereby the seen size is itself "held constant," acquires the appearance of rigidity and is meant as the same seen size as it would be were it already in the core-world. The "ambulatory kinaesthetic compensation" is the phenomenological basis for perceiving phantom unities as the same but with a "smaller" (or: "larger") "format" so that only the visual perspectival "format," for instance, rather than the size (or: shape) is meant as changing: the *third difference* between "depth perception" in here-centered and centerless quasi-objective space.[28]

It is not possible to consider here the many descriptive problems that must be examined for developing the transition in building up from here-centered to centerless quasi-objective space--for instance, the problems of the constituting of "binocular vision," of the constituting of the various constancies of size, shape, color, surface; and so forth.[29] Instead it is necessary to draw a few conclusions concerning the oriented constitution of quasi-objective space to set the phenomenological-philosophical problems of "space perception" of concern in this and the next chapter.

69. FROM PRESPACES TO QUASI-OBJECTIVE CENTERLESS SPACE

To employ a somewhat ambiguous term of Oskar Becker, it may be said that each constitutive substratum of quasi-objective space is "reinterpreted" at the next higher level or stage of oriented constitution. Becker's term may be extended to include the various prespaces as well, so that it can be said that prespace in the narrow meaning of the term, "prespace," is "reinterpreted" such that qualitatively changing segments of the sensum-fields are meant and sensed as changes and non-changes in identical sensa correlative to sensing of certain actualized sequences of kinaesthetic sensa (prespace in the broad meaning of the term, "prespace"). In the further progress of oriented constitution, here-centered quasi-objective space, the sensing of identical sensa is included in the more complex quasi-objective perceiving of perspectival appearances (and, in the case of the animate organism, non-perspectival appearances) of changing and non-changing aspects of phantom unities and of those skiagraphic phantoms themselves. Some of the aspect-changes can be controlled immediately, compensated kinaesthetically, while others cannot. Typical of the changes in the here-centered quasi-objective space are rotational appearances of phantoms the presentation of which is oriented toward a central point in the quasi-objective spatial spread.

In the case of centerless quasi-objective space, here-centered space is "reinterpreted" not only as an appearance of centerless quasi-objective space but as but one among many such actual and possible appearances. In and through progressive ambulatory motion of the phantom organism, centerless quasi-objective space is broadened over the common phantom ground; the centerless quasi-objective spatial spread is "reinterpreted" as an actual quasi-objective appearance of a "space" indefinitely determined and determinable.[30] It is presented as an open continuum in which any place can appear near or far from the place where the phantom organism happens to be actually situated. As a consequence, it is always possible to bring phantoms into optimal presentedness so that, for example, actually unreachable phantoms, appresented as tangible, are brought into tangible presentation in the core-world and co-perceiving of the approach of a phantom over a common ground acquires the appearance of a tactual-visual continuum.

Moreover, only in centerless quasi-objective space is there any basis discriminated for distinguishing between apparent rotation and "real" rotation of phantoms, between apparent size and "real" size of a phantom surface--universally, between "abnormalities" and "normalities"--because only in a centerless spread is there the possibility and horizon of presentation otherwise of phantoms presented as rotating by virtue of strivings to actualize different kinaesthetic sensa-sequences. Finally it should be noted that phantom unities in both here-centered and centerless quasi-objective space are oriented relative to a tactually constituted here-centeredness (the "zero point of orientation") located in and on the phantom organism.

To make a case for this transition from prespaces to here-centered and centerless quasi-objective space we need to turn to a "transcendental deduction" of space.

70. THE TRANSCENDENTAL PHENOMENOLOGICAL "DEDUCTION" OF SPACE

The present investigation of the oriented constitution of the spatial aspect of the real, objective world began by formulating, as historical setting for the investigation, the concept of space as found in modern philosophy. According to that concept, not only is the external world identified with space, but because geometry deals with space, space is said to have mathematical intelligibility.[31] The basic philosophical problem that arises with respect to space so understood is the following: if space defines the real, the relationship of geometrical and spatial properties can be explained. But what, then, defines space? In the light of a discussion of Kant's idea of space it was noted that several specifying assumptions are involved in the modern concept of space that presupposes the difference between positing space in scientific and subscientific experiencing and thinking and that these assumptions are replaced by the Kantian one of the continuity of positing of space.[32] In turn, the continuity of the positing of space in scientific and subscientific experiencing and thinking presupposed that the ground of "space perception" is the "sensation of motion." With respect to this "naturalistic conception of space" we exercised what Husserl calls a "philosophical epoché," and with respect to our experiencing and thinking that posits space in the natural attitude we exercised a transcendental phenomenological epoché (section 34). In so doing the present investigation has arrived at results that are, at times, when made explicit at odds with Husserl as well as with modern philosophic and scientific thinking about space. One of these results concerns the "dimensions" of the various and stratified prespatial and quasi-objective spatial aspects of the one spatiotemporality of the real, objective world. Differently expressed, these results establish the framework for developing the problem of access to the real, objective world in the shape of the problem of "space perception" in general and of the "dimensionality" of space in particular. And it is the question of "dimensionality" of the multi-storied oriented constitution of inherent spationality of transcendental mental living with respect to its mundanization that entails a change in the "context" of the contingent Apriori (section 53). Our building-up analysis has now been carried far enough to permit the philosophical significance of these results to be made explicit with respect to the Kantian and Husserlian ideas of "space perception." The point of departure for formulating the philosophical significance of the phenomenological findings of the previous sections is the "transcendental deduction of space" developed by Oskar Becker.

Like Husserl in, for example, his 1907 lectures published as *Physical Thing and Space*, Becker holds that when we objectivate the stratified spatial aspects of the real, objective world we find that they are what he calls "inhomogeneous," that is, "two dimensional."[33] In his "Contributions Toward the Phenomenological Grounding of Geometry," Becker sets himself the systematic task of relating the phenomenological foundations of geometry back to the various "prespatial" substrata of the constituting of the real, objective world posited, as we may now say, as self-same in the natural attitude of sub-scientific and scientific thinking and experiencing. In carrying out this task he raises again the traditional problem (albeit in altered form) concerning geometry as the definition of space, and of space as the definition of the real. More precisely stated in Becker's terms: In which way does Euclidean geometry as a science produced in the natural attitude relate to the various stratified and hierarchical spatial aspects of the "perceived world"?[34] Or, still more broadly expressed: Can the original Cartesian questions about space still be raised in a meaningful way once continuity of positing belonging to the natural attitude has been made explicit (above, section 41 *in fine*)?

Becker answers this and similar questions with what he calls a "transcendental deduction" of the phenomenological concept of space[35] that he formulates on the occasion of a discussion of the "phenomenological legitimation of Euclidean metrics"[36] subsumed under the broader heading of "the phenomenological foundation of Euclidean geometry for 'actual' space." The discussion of this foundation in the second part of Becker's essay[37] is set, in turn, by the "problem of dimensions" developed by "rationally seizing upon the spatial continuum by means of transition from one limit to another [*Grenzübergange*]." More narrowly, the problem Becker perceives centers on the clarification of two "intuitive" spatial continua: 1) the "inhomogeneous" continua, examples of which are the pitch and intensity of sound, the brightness or saturation of color. And when, in this context, one speaks of the "dimensions of a phenomenon" one means by that the "variable properties <that are> independent of each other." Here the number of dimensions of space is equivalent to the number of independent variable properties.[38] 2) In contrast, the "homogeneous" continua are exemplified by "quasi-spatial (prespatial) sensum-fields: the sight field, the touch field, etc. The continuum now no longer divides into many qualitatively differentiated components. We therefore need a method to ascertain the number of interweaving homogeneous components." To determine a criterion for such a continuum Becker employs a method already employed by Helmholtz and afterwards by Poincaré: "'A continuum is said to be n-dimensional if it can be divided into separate pieces by one or more (n-1)-dimensional continua.' (A point has 0 dimensions.)" Following Brouwer's formulation, Becker refines the formulation of this criterion:

> 'A continuum is said to be n-dimensional if, for each choice of separate pieces σ, σ' there exists a separating continuum π of (n-1)-dimensions, and in contrast <there does> *not* <exist> for *any* choice of σ, σ' a separating continuum of less than (n-1) dimensions.' A criterion of n-dimensionality is therefore acquired since the separation or 'difference' ["*Sonderung*"] is also an operation that can be carried out in morphological continua.[39]

Becker then proceeds to develop and apply this criterion to the various spatial continua and their hierarchical structurization disclosed in the analysis of the constituting of "actual" space belonging to the one real, objective spatiotemporal world posited in the natural attitude. This must not be understood, however, in a metaphysical sense, i.e., as *defining the real* as in the case of modern philosophy generally and, as Becker indicates, in the case of a physics such as that of Hermann Weyl. Rather, Becker says, the criterion of the continuum must be understood only in terms of the "transcendental <phenomenological> legitimation of the three-dimensionality of 'actual' space" which is spelled out in a three-step *transcendental deduction*:

> 1) We show that homogeneous space has *one* dimension *more* than the sensum-field from which the constitution proceeds, i.e., an n-dimensional homogenous space conditions an (n-1)-dimensional sensum-field and conversely. 2) We demonstrated that the sensum-field not only can have *one* dimension, but rather that it must be *many*-dimensional. 3) We give reasons for the fact that the sensum-field *cannot* possess three or more dimensions.--Thus the only possibility that remains is that the sensum-field is *two*-dimensional, and according to (1) this signifies that homogeneous space is *three*-dimensional, which completes the deduction.[40]

Expressed in terms of what has been said about the oriented constitution of space, Becker's *proof* of the deduction consists of showing that the way in which one substratum takes on the appearance of the next highest substratum--the way, that is, the secondarily constituted enters into the primordially constituted--*is by acquiring another spatial dimension*. The gist of the proof, with which, for instance, both Claesges and Mouloud agree,[41] is the following:

1. "The sensum-field possesses one dimension less than homogeneous space." For Becker the sensum-fields in question comprise what I designated as prespatial spreads in the broader meaning of the term, "prespace."[42] That is to say, when prespatial spreads in the narrow meaning of the term, "prespace," coincide in part with prespatial spreads in the broad meaning of the term, "prespace," this can only occur if their dimensions agree in number. In the same manner here-centered (Becker: oriented) quasi-objective space is, as core-world, a genuine piece of centerless (Becker: homogeneous) quasi-objective space; therefore they both have the same number of dimensions. Thus the question of proof boils down to the contrast between here-centered quasi-objective space and prespace in the broader meaning of the term, "prespace." According to Becker, in making the transition from the latter to the former it is clear that but one more dimension is added: *depth*.[43] Accordingly, a saliency spread out in prespace is transformed, "reinterpreted," into a quasi-objective phantom unity, e.g., at some degree or other of distance from the absolute Here located in and on the animate phantom organism--as in the example of the "looming" quasi-

objective appearances of the phantom locomotive approaching Here. Thus one dimension more is added.

2. "The sensum-field has more than one dimension." Were it the case, Becker says, that the sensum-fields (such as the visual pre-spatial spread in the broader meaning of the term, "prespace") were one-dimensional, we would never objectivate, then, a homogeneous, multi-dimensional continuum. In the case of the skiagraphic locomotive spread out in here-centered quasi-objective space, for instance, if the quasi-objective spread were one-dimensional the degrees of depth, of looming, would never be transformed into the series of aspects of increasing seen sizes. For that to occur, more than one dimension must be presented. Or, to take another example--the phantom unity, "whistle" of the skiagraphic locomotive--if there were but one dimension then "in the two-dimensional sound continuum (arrangement of the tones with respect to pitch and intensity) it is only possible, in contrast, to apprehend pure differences of pitch or intensity as 'distances,' but <it is not possible to apprehend> directly the 'distance' of a tone *a* from a second tone *b* which differs at the same time from *a* with respect to pitch and intensity. In particular no constant transition is given from one musical interval into a difference of intensities of the 'same' tone."[44] This is the same situation as objectivated in the case of the transition from depth to, e.g., increase in size, i.e., to breadth. And because it is the essence of space, Becker says, to be a *principium individuationis*, and that means to be a homogeneous multiplicity, its necessary condition is therefore that there be a many-dimensional homogeneous prespatial multiplicity. For otherwise we would not objectivate--as in fact we do--differences in depth and breadth, nor would we objectivate differences between *identical* auditory or visual saliencies in prespace.

3. "The sensum-field has no more than two dimensions." Becker notes that this assertion represents the most difficult part of the deduction, and that while his demonstration of this assertion may not possess the same rigor as do the other demonstrations, he "somehow believes with certainty that it will become capable of a rigorous conception in the future."[45] According to Becker, a given sensum-field exhibits pure "juxtaposition" of elements of such nature that presentation of those field-elements as "behind one another" (e.g. in the visual sensum-field), a "covering over" of one element by another,[46] is impossible because that would also mean

that a corresponding "uncovering" must be possible--a possibility that Becker finds lacking in the prespatial sensum-field. In other words, there is no further course of primary passive visual sensing in which what is "covered over" would be "uncovered." Moreover, from the first proposition we know that presentation of elements, of saliencies, as "behind one another" can only occur by means of a single dimension, namely depth, and that depth--thus Becker--is only found in here-centered quasi-objective space. That is, "in the n-dimensional (oriented <*scl.* here-centered>) space a depth-and (n-1) surface (or breadth-) dimension is given. The prespatial fields would be spread out through (n-1) dimensions. In a four-dimensional space the sight-field would, e.g., possess three (breadth) dimensions arranged 'next to one another.' It would thus seem as though a three-dimensional figure <in the sight-field> (with a two-dimensional delimitation of its surface) would possess the same degree of intuitiveness as does a two-dimensional figure (with a one-dimensional delimitation) in our visual sight-field. But this is an error and precisely on this our deduction is grounded."[47]

Becker illustrates this phenomenological situation with several examples. Suppose that we seize upon a sensuous configuration by means of "sensuous comparison"--upon a circular configuration by comparison with its radii. The complete intuitiveness of a line is given to us by a continual one-dimensional multiplicity of acts of comparison which we perform successively over a period of time. Just this sort of intuitiveness, Becker says, cannot be achieved in three-dimensional "juxtaposition" where the outline would be two-dimensional, a curved surface, and we could never successively seize upon it without ambiguity or with cogency. If we think of a spherical surface, parts of which are spread out in pure juxtaposition, we can then only intuit it along a line, along a meridian for instance. But we can never intuitively appropriate the *whole* curved surface. Infinitely many directions issue from a given point which one can follow, and it is a purely arbitrary choice as to which we actually follow. The full intuitiveness can only be achieved in the case of the *two*-dimensional fields, and this phenomenological finding points to the three-dimensionality of quasi-objective space (and hence of space proper).[48] Indeed, because the intuiting of quasi-objective or of objective physical things is of necessity "incomplete," "presumptive," three-dimensional space "requires" a possible intersubjective world for the possibility of "completion" of the intuition of spatial figures: "It is now indeed justified to regard Euclidean space as the intuitive frame of any possible intersubjective world."[49]

71. CRITICAL REFLECTION ON THE TRANSCENDENTAL PHENOMENOLOGICAL DEDUCTION OF SPACE. THE TRANSCENDENTAL PHENOMENOLOGICAL PROBLEM OF DEPTH

It is not possible to consider here the whole question of the phenomenological legitimation of "Euclidean space"[50] which is of central concern to Becker. What is of interest at this stage of building-up analysis are certain aspects of Becker's deduction which, in many respects, run counter to the phenomenological findings reported in earlier parts of this chapter. More particularly, there are three aspects which can be mentioned here:

1. A basic premise of Becker's transcendental deduction is that the various prespatial sensum-fields have but two dimensions, i.e., that they lack the dimension of depth and comprise "lines" and optimal "breadth." Both the unbuilding reductions and their building-up analysis that I described before suggest, however, that Becker's findings are open to question. Rather than restate what has been objectivated (especially in sections 44, 47f., 51, 66ff.) the whole matter may be expressed as follows: Becker's phenomenological findings are, rather than phenomenological, "logical." By this I mean that, following Husserl, Becker uncritically assumes the doctrine of hyletic data. If that doctrine is true, then such data would, when animated by certain construings, adumbrate the quasi-objective and objective appearances of the surfaces of phantom and real physical things.[51] But because the hyletic data are "subspatial" data by virtue of being "non-intentive" components of the noesis, i.e., as sharing the same time-form as the intentive components of the noesis,[52] they cannot *per definitionem* themselves be spatial. How can something subspatial adumbrate something spatial and transcendent to the stream of mental life-processes (the intentive components of the noesis)? The historical antecedents of this view have been noted earlier along with the philosophical (and psychological) problem of space which those antecedents seek to resolve:[53] the sense-elements, saliencies and unities of prespatial

fields are *aspatial* (or: subspatial) in contrast to the sensuous qualities of quasi-objective and objective physical things and their spatial spreads. In other words, spatial characteristics of physical things can only be explained if space defines reality, and what defines space are aspatial "sensations."

In this context the novelty of Becker's account is three-fold in that, first, the aspatial data are not as, e.g., in Lotze or Helmholtz, "unnoticed" data but rather construable data subject to animation by mental life-processes; and, second, that the aspatial data are "reinterpreted" in the first place as two-dimensional, as having "direction" and "breadth" as a consequence of certain operations of mental life-processes (Becker: the ego); and, third, that rather than being "parts" of quasi-objective and objective space, such data, *in so far as they are seized upon in their adumbrating function*, condition and necessitate an additional dimension, depth, by virtue of motivating strivings to actualize kinaesthetic sensa-sequences which are likewise components of the noesis and hence have the same time-form *as other intentive and non-intentive components of the noesis*. And because the time-form is the same, "the full intuitiveness of a line is given to us by a continual, one-dimensional multiplicity of *acts* of comparison which we can perform successively in the likewise one-dimensional time" of the stream of mental life-processes.[54]

2. Thus a second premise follows from the first: the full intuitiveness of something spatial involves and can only involve two dimensions. But all of this is logical construction if it is the case that hyletic data and, correspondingly, kinaesthetic data are supervenient notions without phenomenological grounding. At the very least, that grounding is not obvious; and I have tried to show that it is non-existant. Nevertheless, Becker insists that the percipient organism is constituted in strivings to actualize various sensa-sequences of various kinaesthesias. Yet the results of consistent unbuilding reductions and discriminations would suggest that included as foundational substrata of "space perception" are substrata of intendings to, and positings of, space and the spatial correlative to accompanying sensings of kinaesthetic sensa-sequences "prior" to any constituting of the animate organism.

Moreover, in working out in detail Husserl's idea of oriented constitution it becomes clear that it is not a question at all of sensum-fields being "reinterpreted" so that they "coincide" with quasi-objective space nor that that space, in turn, is a "part" of a real objective space--the third premise of Becker's transcendental deduction (see above, section 62). It seems to me, instead, that regarded simply and only as they present themselves more originally in corresponding primary passive synthesizing processes, sensum-fields acquire the appearance of quasi-objective space *by being included as foundations* in the more complex intentionalities of perceivings and positings of the phantom "world." And the *reason* why I believe it is not possible to accept Becker's findings as valid is because it does not follow from the exercise of the transcendental phenomenological epoché that *ab initio* the various sensum-field spreads are not "spatial" and transcendent to mental life-processes proceeding in primary passivity.[55] But this signifies that one of Becker's criteria of dimensionality, indeed, of the third dimension, depth, namely "full intuitiveness" and "imperfect intuitiveness" already presupposes three-dimensionality.

As a consequence, Becker's transcendental deduction not only rests upon a logical construct, hyletic and kinaesthetic data as noetic components, but also on the idea that owing to the "intervention" of mental life-processes, and by virtue of its same time-form, spreads and spaces coincide, spaces become "parts" of the one, real objective spatio-temporal world. In short, Becker offers still another version of the Kantian idea of space.[56] But as suggested in the previous sections of this chapter, here-centered quasi-objective space is not a "part" or segment of a whole, namely centerless quasi-objective space; it is, rather, another, and higher, founded substratum of constituting. Nor do the prespatial spreads in any sense coincide with here-cetered quasi-objective space; they are instead more primitive and foundational substrata, stories, of the oriented constitution of space. Clearly the divergence between what I report in the light of unbuilding reductions and building-up analysis consistent with the formulation of the transcendental phenomenological reductions in Chapter One, and what Becker reports is great. Is there any way in which that divergence can be resolved? This question leads to a third feature of the inquiry into "space perception" which must be mentioned here.

3. According to Becker, depth is only presented as a third dimension in the characterization of quasi-objective, phantom space and not of prespaces. This assertion requires clarification. When discussing some of the reasons for rejecting Husserl's doctrine of

hyletic data an example was introduced to test that doctrine and which may be employed to the same purpose here:[57]

> In an unfamiliar mountainous country we see at some altitude a bluish grey formation appearing at one moment as a cloud, at others as the sky-line of mountains. ...As long as the sky-line of mountains appears, the bluish grey color is consolidated and attached to a bodily surface. However, when the cloud is perceived, the spatial localization of the color becomes indeterminate, the color appears to float in the air. Obviously the inner horizon pertaining to the perceptual appearance of the cloud differs profoundly from that belonging to the perceptual presentation of the sky-line....[58]

If, as before, we reduce our phenomenological theme to visual prespace in the narrower meaning of the term, "prespace,"[59] then this example of Aron Gurwitsch can be considered to be a case of what we called "fading" in the prespatial spread where there is presented visually sensed change from a high degree of distinctness (the "sky-line") to a lesser degree of distinctness (the "cloud"). And as the change in the degree of distinctness occurs in the spread there is a sensing of a correlative change in the determinateness or indeterminateness of spread-*localization and therefore of individuation* in the spread "at a distance" *regardless of whether concomitant alterations in "direction" and variations in "breadth" are presented.*[60] "Depth"is presented originaliter therefore even in the most primitive substrata of "space perception." Moreover, because in the case of "surfaces," of "lines" and "breadth," it is possible for "surface" changes to be sensed without there necessarily being presentation of corresponding changes in "vertical" or "horizontal" directions, "depth" is not to be equated with "visually sensed" (*scl.* "seen") direction.[61] As it were, "depth" is already presented in its "own right" in its localizing and individuating functions no matter whether any visual saliencies (e.g. "sky-line" or "cloud") are constituted in founded, assimilative syntheses.

But there is also another side of this phenomenological coin. If we reconsider the examples described concerning the sensing of qualitative change of saliencies in the prespatial visual field intended to as becoming "dark," as becoming chromatic without "mass,"[62] we find that they all point to a correlative seeing of prespatial spreadoutness as "voluminous" or "superficial" which, in turn, is the basis for mutual transfer of sense of the colored prespatial saliency, for instance, and the spread in and on which it "appears." There can be, for example, mutual cancellation of the noematic sense "colored;" but even in this limiting case there is partial assimilation of the sense, "qualitied," which belongs to both saliency and its spread; that is to say, a partial assimilative and differentiative synthesis is constituted with respect to "quality."[63] They are presented as similar to one another with respect to their "qualitied" spreadoutness, but also as dissimilar in that "space" itself is presented as "spread in," i.e., as "empty space."[64] "Empty space" is still presented as "space" but as sui generically "qualitied" because it is so presented regardless of whether anything salient in the spread is presented as "qualitied" ("colored" to some degree or other, in one "direction" or another). "Empty space," in addition, is not presented simply as some sort of primitive "distance" from Here; as noted before, the "center" of visual sensing "retracts" into itself, and hence the "space between" Here and There must not be confused--like Morgenstern's picket fence--with the "space between" the different saliencies standing out "in depth" relative to and "at a distance" from one another.

But this signifies that of necessity there is no actual or possible case where the "color quality" of a sensed saliency, for instance, *would be confused with the sensed "quality" of the prespatial spread itself.* Indeed, *precisely this "quality-difference" allows for the distinction between the two meanings of prespace. "Space" itself can be seen, touched,*[65] *heard,*[66] *and objectivated without visually, or tactually, or auditorily sensing any other sort of "quality"--although the converse is not the case.*

So far, then, there would seem to be no way to resolve the divergence between the phenomenological findings here and those of Becker (and Husserl to the extent that Becker includes Husserl's results as well). There is, however, one other approach possible which, if it does not resolve the divergence, can at least take advantage of it: Restatement of the transcendental deduction of space to include the phenomenological results of the present stage of building-up analysis of the space of the real, objective world.

72. CONCLUSION TO CHAPTER SEVEN: REFORMULATION OF THE TRANSCENDENTAL PHENOMENOLOGICAL DEDUCTION OF SPACE AS THE GENESIS OF THE "IDEA OF SPACE"

There is another angle to the problem of the constitution of depth which has so far gone unmentioned. In developing his account of the transcendental deduction of space, Becker uses examples drawn chiefly from the constituting of the visual prespatial spread and, in one case, from the auditory prespatial spread. But while it makes no difference which prepatial fields are first objectivated in building-up analysis,[67] on the one hand, it is significant that, on the other hand, that one of the first things one discovers as transcendental on-looker is that there is no analogy between tactual qualia in the prespatial tactual sensum-field and the tangible qualities of quasi-objective phantom unities.[68] Thus it is not the case, then, that depth is not presented as dimensional (Becker's proof of the deduction rests on the assertion of the non-existent analogy) in prespatial substrata, but it is also not the case that there is "coincidence" of the spreads of prespatial sensum-fields and here-centered, quasi-objective space. The same point, although also on different grounds, may be made with respect to the prespatial visual spread: what is therein constituted is anomalous in comparison with more complex ("higher") constitutings.[69] Thus the problem is not to see how the prespatial spreads "coincide" with parts of quasi-objective space--an idea that, despite all other differences, is a problem of a distinct Kantian flavor entailed by Kant's notion of "complete space." The problem is rather how prespatial spreads acquire the appearance of something which is non-anomalous.

In this connection, it is interesting that Claesges, while admitting that the results of Becker's transcendental deduction of space are "doubtless correct," nevertheless is sufficiently bothered by it to take the trouble to distinguish between the problem of space peculiar to a transcendental (phenomenological) theory of space-constitution and another, though not unrelated, problem peculiar to a "mundane ontology of space"--namely, the "mathematical determination of space," which is a secondary problem and not to be confused with the transcendental one.[70] Although I find myself in disagreement with Claesges's partial endorsement of Becker and with his division between transcendental phenomenology and mundane ontology,[71] his account based on Husserl's unpublished manuscripts does provide, it seems to me, a phenomenological alternative to Becker's constructive account by raising the important question about the phenomenological origin of idealization in Euclidean geometry.[72]

Like Claesges, Noel Mouloud, in a still valuable essay, finds himself in agreement with Becker's transcendental deduction of space (but he develops it in a way quite distinct from Claesges).[73] After rehearsing the deduction first with respect to the way in which a Euclidean metrics is based on and legitimated by "the structures and schemes of action of phenomenological space," then with respect to the *a priori* justification of the "tridimensionality of natural <*scl.* Euclidean> geometry," Mouloud also shows how it can be applied to account for space as the principle of individuation. While the deduction proves to be valid as a general principle,[74] it also raises important questions concerning its range of application:

> If, in particular, one admits, as we have done, that the schema of a metrics of repeatable and transitive congruences cannot be founded on a determinable property of perceptual-motoric space, but rather to the contrary, that the purely qualitative experience of equivalences and 'congruences' offered by sensuous intuition can only receive its quantitative structure by means of the schema of repeatable metric operations, then the argument which consists of invoking the properties of the principle of sensuous individuation as the ultimate foundation of Euclidean metrics loses its weight.[75]

In other words, there is an "epistemological priority"[76] which must be justified by a set of principles different from those of Becker's transcendental deduction. Indeed, the very question of foundation proves to be altered and "will rest less on the considerations of conditions of coherence pertaining to sensuous intuition than on considerations of conditions of coherence pertaining to the idea of the world and to the system of relations which can be applied <to the world> under the form of direct operation."[77]

Thus at least in this instance Mouloud has turned the transcendental deduction upside down--a view as far from the one developed here as that of Claesges. But no matter how we might now proceed to accomodate or even resolve the even wider divergences and differences from our own results, the divergences and differences themselves achieve their significance by pointing to a two-fold problem of crucial importance for any phenomenology of "space perception:" First, what is the origin and manner of geometrizing idealization in the prespaces constituted in primary passivity?[78] And, second, can that origin be adequately accounted for within the framework of a transcendental phenomenology?

The attempt to answer these questions is the telos implicit in the building-up analyses carried out so far. The problem of "space perception" as the medium of access to the real, objective world is then put on an entirely different footing than that dominated by the naturalistic conception of space. Moreover, the divergences and differences cannot be resolved because of that different footing: as transcendental on-looker, I have refrained from positing my prespatial sensum-fields as having a place in quasi-objective and real, objective space (above, section 21), although they are still posited as having their own inherent spatial spreadoutness transcendent to the sensing of them in primary passivity. Thus the transcendental deduction of space, whether as formulated by Becker or by Claesges or Mouloud, is impossible as it stands. The deduction--if it can still be regarded as a "deduction" at all--would now look quite different:

1. *The quasi-objective spatial spread (either here-centered or centerless) has no more dimensions than prespatial sensum-field spreads" because, according to the law of oriented constituting, an "n-dimensional space" is the appearance acquired by a foundational n-dimensional prespatial sensum-field spread which has no place in, hence is not part of, quasi-objective or objective space.*
2. *Inherently spreadout prespatial sensum-fields, transcendent to the sensing of them, are each not only one-dimensional but are many dimensional.*
3. *Quasi-objective space must be a three-dimensional 'founded' space" because prespatial sensum-field spreads do not possess less than three dimensions.*

To be sure, clarification and qualification of this "deduction" are required; moreover, the "deduction" is incomplete because it does not include in its formulation, at least explicitly, time (above, section 62). Thus in the next chapter still another formulation of the "deduction" will have to be stated. But even with this brief sketch it can be seen that the whole phenomenological issue of space constitution *is not a question of the addition of dimensions* but instead how prespatial spreadoutness, which does not of necessity have a place in quasi-objective (or objective) space is a foundation for perceiving quasi-objective space and of objective space and that in a certain order. Another way to express the problem is how the inherently prespatial spreads, transcendent to the sensing of them, acquire the appearance of one space, the perceiving of which is not exclusively a sensing but has a foundation which is a sensing.

To clarify this problem further it is necessary to carry out a further stage of building-up analysis, specifically of the oriented constituting of *primordial space*. The direction to be taken by further building-up analysis is suggested by a text of Husserl's, mention of which will provide a larger perspective in which to locate the "origin of the idea of space." The text in question is section 150 if the First Book of *Ideas* in which Husserl characterizes the regional idea, Physical Thing, that "prescribes rules governing mulitplicities of appearances," "self-contained series of appearances progressing *in infinitum* taken as an ideal collectivity." Husserl then notes that while the unity of a mere *res extensa* can be conceived without appealing to the corresponding idea of *res materialis*, the converse is not the case.[79] The series of appearances of physical things, moreover, include in themselves a "stratum that we call the *physical thing-schema*: it is the spatial shape merely filled with 'sensuous' qualities--without any determinateness of 'substantiality' and 'causality'...." And the many affairs that we continually take for granted in daily life such as the "fact" that spatial things always appear in certain orientations, at certain distances, etc.--all these factual contingencies prove to be eidetic necessities "even in the least significant empirical particularities" and hence are intuitable only "by means of appearances in which" a spatial physical thing "is and must be given in multiple but determined changing 'perspectival' manners and accordingly in changing 'orientations'."[80]

As a consequence, "The problem of the '*origin of the idea of space*,' the profoundest phenomenological sense of which has never been grasped, is reduced to the phenomenological analysis of the essence of all the noematic (and noetic) phenomena in which space is intuitively presented and 'constituted' as the unity of appearances, of descriptive modes of presentations of something spatial."[81]

Additionally, the analysis is determined by all of the differing "*levels and strata of physical thing-constitution in the frame of originary experiencing consciousness*," and each level can be characterized by a *unity* which serves "as a *necessary middle member* for the full constitution of the physical thing." Among the levels Husserl mentions here are the "many sorts of unity-strata, *sensuous schemata*, the 'sight things' of a higher and lower order.... At the *highest* strata pertaining to this level there is the *substantial-causal physical thing*...still constitutively restricted to *one* experiencing subject and his ideal perceptual multiplicities. The *next higher level* is then the *intersubjectively identical physical thing*..."[82]

Without endorsing all of Husserl's view expressed in the preceding citations, it is to the "next higher level" of the "intersubjectively identical physical thing" that building-up analysis must proceed.

CHAPTER EIGHT

THE TRANSCENDENTAL PHENOMENOLOGICAL BUILDING-UP OF PRIMORDIAL QUASI-OBJECTIVE SPACE. THE TRANSCENDENTAL PHENOMENOLOGICAL "DEDUCTION" OF TIME

73. "SPACE" AND "SPACES"

It is apparent from the course of our investigation so far that the words, "space" and "spaces," are infected with an ambiguity of almost Pickwickian proportions. Indeed, the ambiguity would seem to be unavoidable for "space" (and "time," as will be seen in later sections) in one meaning of the word, "space," can be intended to and posited without intending to and positing existent, real individual physical things. However, in another meaning of the word, "space," it cannot be intended to and posited without necessarily intending to and positing existent, real individual physical things. And these meanings of "space" have been ascertained by showing that constituting of the spatial aspect of the real, objective world includes constituting of many substrata in primary passivity which are foundations of the perception of the spatial aspect of the real, objective world. Therefore, if we understand by "space" (and "time") *pre*spatio-temporality, and by "existent, real individual physical things" identical sensa, then we may say that "space" (and "time") can be intended to and posited without necessarily intending to and positing existent, real individual physical things because there is a clearly distinguishable substratum of space constitution "below" it ("prespace" in the narrower meaning of the term, "prespace"). But if we understand by "space" (and "time") a *quasi*-objective spatio-temporality, such as was made thematic in the last chapter, and by "existent, real individual physical things" quasiobjective phantom unities, then we can say that "space" (and "time") cannot be intended to and posited without necessarily intending to and positing existent, real individual things.

The ambiguity of the word, "space," is accordingly not a vicious one and can be further clarified by reflecting on another phrase which has cropped up from time to time, namely, "empty space."

"Empty space," phenomenologically, must be distinguished from such phenomena as "space between," "space from here to there," and the like.[1] The task of clarifying the constituting of "empty space" concerns the different, although not unrelated, task of the clarification of the constituting of "space in itself" and the "idea of space" acquired in traversing and "exploring" space. Incorporating here some intriguing suggestions of Hedwig Conrad-Martius,[2] it would seem to be the case that "space" intended to and posited correlative to the meaning and positing of quasi-objective physical things such as phantom unities or even full phantoms is "space" intended to and posited correlative to intending to some sort of motion, such as "turning on the spot," "traversing" or "moving from the spot." Such quasi-objective phantom unities are not presented so much as spread out *in* space, as though space were a container of some sort; they are rather present most originally spread out *as* "spatial" ("*raumhafte*"). In these terms, is there a "space" actually presented empty of phantoms and things? What have been called "prespaces" are examples of just such a "space."[3] This signifies, however, that a "space" presented empty of phantoms and things is the necessary foundational substratum for the constituting of "space" intended to and posited correlative to intendings to and positing phantoms and things.[4] Each of what have been called "prespaces" Conrad-Martius calls, in an apt expression, a "space" that is "absolutely empty of itself," a "nothing" in contrast to the "filledness" of the "space" constituted correlative to the constitution of phantoms and things. Indeed, there is again the temptation to speak of two generically distinct "spaces,"[5] one of which is the necessary phenomenological underpinning of the other. Thus while "prespace" is wholly possible (although improbable) without quasi-objective or objective "space," the converse is not and cannot be the case.

The phenomenological findings so far would certainly seem to lead to such a conclusion. But if that is so, then the question arises as to how to further determine the essential interrelationship between the two distinct meanings of "space" and what meaning can be attached to its apparent ambiguity. One answer lies in the consideration of a phenomenological complexity which can be dealt with under the general heading of "apperception" and "appresentation." The phenomena which fall under this heading were already touched upon earlier when characterizing the idea of intentional synthesis.[6] At times I have just considered spatial things "for themselves," just with respect to their *presentive modes*. But at other times it was necessary to consider them as "paired" or as indicating something other than themselves. Noematically expressed, I considered spatial things at times with respect to their *appresentation*.

To take a prosaic example: I am now looking at the front of a wall; but it is not just a wall, not merely a facade. Rather in looking at the wall I also mean and posit it as having another but unseen side; the front of the wall appresents the other side, indicates the unseen but seeable side. Noetically expressed, apperceiving always involves a "meaning more" than what is presented at any given, actual moment. A number of passages from Husserl can be assembled to express what is at stake here for intentional analysis which

> reaches out beyond the isolated <mental> processes that are to be analyzed. By explicating the correlative horizons, it brings the highly diverse anonymous process into the field comprising those that function 'constitutively' in relation to the objective sense of the cogitatum in question...the 'ready-made' object that confronts us in life as an existent mere physical thing...is given in the synthesis of a passive experience. As such a thing, it is given beforehand to 'spiritual' activities...<We> thus find intentional references leading back to a 'history' and accordingly making these phenomena knowable as formations subsequent to other, essentially antecedent formations...<where we> encounter eidetic laws governing a passive forming of perceptually new syntheses...a passive genesis of the manifold apperceptions....When these habitual apperceptions become actually operative, the already given objects formed for the central Ego appear, affect him, and motivate activities.[7]

As it were, apperceptively we fashion a "history" that provides a spatio-temporal environment of "ready-made" objects. Moreover, every sort of experience has its own sort of confirmation, and apperception is no exception. As Husserl suggests in the passages just cited, not only is there "pointing beyond" by given concatenations of passive mental life-processes at any given substratum of constitution, hence full or partial confirmation of experience peculiar to that substratum, but there is also "pointing beyond" to other substrata and full or partial confirmation of the apperceptual experience. The "pointing beyond" is, so to speak, as much "horizontal" as "vertical." It is no surprise then that intentional analysis reveals one substratum of oriented constitution appresenting another and appresented substratum; thus prespace in the narrower meaning of the term, "prespace," appresents prespace in the broader meaning of the term, "prespace," by virtue of apperceiving ("appersensing, as we said earlier") the various combined sensings of kinaesthetic sensa-sequences. The task of further building-up analysis as one sort of intentional analysis is to make thematic those "habitualities" of apperception, as Husserl says, that "motivate" new syntheses in primary passivity appresenting the next substratum of constitution and in which they find confirmation of one sort or another. To be sure, at the same time limiting cases of primary passive apperceptual experience are singled out where there is neither confirmation nor disconfirmation in the further course of the building up of that experience--the principal example so far being the constitution of the quasi-objective phantom organism. Such a limiting case also serves as motivation for the necessary appresentation of other substrata of constitution.

As a result, to put the matter simply, at each substratum of oriented constitution thematically discriminated in unbuilding there is to be found a complexity of appresentative motivations such that by virtue of one substratum appresenting another "perceptually new syntheses" are formed (thus "prespaces" acquire the appearance of quasi-objective phantom "space"). Or, more generally expressed, "space" without anything spatial appresents "space" with something spatial. The motivation for such apperception arises in consequence of primary passive strivings to transform the appresented at one substratum into the *presented* at another and "higher" substratum. And it is in just this phenomenological situation that we find the point of departure for further building-up of the real, objective world in those "habitual apperceptions" of "essentially antecedent formations" of passive genesis. More particularly, in this case the apperceptions in question are those of a specific kind, namely of "if-then" apperceptual experience. It is that apperceptual experience peculiar to what was earlier objectivated as the quasi-objective primordial world in the wide sense of the term, "primordial" (sections 18, 19, 22).

The most immediate task, now, will be to circumscribe the phenomenological theme of this new stage of building-up analysis which will be seen to include of necessity the presentational and appresentational modes of experiencing "causality" and "other minds" which will form the occasion for restating the transcendental deduction of space in terms of the transcendental deduction of time.

74. SOME APPERCEPTUAL MOTIVATIONS FOR THE ORIENTED CONSTITUTION OF QUASI-OBJECTIVE PRIMORDIAL SPACE. "IF-THEN CAUSALITY"

The phenomenological theme of this *second* stage of building-up analysis is the quasi-objective primordial world in the wide sense of the word, "primordial."[8] What distinguishes this substratum of primary passive constituting from those underlying it is, among other things, a "causal" or "if-then" style of experiencing in which other psychophysical beings are evinced and confirmed as psychophysical in my own primary passive mental life-processes. The most primitive substratum of oriented constitution pertaining to the primordial world and its quasi-objective space is that at which quasi-objective individual physical things are perceived and apperceived as what and how they are under "circumstances." To be sure, even in the case of centerless quasi-objective space of phantoms it is possible to objectivate apperceivings of a dependency of quasi-objective tactual appearances on certain co-perceived "states" of the phantom organism and which Husserl calls a "psychophysical conditionality."[9] Thus not only are quasi-objective phantom appearances presented as relative to somatic states such as motion and rest, but quasi-objective qualities such as warmth and cold, color and size, and the like, are also so presented. Quasi-objective apperceiving of this functional dependency motivates a primary passive preferring and striving to actualize the apperceptual horizons of the phantom organism or its parts "under more favorable circumstances" so that, for instance, the phantom organism is perceived with warm rather than cold "hands," with half-closed instead of open "eyes," or so that what is only apperceived and never perceived in an original way is presented in other nonoriginal ways (e.g., turning so that the back of the "head" can be seen in a mirror). Accordingly, under different circumstances there is a more adequate or evident presenting of certain features of the quasi-objective phantom organism. And it is on this basis that the centerless phantom spatial spread is apperceived as an appearance of and enters into, as secondarily constituted, quasi-objective primordial space where things and events are presented and appresented as having certain properties or qualities "under circumstances" (temperature, illumination, and the like) and which are, in turn, dependent on certain co-perceived somatic states and changes. It is the "first level" of the constitution of the *res materialis*.[10]

Stated universally, in quasi-objective primordial space physical things of no matter what sort are perceived and apperceived as what they are always under circumstances and the appearances of which vary correlative to changes and non-changes in certain co-presented somatic states. Apperceiving of such "functional" correlations serves as the ground for the assimilative transfer of the sense, "psychophysical," to everything else intended to and posited in the quasi-objective spatial field. I objectivate, for instance, mutual transfer of the sense, "my psychophysical phantom organism," to other phantoms encountered in the quasi-objective spatial field. To the extent that there is partial mutual confirmation by their "behavior," the quasi-objective field is appresented and posited in primary passivity as an actually existent quasi-objective world populated by a plurality of other quasi-objective psychophysical beings. In other words, the primary passive mutual transfer of the noematic sense, "psychophysical conditionality" is ipso facto the transfer of the noematic sense, "organism of 'someone else'" to all perceived and apperceived quasi-objective physical things. Changes in circumstances correlated with changes in psychophysical conditionalities are appresented as alterations manifesting "behavior" of a plurality of quasi-objective animate organisms (although, to be sure, the "behavior" is still non-egoic). Shortly (section 76f.) I shall return to this situation in more detail, for its clarification is central to the constitution of the real, objective world.

Yet another motivation for constituting quasi-objective phantom centerless space as appresenting itself as the appearance of quasi-objective primordial space is the constituting of "the same place." More particularly, perceiving "the same place" includes perceiving it as the place previously occupied by the quasi-objective animate organism; "the same place" is presented in the perceiving of it as having the sense, "the same, meant again now from here, just as before," appresented as under like circumstances. In short, "the same place" is presented as a "returnable place." The foundation for this constitution has already been objectivated (section 68) in meaning of the actual standpoint, the actual place of the phantom organism, as one possible place among other possible ones as well as in meaning of a common ground shared and traversed by a plurality of phantom things correlative to meaning of place as "exchange of place." But here to objectivate the perceiving of "the same place" is, in addition, to objectivate co-intending to various phantom aspects (or: schemata) appresented as "parts" of one and the same quasi-objective primordial thing under circumstances. In section 75 some examples will be considered; at the moment it is to be noted that correlative to constituting of "the same place" is apperceiving of quasi-objective centerless space as having a place in quasi-objective primordial space.[11] Before objectivation of this correlative apperception can be clarified, and before specific examples can be studied, something further must be said about the nature of the apperceptual motivations and their transcendental phenomenological significance for this stage of building-up the real, objective world.

The various sorts of apperception mentioned may also be said to be a species of *inferring* in the broadest meaning of the term, "inferring." That is to say, apperceiving involves primary passive believing that one quasi-objective physical thing or event exists "because" another such thing or event is believed to exist under similar circumstances. More precisely stated, it is an expectant believing that one thing is likely to exist "because" another exists under the same or similar circumstances. Or, perhaps, the inferring is made on the basis of uncertainty: there is believing that one thing probably exists "because" another thing exists. Universally stated: whenever, in primary passivity, a noematic sense posited in one or another modality of existence is transferred, "paired," and not entirely cancelled out, there is a basis for inferring in this broadest meaning of the term, "inferring." All apperceptual experience in *primary passivity* is then an inferring. The significance of this phenomenological finding can be established by comparing the broader with a much narrower meaning of the term, "inferring."

An example of the narrower meaning of "inferring" is actional, judicative inferring. Consider a logical form, such as the "therefore-form" of judgment: "therefore" expresses a syntactical form in a manner similar to the way in which "and" expresses such a form. If, now, "spatio-temporal Nature" is understood in an extremely broad meaning of the term, "Nature," as has been done so far--i.e., as being there for us by virtue of primary passive oriented constitution--then it may be said that there are no formal-logical "therefore's" in Nature, just as there are no "ands," "if's," "but's," and "is's" in Nature. The subject-predicate form, like the other forms, is bestowed on Nature in consequence of an activity of judging.[12] In other words, a formal-logical implicative relation such as "therefore" is a syntactical relation rather than a functional or "causal" relation. Phenomenologically, Hume was right in seeing that whatever the inferential or apperceptual "if-then" relation might be, it was certainly not one of formal-logical implication. Even if "inferring" is conceived as an active, judicative "inferring," it is still necessary to distinguish between two kinds: 1) formal inferring, and 2) apperceptual or material inferring.[13] When for example, I objectivate and seize upon a material correlation already constituted, this material inferring is different in kind from that expressed when I say, "no triangle is a square, therefore no square is a triangle"--a case of formal inference. But if, in contrast, I infer that it is cloudy, the barometer is falling, that therefore it is going to snow, then there is no formal-logical implication whatever. The justification of the material inference does not lie at the logical level of constituting as does that of formal inference. The correctness or incorrectness of a material inference is not a consequence, then, of its syntactical form; in contrast, the correctness or incorrectness of a formal inference is precisely the consequence of its syntactical form. Just this distinction sets the problem of clarification of the oriented constitution of the quasi-objective primordial world. For what, then, is the justification of primordial material inference in primary passivity, i.e., of the primordial apperceptual "if-then" style of experience? More broadly expressed, what legitimates primordial "causality" proceeding in primary passivity?

The phenomenological significance of these and similar questions for an inquiry into the constitution of the space and time of the real, objective world is that it is clear that the two "generic" meanings of "space" distinguished in section 73 are likewise not the consequence of their syntactical form, that the apperceiving of the one meaning (prespace) as the appearance of the other (quasi-objective and real, objective space) is likewise not a consequence of syntactical form. As a consequence, "space" in the broadest meaning of the term, "space," has no more to do with "Euclidean" or "nonEuclidean" manifolds than formal-logical inferring has to do with material inferring. Oriented constitution of the quasi-objective primordial "world" in primary passivity, and the identifying and assimilative syntheses (among others) that comprise it, is a material inferring, an apperceptual "causality" or motivation entirely distinct from any other meaning of those terms.[14]

Thus we may speak of "causality" in a two-fold meaning with respect to primordial constitution in the wide meaning of the term, "primordial:"

> 1. The quasi-objective apperceiving of the conditionedness by mental life-processes that pertains to states and events spread out and occurring in quasi-objective space--e.g., primary passive strivings to bring about change of place or to be at rest of the primordial animate organism--*and* the correlatively perceived conditionedness of mental life-processes by circumstances under which physical things are apperceived in the primordial "world."
> 2. Conditionedness not of the quasi-objective physical things and events themselves but rather of their quasi-objective perspectival appearances by co-perceived quasi-objective circumstances--e.g., including states of the quasi-objective primordial animate organism and things and events (such as illumination, cold, and the like).

This two-fold meaning of "causality" is wholly peculiar to the oriented constitution of the quasi-objective primordial "world" in general and to its spatio-temporality in particular. Such "causality," of course, is to be sharply distinguished from that pertaining to the "world" of the physicist, i.e., from mathematically "exact" causality.[15] We shall begin with the second meaning of "causality."

75. QUASI-OBJECTIVE PRIMORDIAL APPEARANCES. LIGHT AND SOLIDS.

When characterizing the prespatial visual sensum-field it was noted that there is no basis for clear differentiation of so-called "film" from "surface" colors.[16] One of the reasons for this is, as Katz indicated,[17] that surface colors may be presented in any orientation while film colors, in contrast, can only be presented parallel to the frontal plane of vision.[18] Such a distinction, then, at least presupposes a primitive constituting of a percipient animate organism. Strictly speaking, the prespatial visual sensum-field in the narrow meaning of the term, "prespace," is characterized only by surface colors. In the case of here-centered quasi-objective space, however, film colors are not only presented but are presented most distinctly when what is objectivated as the moving axis of rotation of an approaching phantom, such as the phantom locomotive, approximates a position perpendicular to the line of vision.[19] Thus especially within the near-sphere a primitive differentiation of film from surface color can be noted[20] such that when surface colors are distinctly presented as surface colors seen phantoms are, at the same time, most originally presented as having a clear superficial structure, as fully determinate visual phantoms with their colors.[21] It is this presentation that provides a phenomenological clue to the constituting of quasi-objective solids spreadout in primordial space.

To clarify the constituting of quasi-objective solids we can adapt an example of Merleau-Ponty:

> Electric lighting, which appears yellow to us at the moment when we leave the daylight, soon ceases to have any definite color for us, and, if some remnant of daylight finds its way into the room, it is this 'objectively neutral' light which seems to have a blue tint about it. We must not say that, the yellow electric lighting being perceived as yellow, we take this fact into account in our appreciation of appearances and thus find ideally the proper color of objects again. Nor must we say that, to the degree that yellow light becomes all-embracing, it is seen as daylight, and that in this way the color of other objects remains really constant. We must say that the yellow light, in taking on the function of lighting, tends to put itself in a position this side of color, tends toward the

> zero of color and that, correspondingly, objects distribute the colors of the spectrum among themselves according to the degree and manner of their resistance to this new atmosphere.[22]

A color is meant and seen as the same under varying circumstances such as changing illuminations, distributed with respect to a penumbra such as Merleau-Ponty describes in his example. Yet the presentational and appresentational nature of such colors is not quite so obvious. In his still important study of perception, Wilhelm Schapp calls such colors "*anhaftende Farbe*,"[23] that is, colors that "cling to" the shapes of things, with which the shapes of things are "endowed." He clarifies this situation by distinguishing such colors from the co-perceived shadows of colored physical things: "The shadow is moveable by itself, the color with which a thing is endowed is unmoveable without the physical thing."[24] As in the case of the yellow light in its function of lighting, so here the "moveable shadow" and "immobile color" present a situation of conditionedness of quasi-objective perspectival appearances. Accordingly, a color posited as the same under varying circumstances of light and shadow is intended to most originally as a surface color of a seen quasi-objective physical thing constituted in an identifying synthesis as the same in and through varying perspectival appearances and quasi-objective spatial orientations (especially of things at a distance where a color-constancy tends to be preserved). Similarly, one and the same surface color is constituted in an identifying synthesis as the same under varying illuminations and as distinct from the visually co-perceived and apperceived lighting. The point is that an identical color and an identical thing "endowed" with that color are constituted correlatively in primordial quasi-objective space in contrast to *phantom* quasi-objective space (here-centered or centerless) where lighting and shadow are but quasi-objective appearances among others and where phantoms do not "distribute the colors of the spectrum among themselves according to the degree of their resistance to this new atmosphere." What in primordial quasi-objective space "tends to put itself in a position this side of color" tends in phantom quasi-objective space to put itself among colors and on an equal footing with them. But in primordial quasi-objective space "if" there is a certain lighting, "then" the solid casts a shadow, then there is resistance to the lighting which tends toward the "zero of color."[25]

It is then apparent that colors peculiar to the phantom visual quasi-objective world do not coincide with those in the primordial one; rather perceiving of some of the quasi-objective phantom visual appearances prove to be non-selfsufficient foundations of perceivings that are more complex: color spread out in phantom quasi-objective space does not so much serve to make quasi-objective phantom things visible as it is a determination itself of those phantom things. In contrast, spread out in quasi-objective primordial space color serves as the medium for seeing things that appear in this or that spatial orientation. What proves to be compatible in quasi-objective phantom space turns out to be incompatible in quasi-objective primordial space in which lighting and color are distinct. In referring to this incompatibility, Schapp notes that one perceives the physical thing as physical thing when the lighting or light-formations ("*Lichtgebilde*") are simultaneously apperceived as in the background but nonetheless as an indispensable circumstance under which the physical thing appears. Light-formations "should 'lead' the regard 'over to' the physical thing; the regard should not be trapped" by the light-formations.[26] Modifying Schapp's account to apply to the constituting of the primordial quasi-objective stratum of experience, we may say that there is a "causal" relationship between the presentation of color and physical thing such that if the color-presentation undergoes change, then there is a change in the manner of quasi-objective perspectival appearance in and through which the physical thing is presented and, indeed, posited as physical thing: "the physical thing," Schapp says, "seems to be neglected...where <the lighting is> so salient that <it> cannot be overlooked; perception of the <physical thing> is disturbed."

This "if-then" relationship provides a further clue to the manner in which quasi-objective primordial space is constituted. It is worth while for our purposes to cite Schapp in full:

> Color is simply that which presents physical things to us. But it is not sufficient for physical things to be presented if colors are given. Rather for that to be the case color is ordered, articulated and enters into forms. ...By means of this color-order space and shape are presented. In relation to color, space is therefore something manifested [*Dargestelltes*].

> Color itself is not manifested, it is directly given but manifests space and forms in space. ...Illumination-colors continually change without changing anything about the physical thing. From any side the illumination-colors look different and do so everyday. Likewise the physical thing can change without change in surface colors; but if the surface colors change, then an alteration of what is inherent [*Innern*] in the physical thing comes to light. The process of change of iron under the influence of heat is accompanied by an alteration of color--from red hot to white hot.[27]

Thus the conditionedness of the quasi-objective appearances of colored physical things is the result of a rather complex passive synthesis of assimilation and difference in which there is not just a negative fulfillment of the mutually transferred color-sense (such as illumination and surface-colors), but also in which there is an incompatibility such that the colors are perceived and apperceived as "resisting" each other. In short, the conditionedness is based on perceptual conflict entirely peculiar to the correlative "manifestation" of primordial space. In other words, at the same time that there is conflict between the mutually transferred color and lighting (or "illumination-colors"), at the same time that colors are distributed among themselves in contrast to light-formations in the background, i.e., when "color is ordered, articulated and enters into forms," *primordial* quasi-objective space and spatial shapes are apperceived. This signifies that primordial, quasi-objective physical things are presented and posited most originally as "solids," as *res extensae* with properties. Presupposed, then, is the fact that the primordial, quasi-objective stratum of experience is already constituted, as Husserl says, "with a changeable section of a 'field of perception,' a physical-thing constellation perceived at a particular time in the unity of one perception, a constellation co-existing in a momentary present and, along with that, in a constellation of solids that present themselves through the medium of space that is empty but perceived in its own fashion."[28] To this "medium of space" there belongs the apperceptual possibility of continuous amplification "precisely as a field perceivable only in a successive synthesis," i.e., always apperceived as potentially perceivable under controllable circumstances such as motion or rest. But this signifies in turn that it is not just a single actual perception that is now objectivated, but instead a whole habitual style of actual and possible perceiving and apperceiving as well, a whole system of interlocking, "conditioned" quasi-objective perspectival appearances.

A phenomenological result is that not only is each single quasi-objective physical-thing solid a unity constituted in and through variation of, e.g., near-far perspectives, but each pair of such things, each group of such things is constituted as a unity linking near and far together such that they co-exist in a "momentary present." One "if-then" or "causal" style governs the whole of the perceptual field regardless of whether the physical-thing solids are actually touched or heard as well as seen. This holds *pari passu* for quasi-objective spatial perspectives and their spread as well. As such, primordial space is open but finite,[29] a unity in which configurations or constellations of quasi-objective solids are linked together under circumstances as a "configurative combination" of things simultaneously experienced. Husserl calls the spatial unity of "if-then" togetherness the "first world as world of experience."[30] Thus in contrast to phantom quasi-objective space, primordial space is constituted not as a field of mere co-existence but instead as a field of quasi-objective things simultaneously experienced as combined together such that if this occurs, then that occurs. And the conditionedness not only links various quasi-objective things together but also various parts of the perceptual field into a unity of quasi-objective spatial togetherness. It is a quasi-objective perceptual and apperceptual field in which there is simultaneity of presentation of things in the unity of one perception as well as "action at a distance."

So far, then, we may say that primordial quasi-objective space intended to in primary passivity is not only a principle of individuation but also of an "invariance-system" of actual and possible places apperceived together (as, for instance, in rest or in motion). And this signifies that primordial quasi-objective space is presented or "manifested" not just as a space traversed and traversable; that it is not just presented as perceivable from here and from there, that any here can be potentially any there and conversely; that not only is any place apperceived as a place which can be returned to under the same or similar circumstances. It signifies in addition that by virtue of the enduring, invariant quasi-objective spatial togetherness quasi-objective things and events are presented so that what is perceived as near is extended into a horizon of unfamiliarity and what is presented as far is apperceived as the same nearby. This constitution, in turn, founds a primitive reciprocity of quasi-objective perspectival appearances here and there, near and far, in whatever quasi-objective spatial orientations.[31]

76. THE PROBLEM OF CONSTITUTION OF THE QUASI-OBJECTIVE PRIMORDIAL ANIMATE ORGANISM

So far the concern has been to make thematic the conditionedness of quasi-objective visual appearances of constellations of primordial solids as one aspect of the quasi-objective primordial world. With the interlocking system of reciprocity of quasi-objective spatial perspectives yet another phenomenological theme arises: the first meaning of primordial conditionedness.[32] I said before that primordial quasi-objective space is most originally presented as open but finite in the sense that, as the examples from the domain of vision suggest, there are optima of presentation, thus cases of phenomena not constituted as solids through exchangeable near-far appearances, things that under no circumstances can be brought into presentation nearby as the same when far away. There are limiting cases, thus anomalies, of the one all-embracing "if-then" style of primordial experience in primary passivity: stars, clouds, rainbows, sun, moon, and the like. But these are not the only sort of limiting cases, not the only sorts of presentational "finitude." There is another, perhaps more basic one, and which now introduces still another phenomenological theme. For I cannot objectivate even the limiting cases of visual presentation without at the same time objectivating consciousness of the quasi-objective, primordial percipient organism. In contrast to the phantom organism, the primordial organism, I find, is constituted like any other "natural body existing and moveable in space," able to "convert any There into Here" (Husserl). But this also signifies that primordial quasi-objective space is therefore the "first" and single "space" for a plurality of animate organisms "functioning perceptually," a plurality of "central Here's" each of which is presented in the mode "There" for the "other."

The phenomenological situation is succinctly expressed by Alfred Schütz in language that sets forth yet "higher" strata of constitution:

> The '*hic*' is the zero point of the system of co-ordinates in terms of which the individual groups the *Lebenswelt* into zones within actual and potential reach, each of which carries along open horizons of undetermined determinability. The Other's system of co-ordinates has a zero point which is, as seen from my 'Here,' a 'There,' but which is a 'Here' to him. The 'reciprocity of perspectives'...is founded on the open possibility of an interchange of the standpoints, that is--metaphysically speaking--on the establishment of a formula of transformation by which the terms of one system of coordinates can be translated into terms of the other.[33]

Schütz's statement is especially useful because it singles out the basis for "normal" perceptual experience of a space of objective physical things as much as the foundation of that experience in the primary passivity of the primordial world. The clue lies in the reciprocity of perspectives and of standpoints--actual and possible--which, as Schütz suggests in agreement with Husserl, is the "first" stratum of "intersubjectivity." We may restate Schütz's constitutive "formula of transformation" in primordial terms this way: Because "my own" standpoint can be exchanged for another, and because it can also be exchanged for that of another percipient organism, it is possible to see, or to hear, or touch things from the "right" perspective and the "right" standpoint. For example, the sound issuing from over there is not a mere sound, but the sound of a violin--and, indeed, not the mere sound of a violin but rather a musical sound. Or, seen from there rather than here, the tree is not a mere thing but a rare species of oak. There is, in other words, the basis for more adequate seeing, touching, hearing, all of which includes as foundation the perceiving of one "space-time" in common, of unitary things and events in common, the perceiving of which includes the perceiving of a common ground of the core-world always zoned within reach. The "formula of transformation" expresses a "world in common," the first instance of which is the quasi-objective primordial world, a conditionedness, as was said before, by mental life-processes proceeding in primary passivity. Along with the simultaneous togetherness produced by the "if-then" style of conditioning primordial quasi-objective appearances of constellations of solids, the conditioning that is the "world in common" brings up the problem of the constitution of the primordial quasi-objective animate organism as the problem of the constitution of "other minds." Moreover it is here that this problem acquires its specifically *transcendental* phenomenological significance. Thus some care must be taken to *set the problem* properly in the first place.

At various times in the course of this investigation we have singled out for attention the fundamental role of primary passive tactual intentionality for the oriented constitution of space in both the generic meanings of the word, "space." Special attention was given the dual nature of tactual intentionality--the "produced unity" of the animate organism, to use Becker's expression--or, more generally, what Katz called the "bipolarity" of the phenomenon of touch.[34] In this connection, on a number of occasions it was noted that, historically, the phenomenon of touch has been regarded as the root of "space perception" in general and of the "perception of motion" in particular[35]--although for many writers in the nineteenth century that need not signify that the "sense of touch" is itself "spatial."[36] To be sure, very different conclusions have been arrived at in the present investigation, although they do not deny the basic role of tactual intentionality when it is a question of uncovering the constitutive substrata presupposed by and fundamental to the "perception of motion." Yet it was also seen that the constitutive substratum of tactual prespace, for instance, is of the same originality as the other prespaces.[37] Cassirer aptly sums up the fundamentalness and equal originality of tactual experience by saying that while it has a "peculiar tendency to objectivation," even so it "remains, nonetheless, so to speak, half way;"[38] even though quasi-objective things and objective things are perceived and apperceived by means of our animate, percipient organisms, those perceptions and apperceptions cannot be "detached from this foundation."[39]

The phenomenological situation suggested by Cassirer has been discussed before in terms of the "incompleteness" of the oriented constitution of the animate organism. Cassirer expresses it this way: "the 'objective' content remains, as it were, at the limits of one's own animate organism instead of becoming truly a 'counterpart' ["*Gegenüber*"], instead of being drawn away into an ideal *distance*. Such being at a distance [*Distanzierung*] is only reached in the highest 'objective' senses, *hearing and sight*."[40] Perhaps the most "objective sense" here is the case of vision, and it is for this reason that chiefly examples of vision were introduced to disclose the constituting of distance peculiar to primordial quasi-objective space. More particularly, in our examples of colors this signifies that not only the differentiation of film and surface colors but, correspondingly, the persisting difference between foreground and background, is instituted. At the same time, however, that sight and hearing, we must add, acquire and fulfill the "objectivating function" (Cassirer), the animate organism, the quasi-objective tactual perceiving of which falls short of fulfilling that function, is perceived "as a natural body existing and moveable in space like any other."

To express the situation in rather different terms, it may be said that the quasi-objective primordial animate organism always only *appresents* itself as There, at a distance, but of eidetic necessity *never so presents* itself There. In what sense, then, is the animate organism perceived as a "natural body" like any other? The answer, I shall try to show in the next sections, is that only "on condition" that "my" animate organism is perceived by another, indeed, a plurality, of animate organisms. In other words, correlative to the oriented constitution of quasi-objective physical things at a distance is the constitution of an "intersubjective" primordial quasi-objective space. Yet with these statements the problem of the constitution of the primordial quasi-objective animate organism is not yet properly set. The reason is that the statements made so far harbor not simply a phenomenological problem, but a *transcendental* phenomenological problem. And it is the transcendentality that must be made explicit to set the problem in the context of the oriented constitution of the primordial quasi-objective world. In other words, owing to "if-then" apperceptual conditionedness I no longer objectivate, as before, "my own" transcendently inherent spatial spreads in primary passivity. To be sure, I still find the spatial spread as constituted in primary passivity, but not as own-inherently transcendent to those mental life-processes in and through which it is presented. Rather, owing to "if-then" conditionedness, I find the spatial spread and things in it as "intersubjectively" *transcendent to* my mental life-processes, as, it might be said, *co-herently transcendent to* my mental life-processes. This is the *transcendental* phenomenological meaning of the "formula of transformation" as the way in which the "own-inherently transcendent" acquires the appearance of the "own-coherently (or: intersubjectively) transcendent" peculiar to the secondarily constituted quasi-objective primordial world. This, now, is the proper setting of the oriented constituting of the quasi-objective primordial animate organism which will be expressed in the following sections by the heading, "mundanization of a plurality of mental lives in the transcendental natural attitude."[41] The transcendental phenomenological task will be to show, in short, that the "own-*coherently* transcendent" contains the "own-*inherently* transcendent" as its necessary foundation.

77. THE TRANSCENDENTAL PHENOMENOLOGICAL SIGNIFICANCE OF FURTHER BUILDING-UP ANALYSIS AND THE PLACE OF THE "TRANSCENDENTAL DEDUCTION OF TIME."

The sketch of the two-fold conditionedness of quasi-objective primordial experiencing suggests that a basic ingredient of the "highest" substratum is the quasi-objective apperceiving of states, things and events conditioned by a plurality of transcendental lives in the natural transcendental attitude, i.e., mundanized lives intending to and positing "their own" and a plurality of other quasi-objective primordial animate organisms at rest or in motion in quasi-objective primordial space and time (above, sections 19, 20). At the same time, phenomenological objectivation of this situation of correlative constituting makes explicit a feature unique to the quasi-objective primordial organism, namely the "upright posture."[42] The spatial orientations in quasi-objective primordial space include as foundation of their perception co-intending to the organism presented most originally as having an "upright posture" "under circumstances."[43] To be sure, and with respect to the constituting of the quasi-objective primordial world, this determination of the quasi-objective organism must be considered with respect to locomotion and "resistance." It includes as foundation the constituting of a homogeneous quasi-objective space that is traversed and traversable over a "ground" appresented as here and there, as up and down, as at a distance and the "resistance" of which has been "overcome" in the sense, for instance, that "kinaesthesias of falling" are, as a rule, controlled by the will in correlation with actualization of the kinaesthesias involved in "walking," "running," "jumping."

In other words, the "center of gravity," the "rootedness to the spot," of the quasi-objective primordial organism has been supplanted by a "center of doing and going." Other quasi-objective, co-perceived and apperceived motions of the organism, e.g., arm-motions, hand-motions or motions of the fingers (e.g., pointing) all include as foundation the constituting of "lateral space" which, in turn, is a foundation of the constituting of "use space" and "tool space" which, to be sure, already involve the positing, even in primary passivity, of "means and ends." These are, of course, only some of the primordial phenomena that must be taken into consideration. Their constitution, however, directly involves something else for I cannot objectivate them, as transcendental on-looker, without at the same time objectivating the mundanization not just of my own, but other transcendental mental lives as well.

But, as noted before,[44] my transcendental mental life-processes are most originally presented as a protentive-retrotentive stream. And in their temporal structure, "as the all-embracing cogitatum <of the reflection by the transcendental on-looker,>" they are posited in primary passivity as the "all-embracing life itself, with its openly endless unity and wholeness." Thus Husserl says, rightly I believe, that only

> because it already appears as a total unity can it also be 'contemplated,' in the pre-eminent manner characterizing acts of paying attention and grasping, and be made the theme for an all-embracing cognition. The *fundamental form* of this universal synthesis, the form that makes all other syntheses of consciousness possible, is the all-embracing *consciousness of internal time*. The correlate of this consciousness is immanental temporality itself....The distinction between <internal> time itself and the consciousness of <internal> time can be expressed also as that between the mental process in internal time, or the temporal form of this process, and the *modes of its temporal* appearance, as the corresponding 'multiplicities.' As these modes of appearance, which make up the consciousness of internal time, are themselves 'intentive components of conscious life' and must in turn be given in reflection as temporalities, we encounter here a paradoxical fundamental property of conscious life, which seems thus to be infected with an infinite regress. ...the fact is evident, even apodic-

> tically evident, and indicates one aspect of the ego's marvelous being-for-himself: here, in the first place, the being of his conscious life in the form of reflexive intentional relatedness to itself.[45]

I cite this statement of Husserl at length because it establishes in his thought a basis for the formulation of three steps of what we may call a "*transcendental deduction of time*" that parallels the transcendental deduction of space:[46]

> *First*, owing to the protentive and retrotentive structure, mental life-processes are self-intentive and self-positing no matter what else they intend to and posit; they are constitutive of themselves not only as transcendental but also as self-apperceptive as in and part of the world.
>
> *Second*, as a consequence, determinations of mental life-processes and world in their primary "status" are ipso facto determinations of them in their secondary "status."[47]
>
> *Third*, mental life-processes are self-intentive and self-positing at each substratum of oriented constitution.[48]

It is worth emphasizing again that here, as elsewhere in this investigation, by "mental life-process" is meant "transcendentally reduced mental life-process," and that by "world" is meant "transcendentally reduced world." For the moment, it is the third step of the "deduction" which is of immediate interest because it is at the root of the meaning of "if-then" conditionedness characterizing the quasi-objective primordial world and its spatial dimension. In section 82ff. we shall consider the "deduction" itself and the spatio-*temporal* dimension of the quasi-objective primordial world.

78. SPACE AND OTHER

A basic idea of quasi-objective primordial space is that which we called, with Alfred Schütz, the "formula of transformation;" in Husserl's words it is the idea that "not only the systems of appearance that pertain to my current perceiving 'from here,' but other quite determinate systems, corresponding to the change of position that puts me 'there,' belong constitutively to each physical thing. *And the same in the case of every other 'there'*."[49] Clarification of this idea of "space perception" is clarification of the way in which "my own inherently transcendent world" acquires the appearance of "our own co-herently transcendent world," that is to say, the quasi-objective primordial world. For this another example is required. Perhaps the most convenient example is that of Dorion Cairns mentioned earlier where, "in the mental process, perceiving this white sheet of paper...I myself find, by a suitable turn of mental regard, not the sensation datum white, but an intentive sensing of the sensation datum, white. And it is not the sensation datum, white, but rather the intentive sensing of it, which is something belonging inseparably to the proper essence of the concrete perceiving of this white sheet of paper...."[50] Although clearly distinct in every generic respect from the immediately presented white piece of paper in front of me, my own intentive sensing and perceiving are at least as accessible to me as is the piece of paper. Apprehended by a "suitable turn of mental regard" specifically distinct from the mental life-processes by which there is intending to the piece of paper, my mental life-processes, such as the sensing and the perceiving, are nevertheless laid out for me, so to speak, as "objectively" and indeed as publicly as is the white piece of paper in front of me. *For me*: My own mental life-processes (sensing, seeing, hearing, remembering, and so forth) are presented as publicly as is the piece of paper or the "sensation datum, white," although their presentive modes, while immediate, are yet radically distinct. Thus for me alone to speak of my own mental life-processes, for example, as private and inward (or "subjective"), and of the piece of paper or the white datum as public and outward (or "objective"), is absurd. It is rather the case that *all are presented in optimal public obviousness, meant and believed in as transcendent to the other, as obvious and public for me at no matter what level of oriented constituting.*

Now, however, in so far as you are concerned, your mental life-processes are not presented as public and obvious to me, although the piece of paper you perceive is. To be sure, it is taken for granted by me that your mental life-processes are immediately presented to you and, to you, as obviously and publicly as the piece of paper is to us both. In the primary passivity of my mental living objectivated in the primordial world, there is an assumptive taking-for-granted that the "suitable turn of mental regard" by which you apprehend your immediately presented mental life-processes is not only distinct from the processes by which you perceive the piece of paper, but that it is a case of a turn of mental regard wholly unsuitable to me, i.e., which I cannot perform and still find concrete, actual mental life-processes--despite the fact that the turn of regard is surely similar to the one on the basis of which I likewise seize upon my own mental life-processes. In contrast, a perceiving is performed by me (seeing the piece of paper) which is not only similar to yours, but is also one in which the self-same, concrete actual content is found by me: the public piece of paper, immediately presented and meant as transcendent to my own mental life-processes *and to yours*. Yet that is to assume and take for granted that your mental life-processes are to and for me not only transcendent, as is the piece of paper seen by us, but also to and for me transcendently "inward," "private"--although your mental life-processes are not transcendently private for you but rather transcendently public for you, yet not for me.

Precisely the "not for me" that is correlative to the "not for you" expresses what may be called--seeing the piece of paper immediately presented in optimal obviousness and publicly transcendent to both of us--the incompatibility of "public inwardness," of own-inherent transcendents. The language is admittedly ponderous, and perhaps a less exotic way of expressing the situation would be to speak of the *mutual incompatibility of otherness*. It is worth noting, too, that in order to avoid even greater difficulty in expression, I have used the words "I," "me," "you," "yours," and the like instead of the non-egoic locutions which properly belong to this level of oriented constituting in primary passivity. "I," "me," "you," "yours," even at this highest stratum of the primordial quasi-objective world, are then at best metaphorical expressions for our purposes at the moment.

At this juncture a few conclusions may be drawn. The first is that "inward" and "outward," "public" and "private," are not only mutually correlative but require the qualifier, "transcendent." The second is that only "on condition" of meeting you, as it were, at the perceived piece of paper, is there an inward for me which, however, is never of essential necessity mine but for me always yours, and only yours in so far as it is for me. The third conclusion is that our quasi-objective, mutually incompatible publicity of mental life-processes is not one of empathic let alone expressive or even infectious coincidence of real identity among our mental life-processes. Accordingly, a fourth conclusion may be drawn: "private" or "inward" are words that can only be employed in the second and never in the first person so that--returning to Husserl's statement that began this section--just as you are always there for me I am always here. This signifies that even if we exchange places I can never be there, only here--and not here, but always only here for me. It is not so much that I cannot be your inwardness, but rather that then there would be no inwardness whatever were I you.[51]

These conclusions are of course tentative. Yet they contain a number of distinctions that may be made in the quasi-objective presentational modes that characterize this stage of building-up analysis of the oriented constitution of the quasi-objective primordial world. Thus in the case of "public," it is necessary to distinguish between what is immediately presented as transcendently public, e.g., the seen piece of paper and the sensed datum, white, and what is immediately presented as immanently public, such as my own perceiving of the piece of paper or my own sensing of the datum, white. Nor would there seem to be any actual or possible case where the one presentive mode in its full noematic sense would coincide with or be confused with the other. Likewise it is necessary to distinguish between what is immediately presented as transcendently private (e.g., for me your perceiving of the piece of paper) and what is immediately presented as transcendently public (e.g., the piece of paper perceived by you); again, there would seem to be no actual or possible case where the one presentive mode would overlap or be confused with the other.

Finally, it is necessary to distinguish between what is immediately presented as public and as transcendently private: e.g., my seeing and your seeing the piece of paper, or my sensing and your sensing the datum, white. Not only would there seem to be no actual or possible case where the one presentive mode coincides with the other, but the two cases are also mutually incompatible. To be sure, all of the distinctions just mentioned are distinctions only in so far as the presentive modes are co-presentive modes, paired presentive modes, as the basis of self-identifying and numerically differentiating syntheses. Another way of expressing the same thing would be to say that what my mental living shares is a quasi-objective transcendently public world on the primary passive assumption by me of your transcendent privacy in which you seize upon my immanent or own-immanent public mental living; and conversely in which your public mental living is seized upon by me.

The very distinctions among presentive modes and the very seeing of those distinguished modes already presupposes the mutual transfer of noematic senses of "transcendently public" and "transcendently private" resulting in their partial or full confirmation or disconfirmation. For example, the noematic sense of my immediately presented public mental life-processes for me is paired with the co-presented public piece of paper for me: my immediately presented mental life-processes and the immediately presented piece of paper, as noematic senses, are mutually paired in their very publicity; to be sure, here there is only partial confirmation in the pairing--the one noematic sense of "public" coincides with the noematic sense of the other. But the immediately presented "public immanent" and the immediately presented "public transcendent" cancel each other out, so to speak. They disconfirm each other. And this latter cancellation is, in addition, based on a numerical differentiation in the sense of an identifying of what is immediately presented as immanently public with itself correlative to what is immediately presented as transcendently public with itself.

These distinctions also express the "full fledged" intersubjective, i.e., specifically *social*, relationships of the quasi-objective primordial world. For it follows from the constituting of the mutually correlative immanent and transcendent publicity of our mental life-processes that social relationships are presented immediately as full fledged in primary passivity even if members of the relationships do not share common feelings, beliefs, desires. In so far as a necessary "circumstance" of the "if-then" style of causality is "interpublicity" and "interprivacy" ("intersubjectivity"), this highest substratum of the quasi-objective primordial world is the phenomenological origin of praxis, of mutual exchange of places and encounter pertaining to the "doing of things together." It is the quite specific content of the contingent apriori expressed by the "formula of transformation."

Two significant consequences may be drawn from this "idea of space perception:"

First, the foundation for the oriented constitution of a social relationship--no matter how "primitive"--depends rather on the sharing of transcendently public, own-coherent or common quasi-objective spatio-temporal primordial world; that is to say, each member of the quasi-objective community intends to and posits the quasi-objective spatio-temporality as identically the same public world intended to and posited by others.

Second, each quasi-objective, primordial mental living intends to and posits itself as, publicly, identically the same mental living intended to and posited (*scl.* "experienced," "addressed") by another mental living. Hence the community of mutually incompatible public inwardnesses necessary for any quasi-objective social relationship consists of, *inter alia*, common quasi-objective things and events appresented under circumstances publicly transcendent to any mental livings. In other words, the question about the constituting of community is not one about the coincidence of the "real identity" among mental livings;[52] it is rather about intendings to and positings of a common quasi-objective space and time that include, as necessary foundation, intendings to and positings of public things transcendent to a plurality of quasi-objective mental livings. In short, "space perception" is the phenomenologically fundamental determination of and motivation for the quasi-objective primordial world acquiring the "appearance" of a social world constituted in primary passivity. The words, "social," "social world," "social relationship," and the like, must now always be understood with the qualification of quasi-objective or objective.

79. THE PHENOMENOLOGICAL DATUM OF THE SECOND STAGE OF BUILDING-UP ANALYSIS: PRESENTATIONAL IMMEDIACY AND IMMANENCE

To clarify further this conclusion it is necessary to formulate with greater precision the phenomenological datum of "intersubjectivity." But this clarification is also a critical reflection for it will allow of establishing the correct transcendental phenomenological question about "intersubjectivity."

a. The Basic Assumption

To express the matter in another way it may be said that a basic assumption objectivated as peculiar to the essence of my transcendental mental living in the natural attitude is that it constitutes for itself a quasi-objective primordial world (and, by extension, a real, objective world) presented and posited as having a valid meaning of being such a world for others regardless of whether any individual other mental living may be illusory.[53] As a consequence, it is essential to my world that it be a primordial, quasi-objective world for some experienceable others who are "real" in the same sense that I am.[54] As truly, therefore, as my purely passively constituted quasi-objective mental living is in the world, so are purely passively constituted other mental livings in the quasi-objective primordial world; and as truly as this is a quasi-objective primordial world for my own transcendental mental living in the natural attitude, so it is for other transcendental mental livings.

b. Presentational Immediacy

A consequence of the discussion of the constituting of "public" and "private" is that something may be transcendent to mental life-processes and yet be *immediately* presented. And to be immediately (or: directly) presented does not, of course, necessarily signify apodictically presented. Thus to say that the mental life-processes of someone else, presented and posited as transcendent to my own, as transcendently private, are not apodictically presented, is not equivalent to saying that someone else's mental life-processes are not immediately presented.[55] To be sure, the further course of experience may indeed show that mental life-processes other than mine, just like anything else immediately presented as transcendent, such as the white piece of paper, is illusory. Nonetheless, no such further course of experience can cancel out or provide a rational motive for cancelling out my belief in the existence of my own stream of mental life-processes: a case of immanently public, or apodictic, presentation. Accordingly the question about the constituting of mental life-processes of someone else is not so much a question about the presentational immediacy which distinguishes my own mental life-processes from someone else's; it is instead a question about the immediacy and the public immanence of the presentation which comprises the distinction in the first place.[56] Thus although the mental life-processes of someone else are not apodictically presented to me, they are nonetheless immediately presented to me as the mental life-processes of someone else, as a concatenation of processes belonging to another. And precisely because they are immediately presented as belonging to someone else, I have no apodictically evidential intending to them.

Moreover, along with intending to immediately presented mental life-processes other than mine I find a protodoxic believing in them as having the noematic sense, "existent--but not really inherent constituents of my own mental living." And this belief does not consist of the fact that I do not find them as such constituents. It consists instead of the insight that really inherent immanental public constituents are absolutely incompatible with the mental life-processes of someone else.[57] Just this incompatibility is the prima facie justification for the correct believing in the immediate presentation, "mental life-processes other than and transcendent to my own."

c. The Transcendental Phenomenological Datum

We therefore arrive at the transcendental phenomenological datum: one mental living intends to and posits other mental livings and all intend to and posit one self-same quasi-objective spatio-temporal world in primary passivity. *If* my mental living is what it is immediately and apodictically intended to and posited as being, *then* the mental livings of others are necessarily presented immediately and posited as existing as transcendent to mine and we all intend to and posit one and the same quasi-objective primordial world and each other--immediately.

80. CRITICAL REFLECTION ON THE RIGHT TRANSCENDENTAL PHENOMENOLOGICAL QUESTION OF "OTHER MINDS"

This formulation of the transcendental phenomenological datum expresses that from which discrimination was made in carrying out the unbuilding reductions. The task at hand is the discovery and clarification of the motivations in the quasi-objective primordial world for building up that "highest" substratum at which there is immediate presentation of the mental life-processes of someone else.

At the beginning of his Fifth Cartesian Meditation,[58] Husserl makes the following statement which may serve as point of departure for further analysis: "When I, the meditating I, reduce myself through the phenomenological epoché to my absolute, transcendental ego, have I not then become *solus ipse*; and do I not remain that, as long as I carry on a consistent self-explication under the name of phenomenology?" Husserl's question is applicable to the whole set of steps effected in carrying out an examination of the oriented constitution of the real, objective spatio-temporal world. Indeed, by exercising the psychological phenomenological epoché "the" world itself is made thematic,[59] and by making "the" world thematic it is examined and judged about precisely and just as the noematic correlate of experiencing and thinking about it. As noematic correlate, "the" world is not something lying "behind" or "above and beyond" what we are familiar with in ordinary life. It is just that world and no other; in no way is it impoverished by exercise of phenomenological epoché. But the very form of the question raised by Husserl in the foregoing passage cited (and in the corresponding statement in the First Book of *Ideas*, section 33) points to the fact that he does not plan to stop with the psychological phenomenological epoché; he intends to exercise the transcendental phenomenological epoché. Thus he continues by saying that "Transcendental reduction restricts me to the stream of my pure conscious processes and the unities constituted by their actualities and potentialities." As will be recalled (above, section 21f.), in consequence of this phenomenological epoché a more basic status of my mental life-processes is disclosed which is not itself mundane or intramundane. And it is this more fundamental status that forms the context in which, with Husserl, we undertake the "exposition of the problem of experiencing someone else." This exposition seeks the way from the optimal public obviousness of my own mental-life processes, we may now say, to *our secondarily* constituted, mutually incompatible transcendently public inwardness.

Set in this fashion, the phenomenological question concerns how it is that, at the substratum of primary passivity disclosed by exercising transcendental phenomenological epoché and objectivating the quasi-objective primordial spatio-temporal world, there is acceptance of the fact that the existence and nature of my quasi-objective perceiving and apperceiving is a function of the existence and nature of what is presented and appresented in and through intuiting of them "under circumstances" in primary passivity. And this phenomenological question was arrived at precisely by a series of unbuilding reductions that discriminated the noematic sense, "mental life-processes belonging to someone else," and, correlatively, immediate intuitings of mental-life processes other than my own. By methodically discriminating the fully constituted real, objective spatio-temporal world, a series of substrata are uncovered that are quasi-objective worlds, quasi-objectivities (prespatial, phantom and primordial quasi-objectivities) "through" which the real, objective world is experienced. Even though "myself" is constituted at these substrata as "myself," as *solus ipse*, it is of the utmost significance to realize that this is not a privative meaning of *solus ipse*. To discriminate the foundation of a perception is not to deprive oneself of the perception. There are basically two reasons for this, the second of which becomes thematic, and only can become thematic, at this second stage of building-up analysis of the quasi-objective primordial world.

The first reason is that I, as transcendental phenomenological on-looker, have objectivated and made explicit the noematic sense, "one stream of transcendental mental life-processes among others," by considering in a series of unbuilding reductions the fact that in order to be a mundane, hence public, stream of mental life-processes this is, as contingent apriori, necessarily to be a *transcendently public transcendental* stream of mental life-processes. Thus the *second* reason, for this equally signifies that any other transcendental stream of mental life-processes incompatible with mine constituted in harmonious identifying and numerically differentiating syntheses would, like mine, exercise transcendental phenomenological epoché and, on that basis, objectivate and make explicit the corresponding world-intending and world-positing transcendental mental life-processes. The philosophical-phenomenological significance is that such oriented constitution is presupposed by any attempt to "communicate" a transcendental-phenomenological conviction or opinion; in so "communicating," to be sure, I, the transcendental on-looker, address myself not as a "human being" to other "human beings," but rather as a transcendental-phenomenologizing mental living to other such mental livings whom I mean as at least possibly exercising transcendental-phenomenological epoché, as possible transcendental empirical exemplifications (above, section 9), and, as non-participating on-lookers, each observing their own transcendental stream of mental life-processes.[61] Included, then, in this specific oriented constitution in primary passivity at the stage of building-up the quasi-objective primordial world is each transcendental stream of mental life-processes meaning itself as identically the same stream also meant and addressed by any other transcendental streams of mental life-processes that include phenomenologizing on the basis of their world-meaning and world-positing.

It is necessary then to define "transcendental intersubjectivity" as a transcendental association of transcendental phenomenologizing mental livings, each--actually and potentially--objectivating itself as mundanized stream of transcendental mental life-processes *transcendently public and correlatively but incompatibly transcendently inward*, appresented in the primary passivity of the quasi-objective primordial world included as foundation of perceiving and apperceiving the real, objective world as real, objective world for "everyone." The phenomenological datum which must then be further clarified is, "intending to and positing the mundanized plurality of transcendental mental livings constituted in primary passivity as mundanized plurality of transcendental mental livings, of which my own quasi-objective primordial mental living is a member."

81. THE INTENTIVE FOUNDATION OF QUASI-OBJECTIVE PRIMORDIAL "SPACE PERCEPTION"

Earlier in this chapter (section 74) objectivation of the "lowest" substratum of the quasi-objective primordial world disclosed my animate organism presented originaliter as the "only" quasi-objective physical thing posited with the sense of something component: both physical and mental components constituted correlative to the constituting of the reciprocity of quasi-objective perspectival appearances intending to which includes intending to the exchange and repetition of the quasi-objective spatial orientations of here and there, near and far. This is one of several appresentational factors grounding intendings to unpresented but apperceptively imputed determinations of immediately presented things. Yet another factor of importance here is the presented "incompleteness" of the constituting of my animate organism. Together these two factors motivate the transfer of the noematic sense, "animate organism of my mental life-processes" to immediately presented and appresented things constituted in identifying and numerically differentiating syntheses. What, now, is the "fate" of this noetic-noematic transfer of sense?

Unlike the cases of assimilative synthesis examined before where at least partial similarity was constituted, here there is an assimilative synthesis of annulment, of complete disconfirmation of the transferred sense, "organism of my mental living." More particularly, the sense, "immediately and apodictically presented immanental mental living," is transferred to and overlays the sense, "immediately and transcendently presented mental living apperceived as manifesting itself in something other than my organism;" apodicticity and immanence are completely cancelled out in the overlay of mutually and immediately presented senses, by virtue of which there is constituted an annulling synthesis. But there is also a synthesis of partial similarity in that immanent and transcendent mental livings are immediately presented on the basis of that annullment as intending to and positing one and the same space in which Here can be converted into There and There into Here, and in which a reciprocity of perspectival appearances is operative.

To express the same phenomenological situation in another way, we may say that the other thing paired in the mutual transfer of sense is apperceived not as, e.g., a quasi-objective physical component of a psychophysical thing like my animate physical organism. Instead, on the basis of the complete disconfirmation and the consequent assimilative synthesis of annulment, the other thing is perceived as a physical component like a psychophysical thing so that the mental living belonging to someone else, which is intuited by me as immediately presented in primary passivity as transcendent to my mental living, is apperceived as manifesting itself in another quasi-objective organism.

By no means, accordingly, is the mutual transfer or overlaying of sense to be confused with an "empathic" perceiving of someone else (as in Lipps and Scheler), nor with an inference by analogy (in either of the two meanings of the term, "inference," distinguished earlier[62]) from my own quasi-objective primordial organism presented with the sense of something component to someone else's organism (Hume, Russell, Carnap), nor with the spontaneous bonding together of individual wills (Marx, Tönnies, Heidegger). It is rather the case that "empathy" and inference as well as the bond of individual wills are based upon and presuppose the constituting of transcendently public mental life-processes correlative to and incompatible with transcendently inward mental life-processes in a transcendently public and common world. It is this constituting which, as it were, allows me to expect other organisms to "behave" in general as mine would were my mental life-processes components of those organisms; when there is "behavior" contrary to such expectation, then the transferred sense, "organism of someone else's mental life-processes," is cancelled for me by virtue of which it is experienced as "mere thing" rather than someone else's organism. And that expectation is based on mental life-processes immediately presented as transcendent, hence as absolutely other than mine.[63]

Whatever further meaning can be drawn from the foregoing conclusions, it is clear that "intersubjectivity," even at the constitutive level of the quasi-objective primordial world, cannot be accounted for by starting out with the assumption of a world with a plurality of "subjects" in it, and then asking, as do, for example, Scheler and Schütz,[64] how it is that we mean others in it. Instead, it is necessary to proceed, no matter what the imaginational strain may be, by first isolating in an unbuilding reduction all the substrata underlying the stratum in which an "intersubjective" quasi-objective world is constituted in primary passivity; and then, second, by describing the residual substrata, it is possible to ask about the motivations by virtue of which a mental living characterized only by primary passivity does not rest with the intending to a quasi-objective world in which it is aware of itself as the only mental living and its organism as the only animate organism, but instead necessarily goes beyond them to constitute one primary passive quasi-objective world as containing a plurality of psychophysical beings intending to and positing the same quasi-objective world.

There is then no need to presuppose a world with a plurality of "subjects," nor to introduce, like Leibniz, the *deus ex machina* of pre-established harmony to account for the experience of an identically common world.[65] Rather *if* my own mental living is what it is, *then* other mental livings necessarily exist "outside" and "alongside" it and "we" experience one and the same world and each other immediately: the basic meaning of "causality" and "objectivity," that is to say, of "mundanization."

More broadly expressed: Intuiting of a plurality of mental livings pertaining to others not only goes on regardless of whether I am engaged in that intuiting, but also there is only such intuiting, in turn, on condition of intending to and positing one quasi-objective primordial surrounding space which remains the same, as it were, whether intended to by me or by others as well. And because quasi-objective primordial space, constituted correlative to the constitution of "other minds," is most originally presented as a traversable one of "exchanged" and "exchangeable" places, further exhibiting the "formula of transformation" (above, section 76), it is equally the constitutive origin of a "praxis space" or "action-centers" enabling us to work together in a common situation in the real, objective world.[66] Important as this constitution is for tracing the foundation of the constitution of the space of the real, objective world, it is necessary here to briefly examine the intentive foundation of "exchange of place" and the reciprocity of perspectives. The constituting of one quasi-objective primordial space and of "other minds" are strictly correlative and include as foundation the constituting of my own quasi-objective primordial organism and of a quasi-objective primordial earth-basis. As in the case of the constituting of my own quasi-objective space, so here organism and earth-basis are anomalies--a peculiarity of primordial constitution noted much earlier (above, section 21). More particularly, as anomalies, my primordial organism and earth-basis are "secondarily" constituted quasi-objective, centerless phantom organism and ground (above, sections 65, 68). They are anomalies because, even here, organism and earth-basis are not presented as bodies among other bodies. What, then, are their anomalous presentational modes peculiar to the primordial world by virtue of the "perception of other minds"?

Perhaps the first thing to note is that, in consequence of the "transformation formula," the quasi-objective phantom core-world in centerless space acquires the appearance of one actual among many possible "world-segments" near and far relative to my quasi-objective organism as central Here and relative to which the rest and motion of all other quasi-objective things are co-perceived. The "relativity" in question also signifies, however, that world-*segments*, unlike the *core*-world, are not absolute optima; under circumstances, world-segments always are presented as allowing for "experiencing more" so that if motion ends, rest or termination is appresented with the possibility of becoming motion again in the same or other directions under the same or even differing circumstances. Thus for this, if for no other reason, quasi-objective primordial space requires the qualifying epithet, "surrounding." In short, world-segments are presented and appresented as "open possibilities." So perceiving my own organism as determined and further determinable in that double meaning of relativity, there is necessarily included as foundation an apperceiving of my quasi-objective organism and of other quasi-objective things actually or possibly pertaining to a given world-segment as relative in another respect, namely relative to what Husserl called a "basis-body," i.e., relative to a common ground as at rest--be it a particular piece of the earth or of a vehicle which carries us (Husserl's examples range from railway carriages to space ships to his toy wagon on and off which he jumped as a child). To take just one of his examples: "Consider the relative 'basis-body': I can be in a moving vehicle which is then my basis-body; I can thus be borne by a railway carriage, in which case my basis-body is first of all the body carrying me while moving. ...The vehicle is experienced as at rest. But when I look out the window I see that the countryside is in motion."[67] Thus under one set of circumstances, the relative basis-body is apperceived at rest, but under another it is perceived as in motion. Ultimately, however, all such relative basis-bodies are at least implicitly apperceived as having the sense of being relative to an archontic basis, the "earth-basis as a whole."

Although, as the example suggests, relative to the earth-basis a basis-body may be perceived as in motion or at rest, depending on the circumstances under which it is perceived (the train stopping, getting on or off, etc.), the earth-basis is apperceived neither as at rest nor as in motion: "It cannot be experienced as 'a body' which not only has its extension and its qualitiedness but also its 'place' in space, and which can possibly exchange its place and be at rest or in motion."[68] As transcendental on-looker, I cannot objectivate the apperceiving of the earth-basis as a whole without, however, objectivating the perceiving of the surrounding quasi-objective spatial world-segments of bodies in near-far fields, presented as having places that are not only exchangeable but also as possible terminations of motion, but with the sense of further or other motion. As a consequence, quasi-objective phantom centerless space and phantom things acquire the appearance of being "earthly bodies and <quasi-objective primordial> space is given as having the sense of being earth-space" so that the interpublic community of "animate beings" is presented as "earthly," there being "no contrary in the nonearthly."[69] Thus "earth," as earth-basis as a whole, remains an anomaly in quasi-objective surrounding space of the primordial world; there is "not yet" a foundation for its constitution as having a place, for its being presented as one body among other actual and possible bodies.[70] To be sure, earth-basis as a whole may be said to be appresented as at rest, except that "rest" in this case is not a mode of motion.[71]

The correlative constituting of earth-basis and quasi-objective primordial surrounding space, determined and further determinable in its segmented near-far fields of experience, presupposes, as mentioned, the constituting of my own animate organism as central orientational Here. A few more comments are in order in addition to what has been said so far about the quasi-objective primordial organism. Presented always as central, my animate organism, under no matter what circumstances, either goes or is still, but is never perceived as in motion away from or as approaching itself relative to the "immobile" earth-basis as a whole: "...the basis on which my quasi-objective primordial animate organism goes or does not go," Husserl notes, "is also not experienced as a body, as *wholly* to be moving away or not moving away. Animate organisms of others are bodies at rest and in motion (always: in the sense of approaching or receding from me). But they are animate organisms in the form of 'I move,' whereby the ego is an 'alter ego' for which my animate organism is a body, and for which all outer bodies which are not animate organisms for it, are the same outer bodies that I have."[72] For me, those bodies appresented as manifesting mental livings transcendent to mine, immediately presented as transcendently inward, are presented then not as transcendently public "outer bodies" or "mere things" but instead as transcendently public "body-organisms" which not only move or are at rest but go and are still relative to my animate organism. For me, each transcendently public body and body organism is presented as having locatedness, extensity, as being at rest or in motion approaching or receding relative to Here. Likewise, my own animate organism is perceived by others as for them a body, more specifically, as a body-organism, although not for me. Here again, in the mutual incompatibility of our transcendent inwardness, "for me" and "for you" are the intentive foundation for apprehending each other as manifest in transcendently public body-organisms excepting our own, which is only an animate organism and for each of us never a body--and that for all of us directly or indirectly necessarily relative to "earth" presented originaliter always only as transcendently public basis and not as body, as basis that neither goes nor is still, as neither in motion nor as at rest but rather as "immobile."[73]

To express the same phenomenological situation differently, it may be said that by virtue of the correlative anomaly of "earth" and "organism" an interlocking system of quasi-objective primordial perspectival and non-perspectival appearances is constituted in the quasi-objective surrounding space of the primordial world. The "mutual perception of other minds" is the intentive foundation for those syntheses of similarity and difference in which body-organisms, bodies ("outer bodies") and earth-basis as a whole are constituted most originally in indefinite, open surrounding primordial space under circumstances as a quasi-objective public world.[74]

Before leaving this sketch of the distinction between body, body-organism and organism, it must also be mentioned that just as "earth" remains anomalous in quasi-objective primordial surrounding space, so does the sky, "constituted as the field of what is outermost, yet which can be spatially experienced for me and all of us--with respect to the earth-basis."[75] Presented as the outmost horizon of reachable distance, stars are "limits" (though not optima) "in which what can still be experienced as a distant physical thing finally disappears by moving away from me."[76] And here problems of further building-up analysis arise: Upon what intentive foundation do earth and stars acquire the appearance of "normal bodies" rather than that of basis and field of what is outermost? Upon what foundation are my own organism and body-organism of others constituted as "normal bodies"? These and similar questions must be left for another occasion for it is necessary to turn now to the fact that, like the other "spaces" considered so far, primordial space is space-*time*.

82. THE SETTING FOR FURTHER BUILDING-UP ANALYSIS OF "INTERNAL TIME"

The preceding sections have examined in some detail the third step of the "transcendental deduction of time," especially as it pertains to the oriented constitution of the *space*-time of the quasi-objective primordial world. The task remains, now, of examining all three steps of the "deduction" as they apply to any space-*time* and to any temporality whatever. To be sure, in previous chapters it was necessary on occasion to refer to the largely presupposed temporal dimension of the various prespaces and quasi-objective spaces examined. Indeed, in retrospect it is clear that essential to phenomenological practice is the taking of the present or current phase of the stream of transcendentally reduced mental life-processes as transcendental clue for the building-up analysis of the quasi-objective spatial dimension of the real, objective world (above, section 61). Both in that connection and elsewhere (section 10f.) it was noted that the protendings and retrotendings (Husserl: "retentions") comprising the stream of mental life-processes proceed, and can only proceed, in primary passivity. Although what has been said so far about the "consciousness of internal time" has taken issue with, and revised, Husserl's account (especially with respect to hyletic and other sensuous data, sections 25f., 56, 61), nevertheless Husserl's setting of the analysis of the "consciousness of internal time" has been retained as he established it in his 1905 lectures.[77] The setting is expressed in the following words of Husserl:

> As soon as we attempt to render an account of consciousness of time, to set consciousness of objective time and consciousness of subjective time in the right relationship, and make comprehensible to ourselves how temporal Objectivity, thus any Objectivity whatever, can be constituted in consciousness of subjective time, indeed, just as we also only make the attempt to draw into analysis the consciousness of purely subjective time, the phenomenological content of temporal mental processes, we put ourselves in the greatest difficulties, contradictions and confusions.[78]

To resolve those difficulties, contradictions and confusions, Husserl deals with "subjective" and "objective time," with "time as constituted" and "time in itself." To do this, Husserl raises two basic questions: 1) How is time constituted in the "consciousness" of it? 2) How is something temporal constituted in the "consciousness" of something temporal? In other words, it is presupposed that not only are "time" and the "temporal" constituted, but that they have their origin in the "consciousness" of them.[79] The immediate question is how they are constituted, and to ask such a question signifies that the phenomenological epoché has already been affected to render an account of "subjective" and "objective time." To effect phenomenological epoché with respect to our "natural" believing in the real, objective spatio-temporal world as existent, and in ourselves as part of that world, is to parenthesize world and therefore world-time and world-space. Likewise, the sense of my own mental-life processes as "in the world" is parenthesized. Nevertheless, my own mental life-processes are still meant and posited as temporal, as comprising a stream in which each mental life-process has its duration and relationships to other enduring, flowing mental life-processes. Similarly, the transcendentally reduced world is still intended to and posited as having its own inherent extensity in space and time.[80] As suggested by the formulation of the three steps of the transcendental deduction of time,[81] the mental life-processes at issue here are "transcendental" in that not only are they temporal in themselves and intentive to "world-time" and "world-space," but also in that they are self-intentive as temporal quite apart from whether their own "time" is apperceived as in the world. Thus the "time" of mental life-processes is constituted "time," and constituted precisely in the intentiveness to time regardless of whether I am engaged in that intentiveness, busied with it or with things other than it.

83. THE RIGHT ORDER OF ANALYSIS OF "INTERNAL TIME." MODES OF TEMPORAL ORIENTATION

A result of Husserl's investigations of the "consciousness of internal time" reported in his 1905 lectures was that protendings and retrotendings not only constitute an enduring mental life-process, e.g., an "impressional" process, but that they are also themselves constituted as parts of that process. Husserl was then led to the conclusion that the mental life-processes in question are "pre-temporal,"[82] which is to say, "non-temporal." Realizing by the time of his 1929 work, such as *Cartesian Meditations* and *Formal and Transcendental Logic*, that this was not a satisfactory conclusion, Husserl further distinguished between a genuinely mundane and genuinely non-mundane status of mental life-processes and, correlatively, between the latter's genuine temporal and genuinely non-temporal status constitutive of an enduring mental life-process. Thus while we shall retain for the most part the framework of Husserl's earlier discussion of "time" in developing its bearing on the oriented constitution of the spatial dimension of the real, objective world, it will be necessary to take into account Husserl's later reformulations.

What I mean is this: Husserl's various phenomenological objectivations of "consciousness of internal time," despite all differences from 1905 to 1929, make explicit two eidetically necessary facts pertaining to that "consciousness:" first, the fact that each extent of a mental life-process, any mental life-process, is immediately presented in full concreteness as protentive and retrotentive; and, second, the fact that the only noematic objects protended to and retrotended to are immanental ones consisting of other intentive extents or phases of mental life-processes. Together, these two eidetic facts raise a critically important question of the right order of analysis of the "consciousness of internal time."

In section 8 of the 1905 lectures on the "consciousness of internal time," Husserl begins his account with an examination of a "tone purely as hyletic datum." According to his theory of hyletic data, however, there is no single thing, "the tone" for there is either tone as an immanental hyletic datum or tone as a transcendent object as a violin tone. And the use of the same word, "tone," to designate both is equivocal because the relation between them is not one of identity but rather of adumbrating to adumbrated. Tacitly it would seem then that Husserl assumes that the "tone purely as hyletic datum" is something non-transcendent to the stream of auditory mental life-processes and is thus something really inherent in the stream of consciousness. As a consequence, Husserl also assumes that consciousness flows as two currents--a noetic current and a hyletic current.[83] Thus when Husserl refers to the excluding of "construing and positing of something transcendent" (which are moments of the noetic current) he means an excluding of every noematic object that has the sense of transcending the stream of "internal time." But Husserl also exercises a reduction of the really inherent hyletic datum to the status of a noematic objective sense of the flowing consciousness-of "the tone" in its manners of appearance of actually now and just past. Relative to this consciousness-of, this hearing, the hyletic datum and its duration are "transcendent" in the sense in which anything noematic is transcendent to its corresponding noesis, although the hyletic datum does not have the sense of being transcendant to the stream of "internal time." Thus if one were to try to follow Husserl's account and retain his theory of hyletic data, it would be necessary to consider retrotendings, for instance, as retrotendings *first* to particular sensings of sensa (e.g., auditory sensings of tones, to use Husserl's example), and, *second*, as retrotendings to the hyletic data in question which adumbrate the appearances in and through which the tone is presented. However, if the theory of hyletic data has been rejected along with the theory of consciousness, as has been done in this book, then the order of analysis is quite distinct. So as not to impede the discussion of what I believe to be the right order of analysis, the necessarily rather elaborate graphic representation of "consciousness of internal time" is given as an appendix to this section (section 86).

Contrary to Husserl, and *in the first place*, protendings and retrotendings must be examined as protendings and retrotendings to *intendings of any kind whatever. In the second place*, retrotendings, for example, must be analysed as retrotendings to retrotendings and protendings. Therefore retrotendings to, e.g., *sensings of* prespatial sensa are analyzed under the heading of "retrotendings to intendings of any kind whatever." *Sensings themselves* of prespatial sensa and sensum-fields are retrotended to and protended to. But the sensum-fields themselves as wholes and the particular prespatial sensa salient in them are only intended to as objects of, and inherently transcendent to, those sensings. In this light, we examined (in section 62 above) what Husserl called the "continua of recession-modes," using a graphic representation based on section 10 of his 1905 lectures. In more general terms than before, we may restate our findings as follows:

A diversity of presentational modes of one or another extent of visual, auditory or tactual sensing was distinguished. For example, there is presentation of an extent of visual sensing as past, or as just past, or perhaps of another extent as yet to come. The future extent, we said, is meant as having been expected in its manner to come, while, simultaneously, there is a continuous retrotending to the present extent of the sensing "flowing off into the past." Accordingly, what was earlier described above (section 62) was, in part, the present extent or phase of sensing immediately presented as a retrotending to itself while protending to more of itself to come. Objectivated in each temporal extent of sensing, as an example of any intending whatever, was a retrotending to its own earlier extents and a protending to its own later extents, regardless of what the sensing is a sensing of. Indeed, precisely this is the *transcendental phenomenological datum*: each extent of a mental-life process is *an intending previously protended to and subsequently retrotended to.*[84] Equivalently stated: Each temporal extent is characterized by a reflexiveness, a self-intentiveness, no matter what the object may be of the mental-life process. Regardless of its object and regardless of any ego-involvement in it, every mental life-process continuously identifies itself as the one extent or phase of sensing (to remain with Husserl's example) previously protended to and subsequently retrotended to; in short, it continuously constitutes itself in a primary passive identifying and differentiating synthesis because, simultaneously, it differentiates the current sensing from that extent now "receded into the past," as well as from all extents yet to come. And the same holds true of any mental-life process whatever.

Accordingly, with respect to temporal orientation I find a continuity of presentational modes of past and future. The "date" in time is kept, so to speak, but the distance in time increases, and it is this which makes up orientational differences.[85] The current intending--be it a sensing of a prespatial sensum (e.g., a "tone sensation"), a quasi-objective perceiving of a phantom, a quasi-objective apperceiving of something primordial under circumstances--has its horizon which is a future intentiveness, and the future extent of the current intending will be intended to, but in a different way. Nonetheless, the current intending is always and continuously an identifying synthesis: an identifying of itself presented and posited as one extent of a mental life-process (be it actional or passive) previously protended to and subsequently retrotended to.

Thus when I say, for example--objectivating the current extent of my own mental living--that the "prespatial tone-sensum begins," I do not only mean that the first "point" of the tone-sensum is presented as having a determinate place in "internal time," but I also mean that the first "point" (with the "point" occupied by the tone-sensum in "internal time") is "now"--i.e., that that "point" now appears at no temporal distance (or, as may also be said, appears in the null-mode of temporal distance) from the mental life-process "eternally" now. Similarly, when, in the further course of objectivation, I say that the "tone-sensum endures," then I mean that the tone-sensum is "now" while the earlier extent of the tone-sensum (along with the extent of "internal time" filled by it) is "past"--i.e., appears in the mode of distance in the past, *scl.* distance from the now-intending to it. And when, finally, I say with Husserl that the "unity of the process in which the tone<-sensum> begins and ends 'sinks back' toward the end in the always more distant past," this is not a description of the process as it is in itself but rather of it with respect to its temporal orientation in relation to the always-now sensing mental life-process: in this sensing, the process-unity appears with its continually growing distance into the past. As a consequence, the word, "past," signifies the universal mode of temporal orientation by which all events appear which are earlier than those which appear in the sensing mental life-process without temporal distance.

It is therefore possible to distinguish between differences in the orientational mode of each extent or phase of mental life-processes: the present of the present becomes identified with the future of the past and the past of the future: the past of the present is identified with the present of the past, while the future of the present is identified with the present of the future.

84. "TIME" AND "TIMES"

If, now, the terms, "time," and "temporal," are employed to include only the "internal" or immanental times of mental life-processes, then one must say that these "times" not only are constitutively prior to all space-times, but that "times" which are abstract parts of space-times are not truly or genuinely times because they require qualifying adjectives such as "prespace-time," or "phantom space-time," or "primordial space-time," or "life-world space-time," or "time" as conceived by Newton or by Einstein, or simply space-time of the real, objective world. In the proper sense of the word, "time," "time" is just what Bergson, for example, called the "*durée*" of mental life-processes.[86] In this connection, we may employ the terms, "space," and "spatial," to include not only space as an abstract part of the real, objective world, but also as abstract parts of what we have called sensum-fields, quasi-objective phantom worlds and quasi-objective primordial world (above, section 73). These usages allow for reformulation of the three steps of the "transcendental deduction of time:"

1. Transcendental mental life-processes in the natural attitude are not only transcendentally temporal and intentive to any mundane space-time whatever, but also to themselves as apperceiving themselves as within space-times having no fewer dimensions than those of sensum-fields.

2. As a consequence, transcendental mental life-processes are objectivated as self-intentive at each intermediary level of the oriented constitution of the one real, objective world.

Addendum 1, 2: Transcendental mental living constitutes itself Now as inherently temporal in a dual status of transcendental and mundane "in the form of reflexive intentional relatedness to itself."[87] We may therefore also say that transcendental mental living is the "product" of its own intentiveness to time, that it is essentially and necessarily an identifying and differentiating synthesis having gone on and yet to come. But this can only be because the retrotentiveprotentive structure constituting time in the proper sense, and mental living as inherently temporal, is of necessity objectivated *as the identical time at each intermediary level of oriented constitution*.[88] Proceeding in primary passivity, the process of "self-temporalization," the process of "self-constituting" of transcendental mental living as past, present, and future in the manner described, does not, however, "reconstitute" itself or "multiply" itself, so to speak, at each foundational level of oriented constitution characterized by one or another sort of intentiveness to transcendent space-time manifolds.[89] That is to say, at the level or substratum of oriented constitution peculiar to each space-time, transcendental mental living is transcendentally spatio-temporalized, *mundanized, with the identical structure of transcendental intentiveness to time.*

3. As a result, "do I not find then that my transcendental life and my psychic, my worldly life have, in each and every respect, a *like content*? How can it be understood that the 'ego' has constituted in himself the whole of what belongs to his own peculiar essence as, at the same time, 'his psyche,' psychophysically objectivated in connection with 'his' bodily organism and as thus woven into the spatial Nature constituted in him qua ego?"[90] The answer must be in the affirmative because, starting with the lowest intermediary substrata or levels of the real, objective world, transcendental mental life-processes are protentive-retrotentively apperceived as spatio-temporalized mental life-processes making up my mental living presented and posited as enduring with the quasi-objective and objective n-dimensional spaces that are themselves constituted as appearances acquired by a founding (n-1) dimensional prespatial sensum field (having more than two dimensions).

Addendum 3: If we consider a given extent of a transcendental mental life-process with respect to that process as a whole, then that extent is objectivated as flowing through a "stationary" or "unflowing" frame, as it were, consisting of future, present, and past. The current extent or phase flows through this frame such that the relation of any portion of the extent to each part of the frame changes continuously. Accordingly, it may be said that the "tense" of the posited characteristic of each portion changes continuously from "will be later" to "will be soon" to "is" to "was recently" to "was earlier" to "was still earlier." That part of the frame which is the future never becomes present, but that part of the process which is future will presumably become present and be present. That part of the frame which is the past was never present; but that part of the process which is past (which has passed through the present and is in the past) was present and became past.

Thus no matter what the specific difference in kind, extents and their portions are identified and differentiated temporally. The change in "tense" of the posited characteristics of the extents is a consequence of the flow out of the future (where the "tense" of posited characteristics is decreasingly future), through the present (where the "tense" is present) into the past (where the "tense" is increasingly past). Were this not the case, mental life-processes would be nothing but continuous recurrence, hence would provide no basis for building up the real, objective world within which mental life-processes find themselves. It is therefore the "condition" for my transcendental life" finding itself "psychophysically objectivated...and as thus woven into the spatial Nature...." But the change in "tense" just described is *only a necessary*, not a sufficient, "condition" for being in the world.

Of what, then, does the sufficient "condition" for mundanization consist?

85. THE SUFFICIENT "CONDITION" FOR THE MUNDANIZATION OF TRANSCENDENTAL MENTAL LIVING IN THE NATURAL ATTITUDE

When the current extent of a mental life-process is presented as an enduring one, that extent is, according to the first two steps of the "deduction," bound up with all the extents of mental life-processes previously protended to and subsequently retrotended to--the current extent meant and posited as the limit of the "sinking back into the past." Moreover, it follows from the first two steps that were nothing identical not presented and posited as it itself throughout the changes in temporal modes of orientation and changes in "tense," i.e., changes in temporal appearings and being, all that would then be objectivated by me as phenomenological on-looker would be a series of "punctiform acts," one of which would bear the stamp of actual presence while the others would be stamped with charateristics different from each other and from the currently actual one.[91] Presented simultaneously, those "acts" or processes would then appear as isolated and lacking any inherent relation to one another. In short, *the unity of an enduring extent of any mental life-process is possible only in so far as it presents itself in correlation with something identical presented in and through a multiplicity of different temporal extents continually changing in orientation and "tense."*

Using a metaphor of Husserl's from the *Cartesian Meditations*, we may also speak here of a "polarity" of identity and time. With respect to the third step of the "deduction," i.e., with respect to the apperceiving of mental life-processes as they themselves enduring with the real, objective world, this signifies that their temporal flow may be described as temporally extended and flowing in world-time and world-space together with mental life-processes of others that are temporally extended and flowing in one and the same or identical world-time and world-space, objectivated with their "bodily organisms." In Husserl's words, "All objectivity...is related back constitutionally to the *first affair that is other than my Ego's own*, the other-than-my-Ego's-own in the form, someone 'else'--that is to say: the non-Ego in the form, 'another Ego';"[92] the result is that the "otherness of 'someone else' becomes extended to the whole world, as its 'Objectivity,' giving it this sense in the first place."[93] And the oriented constituting of such "Objectivity" involves the distinction between the genuine and non-genuine or worldly meaning of "time, "that is, the "time" of the world publicly transcendent and the "time" of a plurality of transcendently inward mental livings. Thus the descriptions in the foregoing paragraphs must be altered slightly to include this distinction. More precisely stated, it would be more accurate to speak of the "pole of identity" of publicly transcendent objects as *contemporaneous* with the corresponding noetic "pole of identity" of those mental life-processes, both actional and passive, intentive to the former "pole of identity."

To try to express the situation in still another way, to assert that one "pole of identity" is transcendent to the other, the noetic "pole of identity," is in at least one meaning of the word, to assert that it is *objective*. That signifies, in turn, that the "pole of identity" of the publicly transcendent is experienceable by someone else and is thus not a constituent of my own transcendental mental living. The guarantee, as it were, that the transcendental mental living of someone else is not a really inherent constituent of my own transcendental mental living does not consist, then, simply of the fact that I do not find it as such a constituent; the guarantee consists instead of the eidetic insight that really inherent "inwardnesses" are mutually but correlatively incompatible. And the mutual incompatibility of immediately presented inwardnesses in experiencing a world experienceable by others is the phenomenological basis for the transfer of "objectivity" to the world in the first place--a noematic sense everywhere and universally confirmed and not cancelled out. Accordingly, it is essential to "my" world that it be a world for others who are ostensibly real in the same sense that I am. As truly as I am in the world, so are others; as truly as this is a world for my transcendental mental living, so it is for other transcendental mental livings and, more particularly, mental livings that are inwardly transcendent to me ("monads," to use Husserl's expression).[94]

Moreover, objectivation of the one "pole of identity" is ipso facto objectivation of the other for both "poles of identity" are constituted as enduring unities throughout a multiplicity of extents of the primary passive protentive-retrotentive flow or stream of mental life-processes. But this also signifies that the founding, temporal identity is that of the flow of those processes and that the correlation of identity peculiar to mental living and world is not a one-to-one correlation. Aron Gurwitsch succinctly expressed this in the following words: "Consciousness is not to be mistaken for a mere uni-dimensional sphere composed of acts, as real psychic events, which coexist with and succeed one another. Rather it ought to be considered as a *correlation, or correspondence, or parallelism between the plane of act...and a second plane which is that of sense* (*noemata*). This correlation is such that to each act its noema corresponds, but the same noema may correspond to an indefinite number of acts. It is then not a one-to-one correspondence."[95]

As "correlatives" or "parallels," in what, then, are the time of mental life-processes and world space-time contemporaneous? The answer can only be that they are contemporaneous "correlatives" and "poles of identity" in a "*third time*" distinct from both time in the proper or genuine meaning of the word, "time," and world space-time in the non-genuine meaning of the word, "time." *It is this "third time" that is the sufficient "condition" for mundanization of transcendental mental living in the natural attitude.*

Whatever the phenomenological characteristics of this "third time" may be, however we may proceed to analyze and clarify it and by what methods--phenomenological or others--and whether indeed the "third time" is a "time" at all in either the genuine or non-genuine meanings of the word, "time," are questions which cannot be explored here. Certainly the phenomenon at which this investigation has now arrived outruns, as it were, the phenomenological method established in the first part of this book. The task remains in the next, and final, chapter to restate the results of the investigation of Space, Time and Other within the framework of a universal phenomenological philosophy.

By way of conclusion to this chapter, it may be sufficient to note that the contemporaneity of time and world space-time is, at every intermediary substratum of the oriented constitution of the real, objective world a mundanization of transcendental mental living. Accordingly, *the "origin," the content of the contingent Apriori, of space-constitution lies in the time of transcendental mental living as a founding identity of extents of its mental life-processes previously protended to and subsequently retrotended to in a "time" with which the latter and world space-time are contemporaneous.* This conclusion takes us to the outermost reaches of the methodic sight of the will to return to the "things themselves," the access to which is secured by phenomenological epoché (above, section 4).

86. APPENDIX. GRAPHIC REPRESENTATION OF "CONSCIOUSNESS OF INTERNAL TIME"

At the beginning of Chapter Seven (section 62), a diagram representing temporal distance and recession-modes was introduced to express the temporal determination, "as present," that pertains to the stream of mental life-processes presented as temporally extended and flowing off in quasi-objective space-time of the phantom world. Despite its particular application, the description may of course be universalized to apply to mental life-processes flowing off in any space-time, be it of a quasi-objective or objective world. The diagram now introduced includes and extensively supplements the earlier one. Both diagrams are based on that found in section 10 of Husserl's 1905 lectures on the "consciousness of internal time." For purposes of illustration Husserl's example of intending to (sensing of) a tone-sensum is retained, although the description of it is considerably altered in accord with the revisions made in section 83 above. In so far as is possible without lengthy exposition of specific texts, I have also tried to express the diagram in such a way that it takes into consideration both Appendix II (section 2, c; section 3, a) of *Formal and Transcendental Logic* and Appendix II of *Experience and Judgement*. In this I have been aided by what I can still reconstruct of Dorion Cairns' lectures and conversations. Unhappily, those lectures still remain unpublished today.

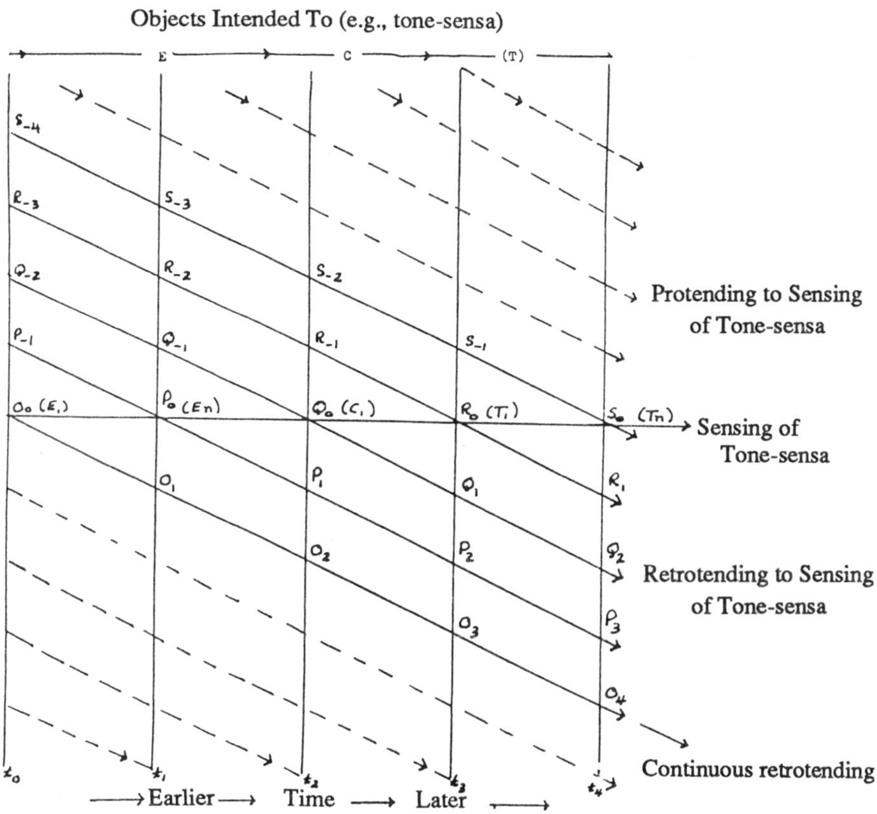

The exposition of the diagram is divided into five brief sections: 1) The "noetic line of description" of intendings and retrotendings to a tone-sensum; 2) the "noematic line of description" of the same; 3) intendings and retrotendings to two enduring tone-sensa; 4) protendings to sensings of tone-sensa; and 5) a case of actional remembering (*Erinnerung*--in contradistinction to an actional recollecting, *Wiederholung*). These sections hardly exhaust the description of time in the genuine or proper meaning of the word, "time." They are meant to be sufficient, however, to graphically represent the statements made in the preceding discussions of the time of mental life-processes of transcendental mental living in the natural attitude.

I. *Noetic Line of Description of Intending to a Tone-sensum That Begins and Stops.*[96]

1. ...O_o to S_o...is a continuous extent of intending to a tone-sensum presented and posited in primary passivity as transcendent to the stream of mental life-processes.
2. O_o, P_o, Q_o, R_o, and S_o are extremely short, non-contiguous extents or phases within ...O_o S_o ...
3. O_o is earlier than P_o; P_o is earlier than Q_o in time.
4. O_o to P_o, P_o to Q_o, Q_o to R_o, and R_o to S_o are all contiguous extents within ...O_o to S_o...
5. Together, those extents described in Proposition 4 make up O_o to S_o.
6. O_o to P_o does not include P_o; P_o to Q_o does not include Q_o; Q_o to R_o does not include R_o; but R_o to S_o does include S_o because it is a terminal point.
7. Because O_o is earlier than P_o, and because P_o is earlier than Q_o (see proposition 3), O_o to P_o is earlier than P_o to Q_o. For similar reasons, P_o to Q_o is earlier than Q_o to R_o, and Q_o to R_o is earlier than R_o to S_o.
8. O_o to O_4 is a continuous extent of mental living which is simultaneous with O_o to S_o; P_o to P_3 is a continuous extent of mental living which is simultaneous with P_o to S_o; and so forth.
9. Therefore, O_1 to O_4 and P_o to P_3 are simultaneous with each other; O_2 to O_4, P_1 to P_3, and Q_o to Q_2 are simultaneous with each other; and so forth.
10. As stated in Propositions 1 and 2, the earliest extent O_o of [O_o to O_4] is an extremely short extent of auditory sensing of a tone-sensum. The subsequent extent of O_o to O_4 is an enduring intentive process immediately retrotentive to O_o. The same holds for the case of P_o to P_3, for Q_o to Q_2, and for R_o to R_1.
11. O_1, O_2, O_3, ... are extremely short non-contiguous extents within [O_o to O_4]; P_1, P_2, ... are extremely short non-contiguous extents within [P_o to P_3]; and so forth.
12. O_1 is simultaneous with P_o; O_2 and P_1 are simultaneous with Q_o; and so forth.
13. [O_1 to P_o] (except p_o) is an immediate retrotending to [O_o to

P_o] that is simultaneous with P_o; [O_2 to Q_o] is that immediate retrotending to [O_o to Q_o] that is simultaneous with Q_o.

II. *Noematic Line of Description of Intending to a Tone-sensum that Begins and Stops.*

14. O_o, P_o, Q_o, R_o, and S_o are intentive to E_1, E_n, C_1, and T_1, T_n, as their respective intentional objects.
15. Each of these intentional objects is presented as simultaneous with the intending to it in time.
16. Each successive phase of intentionality from O_o to Q_o is intentive to a tone-sensum like E.
17. In particular, P_o is intentive to E_n, which is a sensum like E_1.
18. As stated in Proposition 12, O_1 is phase of mental living simultaneous with P_o.
19. Moreover, O_1 is a retrotending to O_o (see Proposition 9) that is intentive to E_1 as its intentional object (see Proposition 14). Equivalently stated, O_1 has its retrotentive object ["O_o, the intending to E_1"]--or, noematically expressed, ["E_1 the intentional object of O_o"].
20. Now, P_o has E_n as its intentional object (see Proposition 14).
21. In addition, the intentional object E_n of P_o is like the retrotended object E_1 of O_1 (see Proposition 16).
22. P_o is a phase within [O_o to Q_o]. Let us assume for the sake of illustration that it be the phase contiguous to O_o and that it is an auditory sensing of E_2.
23. O_1: P_o is then both retrotentive and intentive to retrotended E_1 and E_2 as not only alike, but also as contiguous--thus constituting the enduring object E_1: E_2 in an identifying synthesis.
24. Similarly, at each sucessive phase of [O_o to Q_o] (and simultaneous with [O_o to O_2]) O_1: P_o will be a retrotending and an intending to E_1: E_2: E_3... In short, it is an intending to E as continuously enduring and unchanging tone-sensum, presented as it itself retrotentively and intentively.

III. *Intendings and Retrotendings to Two Enduring Tone-sensa.*

25. Q_o is intentive to the founded object C_1, different from the founding sensum, E.

26. But simultaneous with Q_o there is a retrotending to O_2: Q_o (except Q_o) of [O_o to Q_o] as an intending to E_1: E_2: E_3... which has gone on before.
27. As a consequence, O_2: Q_o is, *first*, a retrotending to the intending to E as continuously enduring and unchanging and, *second*, an intending to C_1. (In non-graphic language, we discriminate the auditory sensing of a tone-sensum as having changed, for instance, in pitch; or a sensing of a new tone-sensum as higher in pitch than the first.)

Addendum I, II, III. In summary, the diagram represents the continuous retrotending to the phases of auditory sensing [O_o to S_o] intentive to their intentional objects [E_1 to T_n]. At time "t_2" there exists an immediate retrotending to all intentionality (intentive, retrotentive, and protentive processes) that occurred at time "t_1." Likewise, there is an immediate retrotending to all the intentionality that occurred at time "t_o." As a consequence, we can not only discriminate immediate retrotendings, but also retrotendings to retrotendings as well. For example, throughout Q_o: O_2 something is retrotended to as P_o: O_1 as immediately preceding Q_o: O_2; and something is retrotended to as O_0 immediately preceding P_o: O_1. But throughout Q_o: O_2, P_o: O_1 is retrotended to as itself retrotending to O_o. Accordingly, throughout Q_o: O_2, O_o as retrotended to throughout Q_o: O_2 is identified with O_o as retrotended to throughout P_o: O_1, as also retrotended to throughout Q_o: O_2.

IV. *Protendings to Sensings of Tone-sensa.*

Very briefly, the protentiveness of mental living may be represented on the diagram as lying above the line, [O_o to S_o].

28. P_{-1} to P_o is a continuous extent of mental living simultaneous with O_o to P_o; Q_{-2} to Q_o is a continuous extent of mental living simultaneous O_o to Q_o; and so forth.
29. P_{-1} to P_o is immediately protentive to P_o; Q_{-2} to Q_o is immediately protentive to Q_o; and so forth.
30. P_o is the fulfillment of P_{-1} to P_o; Q_o is the fulfillment of Q_{-2} to Q_o; ...
31. Q_o (with its intentional object, C), as an intending (an auditory sensing) protended to by an earlier protending, is the positive

fulfillment of [Q_{-2} to Q_o]. But Q_{-2} to Q_o protends Q_o as an intending to an object of a more or less definite and specific kind.

32. Q_o itself, may be, however, more or less different from what it was protended to as being at a particular time in [Q_{-2} to Q_o]. Accordingly, in addition to being a positive fulfilling of [Q_{-2} to Q_o], Q_o itself may be a more or less negative fulfillment of [Q_{-2} to Q_o] and its particular phases.

Addendum I, II, III, IV. Thus one of the principle features of mental living that the account of the stream or flow of time represented by the diagram makes thematic is that any enduring intending is a process characterized by self-intentiveness; in each temporal phase it is possible to objectivate a retrotentiveness to its own earlier phases and a protentiveness to its own later phases. As noted before, it is precisely this retrotentive-protentiveness that is the phenomenological datum which we expressed by saying that each phase distinguished is an intending previously protended to and subsequently retrotended to. Based on this datum are the various syntheses of identification and difference on which are based, in turn, the various syntheses of assimilation and differentiation that comprise the structure of mental living in the world.

The exposition of the diagram may be rounded out by considering an actional remembering.

V. *An Actional Remembering.*

33. In each temporal phase of a remembering (except for the first phase), there is a retrotending to earlier phases and a protending to later phases of the same remembering. To say, then, for instance, that "the remembering ceases," is to say that a particular phase of the remembering is following by a phase of mental living that is not a remembering and that, therefore, is a partially negative fulfillment of the protendings to it implicit in the preceding remembering.
34. A remembering of [O_o to P_o] (with its noematic object, E) proceeds from "t_3" to "t_4."
35. Along with the continuous intending to T during "t_3" to "t_4" there is a continuous retrotending to [O_o to P_o] (with its intentional object, E).

36. At "t_3," then, we not only find R_o (intentive to T_1) but also O_3 as an immediate retrotending to O_o, and P_2 as an immediate retrotending to P_o as well as being an immediate retrotending to all the phases of [O_o to P_o].
37. At "t_3" the initial phase of a remembering is directed to O_o or to its intentional object E_1 as retrotended to in O_3. This remembering continues until "t_4." If "t_4" is contiguous to "t_3," then at time "t_3" the remembering is directed to some later phase of [O_o to P_o], e.g., to P_o or to its intentional object E_n. It is also directed to P_o as retrotended to at "t_4" in P_3.
38. Meanwhile the remembering has been directed to the other phases of [O_o to P_o] and has seized upon them more or less clearly and distinctly. But P_o is retrotended to at "t_4" along with O_1. And, correlatively, at "t_4" the remembering of P_o is accompanied by a retrotending to the remembering of O_o at "t_3."
39. As a consequence, the structure of the remembering from "t_3" to "t_4" is the same as the intending from "t_o" to "t_1." The remembering thus exhibits the self-intentiveness peculiar to mental living, i.e., as previously protended to and subsequently retrotended to in primary passivity--even though the remembering itself is (and only is) actional.

VI. *Concluding Remarks About the Graphic Representation of Time.*

As previously noted, with respect to its self-intentiveness, the current phase of any intending may be described in its modes of temporal orientation; the "date" in time is kept but the distance in time increases and it is this which makes up the difference in the mode of orientation to the present. The present intending to something, be it actional or passive, however narrowly or broadly understood, is nonetheless always an identifying synthesis which itself, as a phase of mental living, is likewise previously protended to and subsequently retrotended to.[97] (So Propositions 10, 13, 19, 24, 27 and 37 may also be read.) This is the basis, again, for the differences mentioned earlier where the present of the present becomes identified with the future of the past and the past of the future, while the past of the present is identified with the present of the past and the future of the present is identified with the present of the future.

The complex structure of self-intentiveness includes the complex intending to other phases of mental living in identifying syntheses, the intending to an enduring object transcendent to the flow of mental living, and the intending to the enduring intending to the enduring object. As Husserl suggests in the 1905 lectures as well as in later writings, to objectivate an intending as self-intentive, no matter if it is intentive to anything else, is to objectivate the stream of mental living in the strict or proper sense (thus, for example, sections 35 and 36 of the 1905 lectures). The phenomenological "clue" for objectivating any intending as self-intentive is the noematic object of the carious syntheses of phases described in their modes of temporal orientation. As transcendental on-looker, I "ask back" from the synthesis to that in which the synthesis is constituted in the most original way (thus Appendix II of *Formal and Transcendental Logic*). *Here, however, the noematic object is precisely the noesis, the intending to itself.* The flow of "temporalized" mental living is then both the constituted and the constituting. With this a limit is reached for transcendental phenomenological method so far as developed in the first part of this book. Whether another further phenomenological or even non-phenomenological, method can be introduced to carry the analysis any further must remain here an open question.

CHAPTER NINE

TIME, SPACE, OTHER

87. THE PHENOMENOLOGICAL THEMES OF TIME AND SPACE IN THEIR HISTORICAL SETTING

Most broadly expressed, the phenomenological theme investigated in the preceding chapters is: "being in the real, objective world pertaining to my mental living as transcendental mental living in the natural attitude."[1] More narrowly expressed, this theme was developed with respect to the "genesis" of the ideas of time and space as products of subscientific-philosophic thinking and experiencing that take for granted the positing of the naturalness belonging to the natural attitude[2]--a development that consciously set aside the "naturalistic conception" of the ideas of time and space with respect to their specifying assumptions exercised in modern philosophy and science.[3] In particular there were two central specifying assumptions singled out for attention and on which the epistemic and ontic claims of modern scientific and philosophic thinking and experiencing rest (section 38ff.). The first is the idea that geometry and kinematics define space, and if space defines reality it would seem to follow that geometry and kinematics define reality.[4] This assumption is linked to another: that the spatio-temporal form of the real, objective world posited in subscientific thinking and experiencing is accepted by the "naturalistic conception" to be the self-same spatio-temporal form posited in scientific thinking and experiencing.[5] With respect to this historical setting, a chief philosophical purpose of this investigation has been to discover whether there is any basis for testing the ostensible legitimation of these specifying assumptions. Phenomenologically, that task was carried out by examining the way in which the one spatio-temporal form of the real, objective world is intended to and posited in subscientific experiencing. Such experiencing proves to be the oriented constituting of the real, objective world in primary passivity. More precisely, then, the phenomenological theme is: "being in the real, objective world with its one spatio-temporal form intended to and posited by my mental living as transcendental mental living positing the naturalness of the natural attitude in primary passivity."

Development of this theme by means of unbuilding and building-up analysis has yielded several phenomenological findings, the most important of which is that the Kantian view of space and time and its specifying assumptions are not obvious. The first reason for this lies in the profound ambiguity expressed by the words, "space" and "time" (sections 18, 22, 24, 73). The spatio-temporal form of the real, objective world *constituted in primary passivity* includes positings of three-dimensional space without of eidetic necessity positing of any things in space (three-dimensional prespace) and of space-time without anything enduring in space (prespatio-temporality); but also included of eidetic necessity are positings of three-dimensional space that entail positings of things in space (quasi-objective three-dimensional space) and of space-time with things enduring (quasi-objective space-time). In addition, at all substrata of oriented constituting, time in the genuine meaning of the word, "time," is posited regardless of whether there is a positing of it as flowing in any prespatio-temporality or quasi-objective spatio-temporality (sections 85, 86). Phenomenologically, it therefore does not follow by necessity from an evidential intuiting of space as three-dimensional that it is true that space is unambiguously presented (although it is immediately presented); nor does it follow that space (prespace or quasi-objective space) can be "abstracted" from space-time--and that in a two-fold meaning: space cannot be "abstracted" from the time of mental living (there is no positing of space in either of the generic meanings of the word, "space," without self-positing of the time of mental living), nor can space be "abstracted" from time in the derivative meaning of the word, "time," as space-time; nor does it follow that space-time cannot be "abstracted" from "things" in space-time. Because none of the foregoing follow from the intuiting of three-dimensional space, there is no basis for asserting that space, prespace, or quasi-objective space, is "Euclidean" or "non-Euclidean," nor, by extension, that objective space is constituted in subscientific experiencing. In short, there is no basis for asserting that "geometry" can define space by rendering it intelligible[6] for none of the intuitions necessary to make that assertion follow from the intuiting of three-dimensional space.

The second reason why the Kantian view of space is not obvious is that central to the oriented constituting of the space-time of the real, objective world is the constitution of Earth and Animate Organism as anomalies. That signifies that included in some cases of positing of them is positing of them "intrinsically spreadout" presented as neither in motion nor at rest, while in the cases where the animate organism is posited and intended to as in motion or at rest there is included positing of earth as neither in motion nor as at rest. Thus it is not obvious that "kinematics" renders space any more intelligible than does "geometry." Again, there would seem to be no basis for confirmation of the "naturalistic conception," of the Kantian view, and its specifying assumptions which remain, phenomenologically, only assumptions.

Expressed more positively, to draw the conclusion that the naturalistic conception of space and time at the heart of the Kantian view is not obvious, that indeed "kinematics" and "geometry" do not without qualification render space and time intelligible, is to conclude that there is no formal necessity that "fields of sensation" have loci in real, objective space-time or that they be located relative to my own animate organism.[7] Only if "fields of sensation" were located in real space and time ascriptively apperceived of necessity relative to my animate organism would the Kantian view be obvious. Phenomenologically, the contrary seems to be the case so that there is no continuity of positing of space in subscientific thinking and experiencing. At best the continuity of positing is a convenient fiction of the naturalistic conception that allows its quite unfounded substitution for "space perception" constituted in primary passivity.[8]

To be sure, this raises the question of whether or not there is any phenomenological legitimation of the Kantian view of space, of whether or not the naturalistic conception and its specifying assumptions must be abandoned altogether. Are the specifying assumptions necessary at all for the Kantian view of space as formulated here? Can it, or a version of it, still be legitimately held even if there is no continuity of positing, even if "kinematics" and "geometry" do not render space intelligible? Such questions for further phenomenological investigation can only be raised here.[9]

Perhaps the best way of concluding this reflection on some phenomenological findings of the present investigation is by stating them in terms of our phenomenological theme: "space perception" signifies access to being in the world in the sense of the mundanization of transcendental mental living in the natural attitude, the naturalness of which consists in temporally extended transcendental mental living apperceiving itself flowing off in one spacetime of the real, objective world and in the spaces and times of each quasi-objective intermediary substrata included in experiencing the real, objective world constituted in primary passivity. Even though the temporal extension, the time, of transcendental mental living is constituted as self-same throughout, there is no question of reductionism[10] because while the "deduction" of space entails that of time, the converse is not the case. The intermediary substrata of oriented constitution, although "worlds," are then not "other worlds" but only the one real, objective world posited in primary passivity with a new noetic-noematic sense for each substratum. The space-time of "fields of sensation" is then not that of phantoms, nor is the space-time of phantoms that of casual things; and so forth. But, by virtue of the self-identical time of mental living apperceived as flowing off in prespace-time and quasi-objective space-time, the former acquires the "appearance" of the latter as something secondarily constituted in primary passivity.

88. EIDETIC TRANSCENDENTAL PHENOMENOLOGICAL PROBLEMS OF SPACE AND TIME

The phenomenological findings of this investigation require that, at the very least, the specifying assumptions of the naturalistic conception of space and time be abandoned. Their abandonment, however, raises a new set of phenomenological problems, namely those that center on clarification of the material essence of space and time.[11] As suggested before (sections 30ff.), this clarification is located within the framework of an eidetic transcendental phenomenology that requires a corresponding thematic alteration. The phenomenological theme may be tentatively expressed as follows: "Any being in the real, objective world, with one spatio-temporal form, posited and presented in any primary passive transcendental mental living that posits the naturalness of the natual attitude." Correspondingly, theoretic interest lies chiefly in "universals," "essences," the seeing and seizing upon of which are of them as exemplified by real, objective things in a real, objective world constituted in primary passivity.[12] The chief example discussed earlier was that of the material "universal" or "essence," Physical Thing.[13] More particularly, theoretic interest is concerned with what we called the "empirical extension" of Real, Objective World, of Physical Thing, of Other as purely possible exemplificatory varieties which happen to be actualized throughout multi-storied oriented constituting in primary passivity. In brief, the preceding chapters of the second part of this book may be regarded, in this connection, as laying the foundation for seeing and seizing upon the purely possible exemplification of the essences, Mind and Nature as the conferred status of any transcendental mental living whatever in the natural attitude which happen to be actualized in primary passivity,[14] that is, without presupposing any antecendent actionality on the part of any transcendental mental living.

But this is only the beginning and not the end of the problem of clarification of the eidetic laws and lawful structure of Space, Time, and Other. For even though what Husserl develops under the heading of the regional apriori of *res materialis* and *res extensa* serves as a guide for what follows, the phenomenological findings of the previous chapters require some rather radical revisions of those all-embracing laws that prescribe the possible sense for "every factual <transcendental> statement" we have sought to make about transcendental mental living in its conferred status in the real, objective world.[15] And the difficulties here are compounded for, as Claesges has noted, "Because in every respect it presupposes the constitution of time, the constitution of space is unambiguously to be designated a static one."[16] Nevertheless, by examining space in terms of oriented constituting it may very well be that an element of the genetic (as defined above, section 13) is involved so that space constitution is not "unambiguously" static.

To shed some light on this situation it is worthwhile returning for a moment to the first chapter of *Ideas*, First Book, where what Husserl said in the Third Logical Investigation about real individuals as parts of a whole is carried over to other sorts of things, some of which exist ideally rather than really.[17] Accordingly a corresponding set of questions is asked not just about the real individual, for example, a mental life-process such as a perceiving, and about its noematic sense, the perceived as perceived, but also about the exemplified essence, Mental Life-process, and the essence, Physical Thing, as discrete parts of the same whole. Are they, for instance, self-sufficient or non-self-sufficient parts relative to each other?

Husserl attempts to answer these and similar questions in section 15 of the First Book of *Ideas*. Because of the importance of the passage, it is worth citing at some length:

> Of particular interest here is the situation with the ultimate substrates and, still more narrowly conceived, with materially filled substrate-essences. Two possibilities arise for them: either one such substrate-essence grounds with another substrate-essence the unity of *one* essence, or it does not do so. In the first case, we arrive at relationships...which are perhaps of unilateral or reciprocal non-self-sufficiency; and with respect to eidetic and individual single particulars falling under united essences, we arrive at the apodictically necessary consequence that single particulars falling under one essence cannot exist without being determined by essences which at least share a generic community with that of the other essences. E.g., sensuous quality necessarily refers to some species or other of spread; spread is, again, necessarily the spread of some quality united with it, 'covering' it. A moment, 'enhancement,' e.g., under the category of intensity, is only possible as immanent in a qualitative content, and a content subsumed under such a <qualitative> genus is, in turn, not thinkable without some degree or other of enhancement. As a mental life-process of a certain generic determinateness, an appearing is impossible except as an appearing of something 'apparent, as apparent;' and, likewise, the converse holds. And so forth.[18]

In the case of mental life-process, what is presented Husserl calls a *concretum*, and the assertion about the *essence*, Mental Life-process, is the *result of an analogy*:

Just as Sensuous Quality and Spreadness make up a *concretum*, so Appearingness and Apparentness make up a *concretum*. This signifies that the essence, Seeing of a Sensuous Quality, *and* the essence, Sensuous Quality Seen, are parts of a whole just as the essences Sensuous Quality and Spreadness are parts of a whole. In addition, Husserl says, Appearingness, like Sensuous Quality, is also an *abstractum*.[19] Appearingness is then both an eidetically discrete self-sufficient part and an eidetically discrete non-self-sufficient part. it comes as no surprise then that in the third part of *Ideas*, First Book, Husserl develops the view that eidetic sciences prove to be either "concrete" or "abstract,"[20] and that, in particular, "pure phenomenology" turns out to be a concrete eidetic science in the sense of an eidetic science of mental living as a concretum having "abstract moments."[21] Of concern, however, is the analogy with the essences, Sensuous Quality and Spread.

There are, I believe, two ways of reading the analogy. On the one hand, a construed hyletic datum is an "*appearing*" through which physical thing-appearances are adumbrated. The adumbrating or discrete, non-self-sufficient part, and likewise the adumbrated appearance, also itself a discrete, non-self-sufficient part, make up a concrete whole just as do sensuous quality and spread; there can be no adumbrating without something adumbrated (the physical thing-appearances) just as there can be sensuous quality without spread. Yet the adumbrating and the adumbrated, like quality and spread, are distinct as species, have nothing in common. On the other hand, and *universally*, noesis and noema are discrete, nonself-sufficient parts of a concrete whole relative to one another, but noesis is also an *abstractum*. In other words, it is equally possible to ignore the analogy and the second formulation holds regardless of the legitimacy of the theory of hyletic data. It is this phenomenological situation that now requires clarification.

89. TRANSCENDENTAL MENTAL LIVING AS WHOLE

Extended to eidetic domains, Husserl's theory of wholes and parts provides both a radical foundation and an analogy for understanding the relationship between Appearingness and Apparentness--universally, between Noesis and Noema. First, just as Sensuous Quality and Spreadness make up a *concretum*, so Noesis and Noema make up a *concretum*. But, second, like Sensuous Quality, Noesis is also an *abstractum*, a non-self-sufficient essence.[22] In turn, this conclusion yields what Husserl later calls in *Ideas* an "entirely fundamental distinction with respect to intentionality, namely the distinction between *components proper* of intentive mental life-processes and their intentional correlates, or of their components."[23] That signifies, he says, "that it is of its essence to include in itself something such as 'sense' and potentially manifold sense..." What Husserl speaks of here is the case where--in the language of the Third Logical Investigation--"pieces" or discrete self-sufficient parts "enter into foundational relationships with respect to a more comprehensive whole in which they all count *as non-self-sufficient 'moments'*." Husserl's description is the second way of reading the analogy, and expresses the following law:

> To each 'piece' in a relative abstraction there corresponds a 'piece' in each of its relative concreta, so that the mutually exclusive 'pieces' of the former serve to ground mutually exclusive pieces in each of the latter. In other words: *the fragmentation of a non-self-sufficient 'moment' conditions a fragmentation of the concrete whole, in so far as the mutually exclusive 'pieces,' without themselves entering into fundamental relation with one another, attract new 'moments' to themselves by virtue of which they are singly distributed to 'pieces' of the whole.*[24]

Taking this law, which may be considered an early formulation of what Husserl later calls "oriented constitution," to apply with eidetic universality and necessity to essences and therefore to transcendental mental living, there is a two-fold result of Husserl's whole-part theory:

On the one hand, Noesis and Noema enter into a whole discretely as non-self-sufficient eidetic parts, as *abstract*: a mental life-process is an *abstractum* that is "impossible except as the appearing of something apparent, as apparent, and the converse holds." Taken as infima species on the analogy with Sensuous Quality and Spread, and in accordance with the aforementioned law and its reformulation in section 15 of *Ideas*, Noesis and Noema cannot be "combined into the unity of *one* essence," hence as having a place in the concrete whole, they lead to a "separated system of species and genera, thus also to separated summa genera." The concrete whole in which the "parts" are so characterized is "Mental Living Awareness in and of the World."

On the other hand, Noesis enters into a whole discretely as *concretum*, as a self-sufficient and self-contained essence: at the end of section 15 of *Ideas* Husserl says that Mental Life-process is an example of an eidetic *concretum*, i.e, of a self-sufficient material essence exemplified by an ideality (exemplified itself by a reality), while examples of eidetic *abstracta* are spatial shape and spread and, by extension, prespatial quale and sensum-fields as well as quasi-objective phantom and primordial shapes and spaces. Thus Husserl says later on in *Ideas* (section 49) that we "see that consciousness (mental life-process) and real being are anything but coordinate kinds of being which dwell peaceably side by side and occasionally become 'related' to or 'connected with' one another. Only things which are essentially akin, things the respective proper essences of which have a like sense can become connected in the true sense of the word, can make up a whole."[25] This is, of course, not the whole story for in the second case Noesis, as *concretum*, enters discretely into the whole, "Mental Living, regardless of whether it is in the world."

However this richly complex phenomenological situation might be further clarified, enough has been said to see that not only is Noesis both *abstractum and concretum*, but that Noema, as the "reality of the whole world, lacks self-sufficiency by virtue of its essence....Reality <of the whole world> is not in itself something absolute, which becomes tied secondarily to something else; rather in the absolute sense it is nothing at all; it has no 'absolute essence' whatever; it has the essentiality of something which, in its fundamental essence, is *only* intentional, *only* an object of consciousness, something presented in the manner peculiar to consciousness, something apparent <as apparent>."[26] To be sure, Mental Living is only presented as *concretum* in so far as it is presented as *abstractum* and, conversely, presented as *abstractum* only in so far as it is presented as *concretum*. Accordingly, of essential necessity, Mental Living is not presented concretely as apart from World even were the essence, "reality of the whole world," not to have any actual or possible exemplifications. Such formulation, it must be noted, is, at the same time, de jure expression and characterization of the *paradox*, transcendental mental living in and of the world, of transcendental mental living being of "two minds" about the world--in short, Mental Living as *abstractum* and *concretum* is nothing else but an expression of the theoretic will and striving required by the return to the things themselves. Formulated *de jure and transcendentally-eidetically*, the paradox now reads: "Thus, *on the one hand consciousness is said to be the absolute* in which everything transcendent and, therefore, ultimately the whole psychophysical world, becomes constituted; and, *on the other hand*, consciousness is said to be a *subordinate real event within that world*."[27] But to carry Husserl's account this far is also to invite a set of difficulties most adequately brought to light by Aron Gurwitsch.

In his "Phenomenology of Thematics and of the Pure Ego," Aron Gurwitsch noted that Husserl's idea of self-sufficient and non-self-sufficient parts assumes as criterion for their distinction and recognition the possibility or impossibility of being singled out as constituents from a context such as a whole, even when the constituent has not yet been actually singled out. If this is true, then Husserl does not seize upon wholes and parts as what they are, just and precisely with respect to how they are presented. Rather what Husserl does, following Stumpf,[28] is to preinterpret them in the light of what is in fact another and quite different phenomenological situation into which it is assumed that whole and part can be transformed. Thus the basic error, according to Gurwitsch, in describing discrete self-sufficient and non-self-sufficient parts is to "foist into a phenomenological datum as already contained in it what results from it by virtue of a modification, but not otherwise <(*scl.* the modification of singling out)>. The possibility or impossibility of singling out a constituent from a contexture is permitted to play a decisive role for the description of the experience of the constituent within the contexture--that is to say, when it is not yet singled out."[29] Although the various examples adduced by Gurwitsch cannot be rehearsed here, his point is immediately applicable to the law of "mutual fragmentation," hence by extension to the very idea of oriented constitution.

Indeed, one implication of this criticism, not explicitly drawn by Gurwitsch, is that it is wrong to conceive a "part" as both *abstractum* and as *concretum*. To be sure, Gurwitsch's discussion is largely confined to his general analysis of the noema and, more specifically, to the question of whether attentional modifications in some manner effect an alteration in the noematic What. Moreover, Gurwitsch eventually reformulates the phenomena of whole and part in Gestalt-theoretical terms according to which "parts" or "themes" are grouped into those items which lend themselves to being "singled out or made independent," and those which do not--"but these terms, obviously, have another sense here than in the writings of Husserl and Stumpf." Nevertheless, it is in terms of items which are and are not susceptible of being singled out that Gurwitsch interprets Husserl's idea of self-sufficient and non-self-sufficient parts. Consequently, Gurwitsch denies that if an item is singled out it remains phenomenally the same, which is the view he imputes to Husserl. Thus, in our example, if a mental life-process is singled out *in its transcendental status*, then it is what Gurwitsch calls an "independent part;" if it cannot be so singled out it is, *in its worldly status*, a non-independent (or abstract) part. To express the matter still somewhat differently, any mental life-process retrotending to itself as previously protended to is then an "independent part;" that same retrotentive-protentive mental life-process flowing off in world-time is a "dependent part." For Gurwitsch it would be illegitimate to speak of this "same" mental life-process discretely entering a whole as an eidetic abstractum and as an eidetic concretum.[30] At the basis of this assertion is the idea that "To every datum correspond other data, to every theme in its thematic field other themes in their fields, in such a way that any given experience motivates certain transitions as possible." Thus if a mental life-process as in and part of the world, as *abstratum*, is superseded by a mental life-process as of the world, no matter if it is in the world, as concretum, then the transition from abstractum to concretum is a possibility inherent in a "system" of concatenated essences each of which contains motivations for possible transitions to certain others. Just such a notion was discussed before (above, section 8f.) in reaching the conclusion that mental living is either "natural" or "unnatural" (abstractum or concretum), the former in its "bestowed" status contains the motivations for the transition to the latter--hence that psychologically reduced phenomena are implicitly transcendentally reduced phenomena. In still other words, the natural attitude contains the possibility of its own (transcendental) illumination, intelligibility by enabling the ego to refrain from those positings in which the naturalness of its life consists (above, section 9). The significance of applying Gurwitsch's criticism of Husserl is that it not only eliminates the need for the analogy with Sensuous Quality and Spread, but validates the *second* way of reading the analogy which allows for the very possibility of the paradox of Noesis presented both as abstractum and concretum in a way that is not counter-sensical.[31]

There is yet another implication of Gurwitsch's critique of Husserl's theory of whole and parts that leads to a somewhat different set of problems. Immediately following the discussion of wholes and parts in "The Phenomenology of Thematics and of the Pure Ego," Gurwitsch proceeds to reject Husserl's notion of the "transcendental ego" as a category of an *abstractum*.[32] In terms of the discussion in this section, what is at issue is not so much the noetic composition but rather the noetic-noematic correlation itself. Even though it is not important here to rehearse and develop Gurwitsch's thought about the problem of the transcendental ego--our theme throughout the second part of this book has been Mental Living constituted in primary passivity--the fact of his rejection of Husserl's account brings to light another dimension, so to speak, in which Husserl considers the paradox of transcendental mental living in the world. More specifically, it is a dimension which issues from the whole-part framework in which the paradox finds expression: eidetic regions and categories.

90. "NATURE" AND "MIND," "SPACE" AND "TIME"

The feeling for the problem of phenomenological genesis, the paradox of "world in itself" and "world as constituted," and now of Mental Living as *concretum* and *abstractum*, are two sides of the will to return to the things themselves. The one side pertains to the transcendental phenomenological reduction, the other to the transcendental phenomenological eidetic reduction (above, section 7). It is as important to see the will to return to the things themselves as it is to see the difference of the sides. The task of critically clarifying both sides in their difference is one that emerges in *Ideas*, Second Book. The setting of the task, however, reaches back at least to section 2 of *Ideas*, First Book, where Husserl says that everything that belongs to the essence of an *individuum* can also have another *individuum* and highest essence so that eidetic "universalities" such as Appearingness and Apparentness as Apparentness embrace *regions or categories of individua. Individua*, in other words, either embrace regions *or* they embrace categories.[33] In the case of categories Husserl gives the example of "any physical thing whatever," "any duration whatever," "any figure whatever," and the like. As a result, Husserl would seem to *confuse the region*, Any Nature Whatever (Any Space-time Whatever), of which there is only *one* example, *with the category*, Any Physical

Thing Whatever, of which there are a *multiplicity* of real exemplifications. If we take Gurwitsch's criticism of Husserl's idea of the transcendental ego to apply as well to Husserl's account of *individua*, then the confusion can be seen to follow from having the essence, Mental Living (or Time), embrace a region, a sui generic "whole," confused with the category, "Ego," of which there are a multiplicity of exemplifications. To confuse a region with a category, in other words, is precisely to allow what is singled out, in this case the *abstractum*, ego, to define ("play a decisive role in") the whole, the contexture or region, when the category is not yet singled out, i.e., when the "part" is not yet singled out.

To express the matter somewhat differently, it may be said that Husserl is led to the result where, in the one case, the "all of reality" (Nature or Space-Time) is found in the quasi-objective or objective real physical thing, and where, in the other case, the "all of reality" (Time, in the proper sense of the word, "time") is found in Mental Living. In turn, this would seem to lead to the metaphysical assertion that Nature is a "product" of Mind, thus that Space-Time is a "product" of Time. The paradox expressed by the will to return to the things themselves, by the feeling for the problem of phenomenological genesis, turns out to be, philosophically, an idealism of a rather Berkeleyian stamp. And just this would seem to be what occurs at the end of *Ideas*, Second Book. The consequence is that there are two, apparently incompatible, meanings of Nature: either Nature is the "all of reality," or Nature is the metaphysical product, the "creation," of Mind.[34] Thus even if a way is found to formulate the two-sided paradox in a non-countersensical way, there still remains the problem of explaining the apparently contradictory regional and categorical account of the really inherent composition of the noesis in so far as it has a bearing on the whole comprised by the noetic-noematic correlation. The gist of Husserl's view of that correlation in the First and Second Books of *Ideas* is that the being of objects in Nature points back to its positing by mental life-processes such as perceivings in which such objects are presented in original ways, and that there must be corresponding kinds of intuitings in which objectivities of a higher order are presented. In other words, for every sort of object there is a corresponding mode of original intending to it.

Now, the set of such objects for a possible eidetic science is called a *region*, and what the objects have "in common" as essential to them is called a *category*, i.e., the fundamental regional concept. Because such categories embrace the structure of being peculiar to objects Husserl calls them eidetic "regional ontologies." And because regional ontologies state what must belong to objects in unconditional universality and necessity, so that they can be the subject matters of definite sciences, such ontologies are said to be eidetic sciences in contrast to sciences of matters of fact (see above, section 5). *Formal ontology* may then be defined as that eidetic science that cuts across all regional ontologies by disregarding, through formalization, all regional differences and hence considers any object whatever of transcendental mental living in the natural attitude. There is but *one formal* ontology in contrast to *many regional* ontologies which, when viewed with respect to the formal ontology, are called "*material ontologies*."

In order to show and legitimate the "distribution" of objects into their corresponding regions, it is necessary to consider them with respect to the most original modes in which they are constituted in transcendental mental life-processes. Precisely here is where the problem arises because Husserl would then seem to proceed to assert that Mental Living is *another region* with respect to which all other regions are relative. He thus moves, as Landgrebe aptly expressed it, from *phenomenology to metaphysics*.[35] Is the move justified? Or is there not again an implicit pre-interpretation--now of a regional "part" of a formal ontological "whole"--of what can possibly be singled out as fundamental: the region, Transcendental Mental Living?[36]

So distributed into their regions, we find, according to Husserl, that one sort of existent object is basic, namely the physical thing of Nature relative to the region, Transcendental Mental Living. Understood in its multi-layered structure sketched at the end of *Ideas*, First Book, the physical thing of Nature is subjected to detailed analysis in Parts I and II of *Ideas*, Second Book. The results of these intentional analyses Landgrebe summarizes as follows: "If one says that the physical thing <of Nature> acquires its objective, natural-scientific determinations <as existing> in itself, that objective Nature exists in itself, then the sense of this last way of speaking of being in itself is inseparable from the correlation to subjectivity as intersubjectivity and its communal cognitive products."[37] This summary takes us to the core of Husserl's distinction in *Ideas*, Second Book (section 53) between "Nature as it stands at the beginning" and "Nature as it now emerges for us in communal relationships."[38]

Corresponding to the distinction are two different attitudes: the naturalistic and the personalistic attitude. The former, a version of which has already been discussed,[39] may be described as follows in the present context: Were it to seem *at the beginning* as though Nature were All of Reality because everything which would be given to us must have its foundation in the presentation of sensuously intuited physical things, as a consequence of which everything that is can be ordered into its contexture as determinable for scientific cognition, it then *now* seems as though this "naturally observed world is still not the world" but rather a configuration in the world for intersubjective mutual understanding.[40] And if, moreover, Nature now appears as the All of Reality, as the most fundamental stratum constitutive of whatever exists, then we can no longer assert that it exists in itself. There is, in other words, an order of the multi-layered, founding-founded structure of objects, and the "having" of things in view of their regional determinations dependent on an "attitude" of the cognizing subject.[41] The first attitude is the "naturalistic" attitude because it is focused on mere "facts," all value predicates ostensibly being left out of its purview. It is in this attitude that the objectivities of mathematical physics are constituted and life itself conceived as an "annex," i.e., as psycho*physical*, thus defined in terms of the region, Material Thing. But this can hardly be regarded as in any sense the "original" constituting of "life," "soul," of the mental world which, after all, is no more than the rule governing one essentially possible course of appearances apprehended as "behaving" in a lifelike way.

In the case of the second, the personalistic, attitude, the mental world is not, as in the naturalistic attitude, simply one among various possible course of appearances of physical things, not simply one more level or stratum built up on lower ones otherwise accessible in the naturalistic attitude. The personalistic attitude is not one chosen for theoretical-methodic reasons. Instead it designates the very way in which ostensibly we are immediately aware of ourselves in the world such that, for example, the encounter with others is not merely a matter of "adding on" another stratum of constitution to a lower and perchance more basic one, not merely the "addition" of Mind to Matter. The focus is rather on our directly confronting others as persons related to the same things and events and the same world, to earth and heavens, to field and forest, to the room in which we pass the time of day, to the picture we see.[42] We are related to a common surrounding world in which we stand in personal associations a common goal of which is to be directed to those quasi-objective and objective determinations of the given presupposed by, yet which are deliberately disregarded in, the naturalistic attitude and conception in focusing only upon "material" determinations by bestowing a primacy on the region, Material Thing, so that the mental or spiritual world disappears as an object of genuine knowledge.

As Landgrebe notes, in this sense Mind, Spirit, becomes a sort of reflection of objectively defined Nature and not a stratum, founded upon it. Mind or Spirit now proves to be a necessary correlate of scientifically objective Nature with respect to which we can only speak of Nature in the first place as the All of Reality. *Together*, both the naturalistic and personalistic attitudes are then implicitly expression of the paradox of being in and of the world. Unless there is recognition and acknowledgement of that, then Nature is nothing else but the mere product of a method; Nature becomes "naturalized," an example of which is what we have called the Kantian view of space and time.

91. REGIONAL LINKAGES

At first glance it would seem as though three regions have been distinguished: Material Nature, Animate Nature, and Mind. However, as Husserl indicates at the end of the Second Book of *Ideas*, there are but two regions at issue: Nature (Space-Time) and Mind (Time), and of the two it is Mind that has a primacy.[43] The line of thought leading to this finding may be expressed as follows: If we cancel out Mind, then there is no Nature as determined and further determinable by natural science; but if we were to cancel out Nature, then there would only be individual Mind but no community, no social relations or personal associations, no intersubjectivity. In short, there would be no being in the world. Were one therefore to try to resolve the phenomenological paradox in favor of one or the other attitudes, one would not be left with a nothing: either one would be left with no world at all determinable scientifically or one would be left with an individual mental living with no personal associations, no culture, and the like. In either case of "resolution," not a further paradox but instead a countersense arises. Being in and part of the world, and being intentive to the world, make up a paradoxical concrete whole.

Husserl's response to this phenomenological situation is best expressed in his own words: "What distinguishes two things is the real causal connection that presupposes the Here and Now, and we are therefore referred back to an individual subjectivity....The only thing uniquely individual originaliter is consciousness concrescent with its Ego. Everything else individual is something which appears and that has the principle of its individuation in the actual appearings which, on their side, refer back to an individual consciousness....Absolute individuation thus enters into the *personal* Ego" because "*minds are not unities of appearances but instead unities of absolute interrelations of consciousness*--stated more precisely, of Ego-unities." In contrast, Nature is the "X and essentially nothing other than the X which is defined by universal determinations. But mind is not the X; it is instead what is given itself in mental experience."[44] These and similar statements by Husserl are not immediately obvious in their meaning. According to Landgrebe who, along with Edith Stein, prepared much of the manuscript of the Second Book of *Ideas*,[45] such statements of Husserl suggest that he has moved from the constitutional concept of consciousness as the correlation of the given and its corresponding modes of awareness of the given, to the metaphysical concept of consciousness. Consciousness then becomes an "absolute region" of being.[46] Thus for Landgrebe the idea Husserl expresses is that Mind, in the sense of the "object" of a positing mental living, is immediately presented for itself in reflection and that nothing is or need be postulated as standing behind or being inaccessibly beyond the self-intentiveness of mental living. Nature, in contrast, is never itself presented except as the X determined and further determinable *ad infinitum*, and into which we cannot further penetrate as we can with Mind. The reason for this is not because Nature is hidden but rather because Nature has no "innerness" of its own (above, section 78ff.); Nature is always and only public.

As interpreted by Landgrebe, two distinct and apparently contradictory meanings of Nature would seem to emerge from Husserl's discussion:

1. Nature in the sense of the All of Reality, along with its founding-founded stratifications of which animate beings and Mind itself comprise the higher layers.

2. Nature in the sense of an undetermined but infinitely determinable formulation or configuration of physical-thing reality produced by Mind. Because, Landgrebe says, the second meaning cancels out the first by relativizing all regions in so far as they are referred back to an "absolute consciousness" constitutive of the All of Reality, the two meanings of Nature are incompatible rather than correlative concepts. As a consequence, it then makes no sense to speak of the various founding-founded stratifications of Nature: "If, however, Mind is absolute being <and> Nature merely relative, constituted 'only' in Mind, then we can no longer speak of the relationship of founding because the absolute cannot be founded on the relative."[47] Mind cannot both be *one* stratification among many (perchance the highest) of Nature and absolute, that is to say, "unstratified." Husserl has thus illicitly introduced, according to Landgrebe, a metaphysics in his constitutional phenomenology.

Yet as seems to be implicit in section 64 of *Ideas*, Second Book, and explicit in what has been seen of Husserl's de facto and de jure course of thought in previous chapters, Landgrebe's conclusion not only need not be the only one but it is also one that overlooks the peculiar paradoxical situation Husserl sought to clarify from early on in his attempt to articulate the philosophic-theoretic will and striving required by the things themselves to produce critical-epistemological understanding. To see that Nature is *not necessarily* a condition of Nature, and to see that this does not necessarily imply that the one can exist concretely apart from the other, hence that the one can be relative to the other: *This is to see that the fact has already been recognized and acknowledged that Mind is in the world only by being aware of the world and itself in the world. Thus the very distinction between the two meanings of Nature presupposes cognitively seizing upon the paradox as having undergone a "reflectional modification:"*[48] *The distinction between the two mutually exclusive meanings of Nature is one between two objects of mental living.*[49]

Both meanings--the one presented originaliter in the naturalistic and the other in the personalistic attitude comprising the general positing that belongs to the naturalness of the natural attitude--are perfectly valid theoretically; and here, as elsewhere in Husserl's writings where the feeling for the problem of genesis holds sway, the response is to have the two meanings undergo the phenomenological epoché so that a decision to choose the relative over the absolute, or the absolute over the relative, is held in abeyance. It is therefore not a question of introducing an illicit metaphysics but rather of exercising transcendental phenomenological epoché such that the paradox is overcome by ignoring, as it were, the theoretical (*scl.* metaphysical) conception of the impossibilty of Mind being "constitutive" of Nature as the All of Reality, as the one spatio-temporal world posited as the *omnitudis realitatis*.

Returning, now, to the first part of *Ideas*, First Book, and retaining the focus of the transcendental phenomenological eidetic attitude, the two meanings of Nature in *Ideas*, Second Book, correspond to Noesis as abstractum and as concretum respectively. But what has just been said above implies that there is only confusion of the region, Mind or Mental Living, with the category, the abstractum, Mind in the World, if one tries to resolve the paradox by choosing one meaning of Nature over the other. The alleged confusion would arise, then, not so much from singling out what has not yet been singled out (Gurwitsch), nor from introducing an illicit metaphysics (Landgrebe), but instead from not remaining faithful to the paradox of being in and part of the world at the same time, of not heeding the "daimonion" of the will to return to the things themselves (above, section 4). Methodologically, the confusion results specifically from trying to proceed from psychological phenomenological eidetic epoché to transcendental phenomenological epoché, which is impossible.[50] Only by following the right order of exercising phenomenological epoché as a response to the will to return to the things themselves can the confusion of region and category be avoided along with the exclusion of one meaning of Nature in favor of the other.

But if that is the case, if that is the paradoxical and protodoxical original presentation of Mental Living and World, of Time and Space-Time-Other, what then is the basic character of the positings in which the naturalness of Mental Living and World consists? More particularly, of what does the equally paradoxical being-in-the-world and being-for-itself of transcendentally, descriptive and eidetically pure mental living consist?[51] Time and Space-Time-Other are, after all, transcendentally and transcendentally-eidetically reduced phenomena the genuine meaning of "reality" of which is their "irreality." Of what does the positing of the "irreality" of the time in the proper sense of mental living consist? Of what does the positing of the "irreality" of mental living flowing off in the space-time of the real, objective world constituted in primary passivity consist?

92. IRREALITY AND THE PROBLEM OF EXISTENCE IN MODERN PHILOSOPHY

The questions about the "irreality" of Mental Living and World, more specifically about the positing of their irreal actuality and possibility as transcendentally and transcendentally-eidetically reduced phenomena, are historically rooted in the *problem of existence* (or: "being") in modern philosophy. The root bears a complexity such that it is no accident that phenomenologically one must resort to speaking of the genuine meaning of reality or existence as its "irreality" and that specified into "irreal actuality and possibility." The root of the problem of existence in this connection is perhaps most conveniently shown in the Third of Descartes' *Meditations on First Philosophy* where he considers the idea of existence in general with respect to the idea of the existence of God. "For example," Descartes says,

> my ideas of cold and heat are so little clear and distinct that I cannot determine from them whether cold is only the absence of heat or heat the absence of cold, or whether both of them are real qualities, or whether neither is such. Besides, since ideas are like pictures, there can be no ideas which do not seem to us to represent objects; and if it is true to say that cold is nothing but an absence of heat, the idea of cold which represents it as something real and positive could, not inappropriately, be called false, and so for similar ideas." However, the idea of the existence of God proves to be dissimilar to the idea of the existence of cold. This is because "even though one might imagine that such a being [*scl.* God] does not exist, nevertheless one cannot imagine that the idea of it does not represent anything real, as I have just said of the idea of cold.[52]

Perhaps the most succinct estimate of Descartes' view is to be found in Leibniz who, in his "Reflections on Knowledge, Truth and Ideas," says that "Some time ago I was moved to a closer examination of this question [*scl.* of existence] by the agreement, long famous among the schoolmen and renewed by Descartes, which proves the existence of God by this syllogism: Whatever follows from the idea of definition of a thing may be predicated of the thing itself. Now, existence follows from the idea of God, since he is defined as the most perfect being or the being than which no greater can be conceived."[53] Leibniz then proceeds to criticize Descartes, as have so many others beginning with Mersenne's collection of the second set of "Objections." But the criticism that is most important historically and philosophically is that of Hume rather than Leibniz, although I cite Leibniz here because the setting of the problem will turn out to be more Leibnizian that Cartesian.[54] In his *Treatise of Human Nature*, Hume says that it would seem that all impressions and ideas are conceived as existent and that from this conception the "perfect idea and assurance of being is deriv'd." But upon closer inspection this conception gives rise to a dilemma: either the idea of existence is derived from an impression constantly conjoined with every object of thought, or the idea of existence is the same as the idea of the object of thought.[55] Thus the idea of existence is not derived from an impression because that would signify that there are cases where two distinct impressions would be inseparably joined; because such inseparable conjunction is impossible, the idea of existence proves to be, Hume says, "the very same with the idea of what we conceive to be existent. To reflect on anything simply, and to reflect on it as existent, are nothing different from each other. That idea, when conjoined, with the idea of any object, *makes no addition to it*. Whatever we conceive, we conceive to be existent. Any idea we please to form is the idea of a being; and the idea of a being is any idea we please to form." Those who do not agree with this view Hume continues, "must necessarily point out that distinct impression from which the idea of entity is deriv'd, and must prove, that this impression is inseparable from any perception we believe to be existent. This we may without hesitation conclude to be impossible."[56]

If, then, existence is anything at all, it must be a property in some sense or other; but because there is no impression of such property the idea of existence and what we conceive to exist are the same. It is therefore not the case that whatever follows from the idea of a thing may be predicated of the thing. The genuine, the most original presentation of the world in and of itself must be said to be existentially mute. The Cartesian view, as reformulated by Leibniz, is thus untenable. Whether or not Hume's own answer resolves the issue is another question; of interest at the moment is the fact that he has laid the foundation for Kant's "Refutation of the Ontological Argument."

In his "Refutation of the Ontological Argument" in the *Critique of Pure Reason* Kant says that through the concept of an object we think the universal conditions of possible empirical knowledge of that object. In contrast, through the existence of an object we think it as "belonging to the context of experience as a whole."[57] Moreover, because the concept of the object is tied in with the context of experience as a whole, it is in no way enlarged by that context. We only have, rather, further possible perceptions of the object to account for. Kant then says that the

> analytic criterion of possibility, as consisting in the principle that bare positives (realities) give rise to no contradiction, cannot be denied to it. But since the realities are not given to us in their specific characters; since even if they were, we should still not be in a position to pass judgement; since the criterion of the possibility of synthetic knowledge is never to be looked for, save in experience....The connection of all real properties in a thing therefore is a synthesis, the possibility of which we are unable to determine *apriori*.

Implicit in Kant's view is the assumption, beginning with Descartes and made explicit by Leibniz, that "existence," "being," signifies nothing else than property, and the originality of Kant's "Refutation" lies in showing that "existence" is not a real predicate because it is "not a concept of something which could be added to the concept of thing." Kant thus follows the Leibnizian and Humean pattern of orienting his discussion around the determination of physical things:[58] As in the case of Leibniz restating Descartes, and likewise but more radically in the case of Hume, even though God is not a physical thing, to say that "God exists" is, following Leibniz's syllogism, *formally akin* to saying that such and such a physical thing exists. The significance of Kant's "Refutation" is that existence or being is not contained analytically in the concept of physical thing; and by extension the existence of God is not contained analytically in the concept of God. Such concepts are merely thought, but not posited; existence is merely (conceptually) presented. To further elaborate Kant in this connection is not a simple matter.[59] It would seem, however, that Kant may also be suggesting that even if a supervenient "something" were added to thought by positing, there still would not be a *new* determination of the thing in question. Existence is not just one property among others which can, for instance, be lost. One cannot say that there are things which are of such and such a size, weight, shape, color, *and* which also, somehow or other, have the posited property of existence or being. In so far as their concept is at issue, they are conceived alike regardless of whether or not they are thought as actually, possibly, or even necessarily, existing.

But if that is the case, how, then, can we adequately account for a justified belief in things as existing? With respect to the external world, existence may even be posited in accordance with the "postulates of empirical thought" in the light of the "analogies of experience." Even so, existence is not a property, hence not a "real predicate," neither a "reality" nor an "unreality." Whatever else it is, or may be, Kant does not say. Indeed, his "Refutation" is silent on the meaning of existence in such a manner that Jules Vuillemin has drawn the remarkable conclusion that the goal of the "Refutation" is to make phenomenology necessary so that the philosopher, having "parenthesized" the existence of physical things, examines them solely with respect to the faculty of knowledge. But this is not quite what Husserl had in mind, as we have seen (sections 5, 8), especially when making his famous distinction between the "absolute being" of mental living and the "presumptive reality" of being in the world.[60] Even though one might be tempted to agree, as Vuillemin says, that the idea of the "postulates of empirical thought" gives rise to the steps of the phenomenological epoché and reduction which are in no way arbitrary in content,[61] the question still remains: How else may existence or being be understood if not as a "real predicate"?

93. THE PROBLEM OF EXISTENCE IN HUSSERL'S CRITIQUE OF BRENTANO

Husserl inherits the problem of existence in modern philosophy through his critique of Brentano in the Fifth Logical Investigation, explicitly returning to it again in the light of Kant's "Refutation" in the Sixth Logical Investigation (e.g., section 43ff.). Moreover, the way in which Husserl inherits the problem brings to light certain of its dimensions which might otherwise have remained hidden, eventually setting the line of thought he develops in the third chapter of the second part of *Ideas*, First Book. The focus of discussion in the critique of Brentano proves to be precisely the problem of existence as a predicate.

According to Brentano, an act of consciousness, a "physical phenomenon," is either a presentation of something or it rests on a presentation of something.[62] For example, an object desired would be inconceivable unless it were either presented in some fashion or based on a presentation. This rule holds for other acts as well, such as love and hate, belief and judgment. But is it the case, Husserl asks, that all other psychical acts presuppose a founding presentation of the sort Brentano describes, namely a "mere" or "neutral" presentation of something as neither believed-in nor disbelieved-in, as neither loved nor hated, as neither desired nor willed? If we take Brentano's account in terms of Kant's "Refutation," then what Brentano suggests is that, if existence is predicated at all, it is predicated of things by a higher order of psychical acts. Predication of existence then necessarily presupposes mere presentation of that whereof predication is made. Like Kant, Brentano would also hold that the predicate, "existence," is not analytically contained in the concept of the physical thing; rather existence is the addition of a "belief characteristic" that transforms the mere presentation of something into a judgement about it as existing. Yet, this does not signify, nor need it signify, that existence is a new predicate in the sense that mere presentation of a physical thing is sufficient to found a judging about it as existent. The judging (or belief, in Brentano's broad meaning of the word, "belief") is "added" on, to be sure, but not as a non-self-sufficient part of the judgmental-presentational complex.[63] As a consequence, when a concatenated set of psychical acts all have the same object, and when they differ only with respect to their "intentional essence," as Husserl calls it (when one act, for instance, is a judging, another a wishing, still another a doubting of the same object), they all have the same mere presentation of the object as foundation.

Brentano would thus seem to offer a compromise of the view of Descartes and Hume from the side of Kant such that it would follow that the identity of the "noematic core" (or "matter") in the variation of its existential quality or belief characteristic rests on the essential identity of the founding presentation. But what can it signify to speak of the "same presentation"? Certainly it cannot signify, Husserl says, that the object of the act of belief and the founding presentation are actually one and the same since the object of the former is but an abstract moment of the act in which it is merely, neutrally, presented. It would rather seem that what Brentano has in mind is that the presentation of one and the same "intentional object" varies in its modes of presentation while the quality remains the same. For example: I believe in the seen table in front of me as existing regardless of whether I proceed to judge about it, desire to use it. As a consequence, it can be said that each and every presentation has the same "intentional essence." The novelty of Brentano's view, then, is that while every other "intentional essence" is a complex of quality and matter, that of presentation is mere matter.[64] It follows from this that only because of the circumstances that all other acts are complex and necessarily comprise, or are founded upon, a mere presentation as a component part, can we rightly speak of a difference between the dimensions of quality and the meant and seen object with its various and varying modes of presentation.

But if this is the case, how are the various "species" of presentation distinguished from one another? The answer which Brentano, apparently, would have to give is that differences among presentations occur only at higher levels by virtue of supervening psychical acts such as believing, desiring, judging, loving, hating. As a result, those differences are not "generic" ones. Yet it must be noted that this is precisely the form of the question which Kant's "Refutation" takes on Brentano's view, namely, how the "species" of presentation are to be differentiated.[65] It must also be noted that the question, as Husserl takes it over in his discussion of Brentano, is already an eidetic phenomenological one.

Husserl observes that even though there is good reason for Brentano's account of the meaning of existence, it does not exclude other accounts. Indeed, it is surprising, Husserl says, that Brentano gives so much weight to the idea of mere presentation in contrast to acts of belief, to positing acts largely left unexamined as much by Brentano as by Kant. Yet if Brentano is considered in the light of the history of the problem of existence it is not surprising that, after Hume and Kant, what he talks about under the heading, "mere presentation," proves to be *the* alleged *generic*, or regional, feature of mental living in the world. For both Kant and Brentano the question remains: How are the specific differences among acts to be understood? As abstracta in the one case, as concreta in the other? Were I to judge that something is the case, is this not a judgmental complex composed of a presentational act that makes the case in question present to me, and an additional complementary act peculiar to the judging in and through which the case in question appears as existent? Does the principle of differentiation lie outside the genus, "Mere Presentation"? What is the origin of the "added" specific difference expressed by the words, "mere" and "belief"? Is the judgmental characteristic the same in all cases of judging? Is it already the ultimate specific difference? Kant says, in effect, that the act of judgment in which existence is posited is not determinable a priori although the judgment is a synthetic one. In other words, even though presentational, the act of judgment is specifically, and not generically, different in its lack of a priori determination. As a result, Kant, therefore, states that existence is not a "new predicate."

With both Kant and Brentano it can be seen that implicit in the very setting of the problem of existence beginning with Descartes is the assumption, first, that presentation (or re-presentation) is the only generic feature of mental living and that, second, more fundamentally it is in terms of some conception of genus and species that mental living is to be understood. *And precisely here we find the root of the assumptions about space and time peculiar to the naturalistic conception of the real, objective world*: *the continuity of presentation demands*, as it were, *a continuity of positing because of the double assumption inherent in the setting of the problem of existence.* These assumptions as assumptions are brought to light by the way in which Husserl inherits the problem of existence in his critique of Brentano.

Remaining with the examples and the line of thought developed by Husserl in the Fifth Logical Investigation, we may say that if the specific difference consists of the addition of a judgmental characteristic to a presentational act, then the difference between the act in and through which "Emperor" is presented and the act in and through which "Pope" is presented does not consist of the generic presentationality of the acts. The difference must lie elsewhere and, more particularly, it must lie in a different genus. "Emperor" and "Pope" are not specific differences, nor are they different degrees of presentation. Nonetheless, the acts are "qualitatively" alike. Perhaps the place to look for the difference is in the "content" of the acts. But to suggest this entails the introduction of an entirely new genus, namely that of "content," and Brentano's axiom that mere presentation is the only generic feature of mental living would have to be rejected.

Two different examples illustrate what Husserl has in mind in this criticism. The first example consists of the phenomenological situation where, in and through one act, "Napoleon--victor at Jena," is presented; while in and through another act, "Napoleon--vanguished at Waterloo," is presented. In both acts, the "matter" or presentational "material" is the same, but the "quality" varies (*scl.* "victor" and "vanguished"). The second example is the one mentioned before: the presentations of "Emperor" and "Pope," where the presentational "material" varies, but the quality remains the same. And the point is that the view according to which presentationality is the only generic feature of mental living cannot account for either set of differences. And if this set of differences cannot be rendered, how, then, does one explain the difference between the mere presentation of something and the belief in that something as existing? Is "existence" a "quality" like "victor" and "vanguished"? Obviously it would be a truism to say that the "content" of the acts distinguished them. What, then, does distinguish them?

The answer to the question, and the avoidance of truisms, Husserl says, lies in clarifying the distinction between "quality" and "matter." There are two alternative routes that can be taken toward this clarification; the one is a reformulation of Brentano and, by implication, Kant; the second is the one taken by Husserl.

94. BRENTANO, KANT, AND HUSSERL

It is possible that what makes up the changing "intentional essence" and, therefore, at the same time, the changing relation to the object, is a "qualitative" facet peculiar to the presentational act itself. That is to say, those acts in and through which, e.g., "Emperor" and "Pope" are presented as distinct from one another are precisely analogous to the difference between the presentation of the colors red and blue. Thus in this case there is to be distinguished the presentation of "any color whatever," the "general," and this or that "particular" presentational shade or nuance of color (red or blue). Similarly, in the case of the presentational acts, "Emperor" and "Pope," they are different "shades" of any presentation whatever; or, in the case of the same "matter," Napoleon "victor at Jena" and "vanguished at Waterloo" are akin, for instance, to the different degrees of brightness of blue. As a consequence, that an act is directed toward a certain object or objects is not because of any activity "outside" the act on the part of the object in and of itself, or because of any manipulation of the object. It is, instead, owing to some inherent determination of the presentational act itself that this rather than that appears in this or that manner.

But, as Husserl notes, this alternative would now put a premium on the "intentional quality" which comprises the uniform species of not only presentation but also of judgment, desire, volition, and so forth. Yet the species, "presentation," somehow is said to undergo further differentiation with respect to this or that object. Differences such as "Emperor" and "Pope" would be based on the "qualitative" difference of presentation. There would seem to be no other way of conceiving the situation: "This is because it is not possible to re-establish the uniformity by conceiving the distinct contents of different judgments, and likewise of different desires, volitions, etc., as themselves being different species of quality which are judgment, desire, volition."[67] Equivalently stated: distinct species cannot have the same ultimate differences because otherwise there would be no specific uniformity. And is not this incompatibility, Husserl asks, simply replaced by another if different species of the same order have to be admitted which have their ultimate differences at a lower level than the rest of the species which are already themselves the ultimate differences?[68]

As a result, there would seem to be good reason for casting doubt on, if not rejecting altogether, the "genus-species" idea of mental living in the world as much in Brentano's account as in those of Descartes, Hume, and Kant. Eidetically-phenomenologically, the least one can say for it is that it is inadequate, and Husserl's rejection of it in the *Logical Investigations* motivates the quite different account in section 15 of *Ideas*, First Book. For already in the first edition of the former work Husserl says that what must be seen is that the relation between "quality" and "matter" is not analogous to the relation between red and blue, or between color and red and blue. It is, instead, similar to the relation between color and extension--if there is an analogy at all. As a consequence, differentiation does not lie within a single genus (to remain with the language used in the Fifth Logical Investigation). Just as there are two different genera in the case of color and extension which cannot be present one without the other, so there are two different facts of mental living: "quality" and "matter," neither of which can be found without the other. They are mutual and correlative complements, not supplements (as in Brentano and, by extension, in Kant).

In the Fifth Logical Investigation Husserl offers a number of examples to illustrate this conclusion, two of which may be briefly rehearsed here. The first example is the following: Suppose, Husserl says, I am walking around in the Panopticum Waxworks where, on the stairs, I meet an unknown but charming woman--a manikin that momentarily deceives me. As long as I am deceived, I have a perceptual process going on like any other; I see the woman and not the manikin. But were I to recognize the illusion, I would then see a manikin which "represents" a woman. To be sure, when I speak of "representation" in this sense I mean something else than when I speak of the wax figures, such as that of Napoleon, which has a very different representational function; the undeceived perceiving of the wax figure, which is the founding perception for the "picture consciousness" of Napoleon, is quite distinct from the illusory perceiving of the manikin as a woman for the illusory perceiving does not serve as a founding act. Instead, the woman appears at one and the same time with the manikin; two interwoven perceptual apprehensions "coincide" with respect to a certain "content" of appearance. Moreover, the "coincidence" occurs by way of opposition so that the one perception tends to cancel out, in part, the other. But one has to say "in part," because the original perceptual act is not entirely nullified. It merges with a new perceiving, namely that of the manikin as a manikin and not as a real woman. However, the original perceiving does not serve as a founding act here for the second and correct perceiving. What is perceived is only the manikin which alone is believed in as existing there; it is not a "picture consciousness" of a real woman. The one perceiving is, as it were, detached from the other the illusion is discovered. Yet, much more is involved. Were it simply a matter of one perception being detached, then the presented object in the second case would be, in actuality, the same object as is the first case. But after the unmasking of the illusion, the perceiving is the perceiving "cancelled out" in the perceptual conflict. And that is a "quality" not characteristic of the first or original act of perception in which the manikin was believed in as an existent, real woman. Both perceivings do, of course, have something in common: the perceptual "matter" or "material." The difference between them would seem to consist of the "quality" of the acts because the same "matter" can both be "matter" for a serious perceiving and for an illusory one. Obviously both cannot be united; a serious perceiving can never, at the same time, be a perceptual fiction.

Husserl's example is important because it makes doubtful the idea that every perceiving, for instance, is a complex in which the "quality" is added on to an act of perceptual presentation, as though the "quality" or "belief characteristic" were therefore simply detachable from the "matter" without the perceptual situation being in any way affected. Were this the case, as Brentano, Kant and Hume and Descartes suggest, then it would not be possible to explain any difference arising in the conflicting perceivings of the manikin and of the woman, nor would it be possible to recognize the one as deceptive, illusory and non-existent.

Husserl's second example concerns judgment. Although the analysis is more complex than that of perception, the problem is essentially the same and depends on the question of whether what determines that the judging is a judging of this rather than that state of affairs lies in a founding act of the mere presentation of something judged about. And if I begin with the assumption of a mere presentation as a founding act, I am then forced to look for something which is "added" on to it and which makes up the specific difference. At least at the beginning, the possibility must not be excluded that the "quality" is already there with the "matter," and that it is there of essential necessity and not accidentally. As a case in point, Husserl considers the case of assent or agreement; more particularly, it is the situation where I undergo experiences of a certain kind when not judgmentally decided about something but am, instead, merely entertaining it and only afterwards assent to the judgment (or, perchance, reject it). The example is especially instructive because it is a situation where a new act is joined to one of mere presentation (merely entertaining the idea of something without assenting to it or rejecting it); the new act, moreover, succeeds and asserts or affirms the mere presentation. But the question is whether the new act simply arises out of the old in such a way that the specific judgment "quality" (e.g., assent) is added on to the mere presentation, to the neither believing nor disbelieving in the state of affairs judged about. Is it the case, then, that in the new act a "quality" in any genuine sense is added on? Clearly something identical is preserved when the new act emerges from the old. But this identical something need not be a full act of presentation, nor does the addition of a new "quality" founded on it need be the only observable change. It is possible, of course, to understand the phenomenological situation in line with Brentano as one in which the "quality" of the first act is detached from the presentation; the judgmental characteristic of assent is withheld, removed, set aside for the moment until I make up my mind. But it might also very well be that something else is involved.

Put in another way, what legitimates or ratifies assent? Suppose that I assent to a judgment made by someone else; what the other says does not immediately awaken a judgment on my part which agrees or concurs with it. To effect that agreement, to receive or adopt the judgment, is not necessarily to assert it. To do that I must directly understand the judgment without judging it myself; I must be aware of what is alleged in the judgment as something "merely proposed." On that basis, I proceed, in a further act, to examine it in some depth, to make it a question for me and strive to arrive at a conclusion. The conclusion arrives, I find it well-founded, and assert it to be the case and then discover that I am in agreement with the other's judgment. Moreover, in the judgment which I now make, according to the view held by Brentano, the previous mere presentation, the "merely proposed," does not enter into the picture nor do the series of acts of proposing and making of the original judgment a question. Instead, a judgment is presented which concurs, on the one hand, with the judgment of someone else and, on the other hand, with the question on which I have reflected and which I have examined. Assent is produced: I judge likewise on the basis of the same "matter" or judgmental "material;" I assert as true what has taken in the question as questionable.

But what distinguishes this judgment, which I now make, from other possible ones about the same "matter"? Looking at the situation more closely, it can be seen that the succession of acts of questioning and examining, and then the concurring judgmental process, do not make up the whole affair; a transitional act binds together the original judgment on the part of the other person as understood by me, and the judgment of assent which I now make. What occurs is that the deliberation and questioning which make up my examination are fulfilled in the concurring conclusion, and in this unity of fulfillment of the answer both acts are not merely successive, but rather interwoven with one another from within, so to speak, in contradistinction to the serious and illusory perceiving of the first example. It is not, then, a question of a "quality" being added on (or, in the case of disagreement, taken away). It is, instead, a situation in which an act is fulfilled (positively or negatively).[69] It is an unfulfilled act which is therefore not to be confused with mere presentation; of essential necessity the "quality" is there from the beginning but not fulfilled. *From this it follows that mere presentation is rather a modification of an act instead of its generic determination.*

As the two examples suggest, mental living is most fundamentally characterized noetically and noematically by one or another mode of belief and of being or existence. Being, existence, is therefore neither genus nor species in any ordinary sense of the words, "genus" and "species," and accordingly has nothing to do with property. To be sure, this is as much a negative conclusion as that of Kant or even of Hume. Indeed, in these terms Kant's answer to Hume would seem to be confirmed by Husserl: Existence or being is not a predicate or a property. This assertion, however, must now be understood in Husserl's terms in the sense that existence is not just another property among others. Phenomenologically expressed, the noematic position is a characteristic rather than a stratum of the noema, distinct from all other characterizing dimensions of intentionality (such as, e.g., "attentional" modifications). Not only must the posited characteristic, "existing," not be confused with the *objectivated form*, Existence or Being, but neither must be confused with a property of a thing or of a predicatively formed affair-complex.[70] But on the positive side, what does it signify to say that existence is a characteristic? It does not seem quite the same thing as something other and different from a property. Husserl's account certainly cannot be construed as a "refutation" of Kant's "Refutation" at all; nor does Husserl's critique of Brentano seem to do more that "correct" and "improve" upon Brentano's view--although in a very profound way--indeed, a way that leads far beyond Brentano to the outer reaches of transcendental phenomenology. Nor does the way in which Husserl deals with the problem of existence as found both in the historical and systematic reflections seem to be a "new" solution at all, one, for example, which would exclude all others. Nothing, in fact, seems to be excluded.

What sense is to be made of Husserl's account? What bearing does it have on the problem of existence or being represented here by Descartes, Leibniz, Hume, Kant, and Brentano? More broadly expressed, what bearing does it have on the theme, "My own transcendental mental life-processes believing in themselves as being in the world even if nothing else in the world exists"?[71] Still more broadly expressed, what bearing does it have on the theme, "being or existing in the real, objective world constituted in primary passivity that pertains to my transcendental mental living in the natural attitude"?[72] In a way, Husserl's answer to these and similar questions has already been presented.[73] It is necessary, now, to make that answer explicit again and draw some conclusions important for understanding "space perception."

95. THE TRANSCENDENTAL PHENOMENOLOGICAL FORMULATION OF THE PROBLEM OF EXISTENCE IN MODERN PHILOSOPHY

Presupposed by the discussion of the problem of existence stemming from Kant is an unusual situation: If Kant's "Refutation" is valid and confirmed by Husserl in his critique of Brentano, it would then seem to follow that it is possible to objectivate something, indeed, anything, without of necessity having to objectivate it as existing. Yet, suppose that one were to objectivate something specifically as non-existing? From where would the objectivating act derive the something objectivated as objectivated? Even were one to answer that the object, for instance, a centaur, is a mere construct or a fiction, a perceptual fiction, produced by the perceptual objectivating act itself, it is still "something" objectivated. Or, to take Kant's example of the dollar. Even if I only perceptually imagine and objectivate the dollar in my pocket, something is nonetheless objectivated, merely thought of as pertaining to the objectivating. That is to say, the objectivating regard, to use Husserl's term, focuses on something, on a noematic sense which is, in some manner, bestowed. But then I have implicitly distinguished with Kant, for example, between something actual, real, and something merely thought of. And precisely such a distinction is implicit and presupposed by Kant's example of a "real dollar" and the "imagined" dollar in my pocket. More generally, Kant not only presupposes this distinction, "the Scholastic distinction between the '*mental*,' '*intentional*,' or '*immanental*' Object on the one hand, the '*actual*' *Object* on the other hand,"[74] but *also* that the real dollar *exists*. Thus even though Kant concludes that existence is not a real predicate, this in no way alters the perceptual situation of the one dollar objectivated as existing. That it exists, there can be no doubt in the one case, that it does not exist there can be no doubt in the other case. And this signifies that Kant has presupposed the distinction discussed and arrived at before, namely the distinction between two paradoxically juxtaposed objects of my own mental living.[75]

Once one arrives at this distinction, and especially if one arrives at it in the manner of Descartes, Hume and Kant, and likewise Brentano, then there is the *temptation* to draw the following conclusion: "In the mental process the intention is given with its intentional Object which, as intentional Object, inseparably belongs to it, therefore itself *inherently* dwells within <the intention>. Indeed, it is and remains its <Object> meant, objectivated, and the like, no matter if the corresponding 'actual Object' precisely is or is not in actuality, if it has been annihilated in the meantime, etc."[76] And, in a certain fashion, this temptation defines the course of thinking about the problem of existence in modern philosophy.

Or, better: the *degree* to which the temptation holds sway defines the course of thought we have been tracing. As a consequence, if one tries to separate "the actual Object...and the intentional Object, including the latter <as> really inherently in the mental process as 'immanent' to the perception, we fall into the difficulty that not *two* realities ought to stand over against one another while only *one* <reality> is found to be present and even possible."[77] Earlier (section 9) this situation was expressed as a transcendental *empirical* problem. Here the same may be expressed as a transcendental *historical* problem:

It is one thing to seize upon the distinction presupposed all along and another to fix and correctly evaluate it. *Thus in Descartes*, on the one hand, I find that the objects of my mental life-processes are nothing other than "ideas," some of which, to be sure, depicture, are "representative of," myself and things other than myself but which nevertheless are nothing more than dependent meant and intended-to objects of my mental life-processes; on the other hand, at the same time I simply posit the world and my mental life-processes in the world as existing. Starting out from, and objectivating *this* straightforward positing presupposition, as in Descartes, I may then set out to elaborate the view that their existential status as objects of my mental life-processes is accidental to the world and its components: *thus Hume and finally Kant*. But, then, if that view is correct, if I have yielded to the temptation of modern philosophy to a still higher degree, the objects that I reflectively seize upon and, in a further act, objectivate, as essentially objects of my mental life-process cannot also be the world posited as existing in and of itself: *thus Brentano*. As a consequence, the conclusion is forced upon me that the presupposed distinction is not a mere distinction but rather a *numerical* distinction between positing of the world in and of itself and the world as object of my mental life-processes, that is, of "idea." Thus I arrive at the form of the root of the historical reflection and at the root of Husserl's critique of Brentano. It is the paradox encountered in striving to return to the "things themselves" and comprises a distinction prior to any philosophical rendering of the confrontation of world and things in it as existing.

Having seized upon the distinction, but also having yielded to the temptation indicated by Husserl, Kant falls into the difficulty of having "now two realities <that> ought to stand over against one another while only one <reality> is found to be present and even possible." Kant's answer is to show that the "real contains no more than the merely possible."[78] The merely thought of or conceived is not itself "in the least increased," Kant says, "through thus acquiring existence outside my concept." Where, in one respect, Husserl and Kant differ lies in Kant's conclusion: "Whatever, there, and however much, our concept of an object may contain, we must go outside it, if we are to ascribe existence to the object."[79] In other words, the two "realities," the "actual object" existing in and of itself (the hundred real dollars) and the object of my mental life-processes, the merely thought of or conceived (the hundred possible dollars, "the mere concept of them (that is, of their possibility)"), are the same in the sense that it makes no difference if I posit the one as existing and the other as not. It is an easy step, then, to Brentano and the notion of mere presentation as the founding stratum, whereby we "go outside" the founding presentation in a further act of positing existence or being. But for Husserl there is no attempt to "go outside;" indeed, there is the deliberate attempt, in striving to return to the "things themselves," to remain with the distinction itself and to resist the temptation to regard the hundred dollars as inherently dwelling within the intention "no matter if the corresponding 'actual Object' precisely is or is not actuality."

In contrast to proceeding *as though* the two realities were *one*, Husserl says that

> we have to abide by what is given in the pure mental process and to take it within the frame of clarity precisely as it is given. The 'actual' Object is then to be 'parenthesized.'...If we now effect the phenomenological reduction, then every positing of something transcendent, thus above all what is inherent to perception itself, receives its excluding parentheses, and this is passed on to all of the founded acts, to every judgment of perception, to the positing of value, and possibly to the value judgment grounded in it....But we do not tolerate any judgment which makes use of the positing of the 'actual' physical thing, nor of the whole 'transcendent' Nature, or which 'joins in' <that positing>. As *phenomenologists* we abstain from all such positings. But on that account we do not reject them by not 'taking them as our basis,' by not 'joining in' them. They are indeed there, they also essentially belong

> to the phenomenon...instead of joining in them, we make them Objects, take them as component parts of the phenomenon....[80]

As a consequence, we find, "figuratively stated: the 'parenthesis' undergone by perception prevents any judgment about perceived actuality (i.e., any <judgment> having its basis in *unmodified perception*, thus taking up into itself its positing). But it does not prevent the judgment about the fact that perception is consciousness *of* an actuality...."[81] By not making any judgment having its basis in "unmodified perception," a generical distinction is discovered which had surfaced in the critique of Brentano, namely between the noematic core and the posited characteristic; more broadly stated, it is the distinction between presentationality and positionality. These are "quite different sorts of characteristics which always attach to the noematic core, and the ways in which they pertain to <the core> are very different. They find their place within *fundamentally different genera*, within, so to speak, fundamentally different *dimensions of characterization*...The situation is such that...always and necessarily a noematic core, an 'object--noema,' is intended to which *must* be characterized in some other manner and, moreover, according to this or that (on its side, exclusive) species in *each* genus."[82]

Not only, then, do Kant and Brentano base their judgments on "modified" perception, but they fail to see the distinction between "matter" and "quality," between presentationality and positionality; more particularly, they fail to see that they are distinct eidetically and that even though they are interrelated as essences, as eidetic abstracta, they are not related as genus and species. Just for this reason it is necessary to speak of a "characterization"--a figurative and descriptive expression--rather then of a "property" or of something postulated in "going beyond" what is given. Indeed, the adherence to precisely what is itself presented vitiates the need and temptation to appeal to something supervenient. "Existence" is an "irreality" then precisely because it is a *modality of positing* and not a metaphysical meaning.[83]

Thus, although it may be pre-interpreted that way, positing is not something founded on, or even a species of, presentation. Rather presentation is always characterized in some manner or other, and this of eidetic necessity so that if there is variation in manner of presentation there is necessarily variation in modality of positing, and conversely. And just as unbuilding and building-up analyses of the real, objective world constituted in primary passivity disclose that there is not a continuity of manners of presentation, so there is not a continuity of positings.[84] Moreover the co-variation of presentation and positionality cuts across all of mental living, whether intentionality is actional or passive (see above, sections 10, 11). A few examples will suffice here (some more obvious than others) to illustrate the irreal actuality of space and time and other.

96. THE POSITING OF SPACE, TIME AND OTHER AS IRREAL ACTUALITIES

In the preceding chapters a wide variety of examples of positing of space, time and other have been provided. It remains to make some features of that positing explicit and in the context of an eidetic transcendental phenomenology of space, time and other. Husserl's, and an Husserlian, response to the problem of existence in modern philosophy is of course the beginning, and not the end, of clarification, and here only a few steps will be taken at the beginning.

Perhaps the simplest example to start with is that of seeing the white piece of paper and being uncertain as to whether or not what I see exists; seeing the piece of paper, I believe in it as existing, posit it as existing, in a dubitative mode. Notwithstanding, I am certain that it is something meant as selfsame throughout the seeing-unbelievingly. That is to say, the identifying synthesis in which this, something-white, is constituted is itself certain even though the extent of the seeing itself is uncertain, dubitative. Or, in the case of seeing of the manikin in the waxworks: at a distance I see something which, at first sight, looks like a woman, but then a moment later like a manikin. I am uncertain as to which it is. But here, again, the synthesis is certain: the selfsame, meant or intended to and believed in now as a woman, now as a manikin. Correspondingly, the identifying synthesis of the seeing, in contrast to that of the seen, is also certain even though I am uncertain as to whether or not I can believe my eyes, whether or not the seeing is a seeing and believing (or positing) a woman or a manikin as existing. Or, perchance, I refuse to believe in the seeing of a woman or of a manikin; I am neither believing nor disbelieving in the one or the other as existing. In any case, whether the believing is undecided or even purely neutral with respect to what is seen, whether or not I am certain which seeing is going on (seeing of woman or seeing of manikin), that perceptual mental life-process is synthetically retrotended and protended to with simple cetainty as existing in primary passivity.

The two examples disclose the phenomenological situation in which, no matter what the positionality of the mental life-process in question, both identifying syntheses (of seeing or of seen) are positings in the mode of simple certainty. Indeed, no matter what the mode of actional or passive positionality of any mental life-process or of any given extent of one, that mental life-process or its extent is always of necessity constituted in primary passivity in an identifying and distinguishing synthesis that is a positing of itself with simple certainty.[85] In this context, with respect to the basic pairing and synthesizing of mental living, the positionality of any given mental life-process or its extent is like an independent variable. Moreover, as noted before (sections 62, 85f.), mental living is constituted in identifying syntheses as an identity posited as irreal actuality throughout a multiplicity of protendings and retrotendings to itself, and each partial extent is constituted as identical and different from every other partial extent, protended to and retrotended to as paired in primary passivity--and that includes *actional* as well as primary passive mental life-processes or their extents, no matter what the independent variable of positing may be. My mental living in its self-positing may then be said to be "absolute." Thus in the case of immediate, immanently public presentation of mental living *in primary passivity* there is always of necessity and *in primary passivity*, without antecedent actionality, *synthetical* self-positing as existent (see above, section 79b), and that in the mode of protodoxic or simply certainty. But the phenomenological situation also reveals a complexity of a rather different sort.

The situation I have in mind concerns the positing of "spaces" reviewed at the beginning of the last chapter (sections 73f.). Presenting and positing of prespaces, for instance, as existent, as irreal actualities, does not of necessity entail positing of them in primary passivity either as extended or as having real things in them; they are presented and posited as spreads, to be sure, but not as extensities. In contrast, presenting and positing of quasi-objective or objective space as irreal actual extensity of necessity includes presenting and positing of them as spreads.[86] These cases, along with the many other similar ones distinguished before, exhibit dependent variability so far as presentationality and positionality of "spaces" and "space-times" are concerned. Their positing is not "absolute." Their immediate presentation is not apodictic but rather transcendently public. And in both sorts of examples just mentioned there is independent as well as dependent variability of positing and immediate presentation in one or another modality of positing. Moreover, in both sets of examples there is the positing in the mode of simple certainty peculiar to identifying syntheses in primary passivity. Perhaps the example that illustrates most dramatically the positional-presentative dependent and independent variability is that of the "perception of other minds" described earlier (section 79, b) for it exhibits the unity of "absolute" and "relative" positing of mental living in the world. It is, in fact, a limiting case for it involves not only immediate presentation but also mutual incompatibility of such presentation of the mental living of others. Here the positings are likewise mutually "incompatible;" they are, more specifically, mutually incompatible self-positings as existent immanently public but non-transcendent and transcendently public but non-immanent mental livings. In still other words, each mental living of the plurality of mental livings is self-positing in the modality of "absolute" in its immediate presentation *and* positing of another mental living in the modality of "relative" in *its* immediate but non-apodictically evidential presentation. The positing is equally "absolute" and "relative" (above, section 80). To be sure, the "equality" of this positing obtains only correlative to the constituting of a public, primordial quasi-objective world in primary passivity (section 78). That is to say, not every self-positing that is also a "relative" positing of a spatiotemporal "world" as existent is a positing of the mental living of someone else, although the self-positing that is positing of the mental living of someone else includes as essential foundation "worlds" presented immediately and posited as existent, but not as publicly "intersubjective." Eidetically expressed, not every self-positing of mental living as abstractum includes positing of other mental livings as concreta that are abstracta. And this holds universally for the plurality of mental livings positing one real, objective spatio-temporal world as existent.

The foregoing considerations provide a clue to the understanding of several famous (or, according to some, infamous) passages in *Ideas*, First Book,[87] where Husserl speaks of a distinction that must be made between "absolute" and "presumptive" (or "relative") positing.[88] Perhaps the least controversial and most cautious reading of these passages is that of Rudolf Boehm who notes that Husserl insists that the "perception of something immanent" guarantees that what "comes to absolute givenness in this perception is an absolutely indubitable being," that what is presented as absolute "necessarily requires the absolute positing of that which is absolutely given as existing," and that the domain of what is immediately presented as absolute is identified with what is posited as absolute.[89] Thus Boehm concludes, as noted before, that the expression, "absolute being" (or "existence") does not refer to a mode of being of something absolute in a metaphysical sense but instead to a modality of positing, to a modality of ontic believedness. Thus what is believed in with indubitable certainty is what is immediately presented as immanently public in contrast to believing in something with dubitable (but also simple) certainty[90] and which is immediately presented as transcendently public. It is this latter positing, *in the context of the contrast*, that is said to be a "presumptive" (or "relative") positing. Boehm notes, however, that this distinction contains quite radical implications drawn out by Husserl further on in section 49, the title of which is "Absolute Consciousness as the Residuum After the Annihilation of the World"--an exuberant title with a distinct Hobbesian nuance.[91] Important for Boehm is Husserl's statement, "that while the being of consciousness...would indeed be necessarily modified by an annihilation of the world of physical things, its own existence would not be touched."[92] The conclusion to be drawn would seem to be obvious: "With this statement Husserl clearly infers relationships of *being* from positional relationships" so that the "statement: 'Therefore no real being is necessary for the being of consciousness' is ambiguous" because in "principle, the possibility of absolute givenness has its foundation in absolute being: *absolute being establishes the possibility of absolute givenness*."[93] "Real being," in contrast, "understood as being for an Ego," is never posited as necessary and in that sense is not necessary for the being of consciousness or mental living.

But before drawing these and perhaps even more radical consequences from Husserl's statements, a more basic question has to be asked, namely what is the basis for Husserl's distinction between immediate presentation and positionality in the first place? To be sure, the distinction, in a way, is not original with Husserl; it is rather the way in which he inherits the problem of existence in modern philosophy. As it were, philosophically, the distinction is ready-made for him. What he does, however, it seems to me, is to establish a phenomenological basis for making that distinction without interpreting away the paradox that emerges in the will to return to the "things themselves." The phenomenological basis, in turn, allows for a radical departure from the significance attached to the distinction before him.

What, then, is the phenomenological basis for the distinction? The distinction in question is one made in the realm of objectivation or thematization, i.e., relative to objectivating or thematizing acts.[94] In *Ideas*, First Book, the chief sort of positionality (although not the only sort) considered is that of "doxic positionality" which includes all believings, disbelievings in something either with simple certainty or with some degree of uncertainty; disbelievings, doubtings, and the like are said to be modifications of believing with simple certainty.[95] Consider an example of something presented as transcendently public: I perceive the ashtray in front of me and believe in it as existing. I may proceed to objectivate the percept, "Ashtray--blue." However, I may instead objectivate the belief characteristic, "Ashtray--existent." Relative to the believing in the ashtray with simple certainty, "existent" is not part of the objective sense intended to.[96] And if, in a further act of objectivating, I proceed to explicate, for instance, the shape and color of the ashtray, this still is not an objectivating, let alone a seizing upon, the ontic characteristic the ashtray is posited as having in the perceiving of it. That occurs only in a new and further act of objectivating in which I make thematic the ashtray seen and believed in as existing.[97] As a consequence, it would seem to make little if any sense to say that being or existence, whether "real" or "absolute," "establishes the possibility" of givenness, whether "presumptive" or "absolute." What meaning, then, can be attached to Husserl's statements?

In the first place, it is necessary to note the context in which Husserl speaks of "absolute" and "real" being or existence. The context is expressed by the title of section 47 of *Ideas*, First Book: "The Natural World as Correlate of Consciousness;" that is to say, what is to be objectivated, made thematic, are those positings in which the naturalness of transcendental mental living in the world consists. In the second place, the meaning of statements expressing that objectivation must be understood as statements about transcendental mental living as transcendentally purified from a specific positing that confers on mental living its characteristic of being an event related to another event in the world (e.g., a percept) in world-time. What remains in consquence of that purification is the objectivational or thematic distinction between immediate presentation on the one hand (specifically, between immediate presentation of something immanently public and something transcendently public) and, on the other hand, "absolute" and "presumptive" positing. It is a contrast in positing objectivated once it is seen methodologically that refraining from accepting the sense, "being in the world of which my mental living apperceives itself as a non-self-sufficient component" is not equivalent to refraining from accepting the sense, "actuality of my mental living for which the world is"--thus the second theme of the transcendental phenomenological epoché and reduction.[98] In the third place, the (at least necessary) motive for this objectivation is to discover how the examination of mental living can be relevant for an examination of the nature of what is, which is to say, of being of "two minds" about myself and my world. The contrast between "absolute" and "presumptive" positing is made while, as meditating phenomenological on-looker, I live simultaneously in the transcendental unnatural and the transcendental natural attitudes.[99] Accordingly, and in consequence of revision of Husserl's methodological procedure, the second phenomenological theme was expressed earlier as "My own transcendental mental living believing in itself as being in the world even if nothing else in the world exists."[100]

Consistent with the results of the investigations carried out in this second part, I understand the second phenomenological theme to signify that even if there is no necessity that my transcendental natural mental living be posited as having a place in the fully constituted, real objective world, it nonetheless posits itself in primary passivity with simple but apodictic certainty, i.e., believes in itself immediately presented with the noematic sense, "immanently public." Although this is more a Husserlian than Husserl's way of expressing the situation, it still would seem to be consistent with the critical-epistemological understanding sought in the will to return to the "things themselves." By "annihilation of the world" I understand the unbuilding of the oriented constitution of the real, objective world to those own inherent quasi-objective and pre-objective "worlds" the intendings to which and positings of which inseparably belong to the proper essence of intending to and positing of the real, objective world as its really immanent part. Thus, to use a later literary expression of Husserl's,[101] "Every existent (that ever had, or can have, sense for us) stands, as intentionally constituted, in a hierarchy of intentional functions <pertaining to the oriented constituting of the real, objective world> and of existents already constituted intentionally, which, for their part, are involved in intentional functions for a new constituting of being" so that, as Husserl says in the *Cartesian Meditations*,[102] "the primordial enters, with a new stratum of sense, into the secondarily constituted world...necessarily given as a horizon of being that is accessible from the primordial and is discoverable in a particular order."

In *Formal and Transcendental Logic* Husserl proceeds to note that not only is every existent relative in "any usual sense, but *every* existent is *relative to transcendental subjectivity*" which "exists, in itself and for itself, in a hierarchical order corresponding to the <oriented> constitution that leads to the different levels of transcendental intersubjectivity...constituted (in me, and hence relatively to me) as a plurality of 'egos'--each of whom is legitimately accepted as intentionally related to the *same* intersubjectivity along with me..."[103] Such positing, as noted before,[104] is equally "absolute" and "relative" so that with Husserl we may say that this intersubjectivity (or interpublicity) *also exists* "with the mode of existence that belongs to something 'absolute'." Something immediately presented as "absolute" can only mean, then, something "existent in the form, an intentional life--which, no matter what else it may be intrinsically conscious of, is, at the same time, consciousness of itself," hence the essence of which "*includes the possibility of 'self-examination'*" or, differently expressed, which includes the possibility of living at the same time in the transcendental natural and the transcendental unnatural attitude. Beyond this, whatever else "absolute being or existence" might mean exceeds the critical-epistemological boundaries of transcendental phenomenology. There is no need, then, to be led off with Landgrebe into metaphysical forests or to find ourselves, like Boehm, led by Husserl into the theological underbrush of "transcendental subjectivity."[105] It does make sense, however, to ask about the extent and nature of those boundaries. The final sections of this chapter, and this book, will be devoted to taking stock of those boundaries.

97. THE LIMIT-PHENOMENA AND BOUNDARIES OF TRANSCENDENTAL PHENOMENOLOGY

A place has been reached in this investigation of Space, Time and Other where it is necessary to draw a number of phenomenological-philosophical conclusions. The conclusions serve as much as the end of this investigation as they do the beginning of new investigations, such as of the way in which primordial quasi-objective space and time acquire the appearance of the fully constituted real, objective world and then of how the spatio-temporal dimension of the real, objective world acquires the "appearance" of space and time determined and further determinable by natural scientific thinking and experiencing.[106] The critical-epistemological understanding emphasized in the Introduction to Part One as the goal of the will to return to the things themselves by now has acquired a much broader scope: by attempting an increasingly self-conscious exercise of transcendental phenomenological epoché an equally increasingly concrete form has been given to the will to return to the things themselves as striving toward the "presumptive end, genuine science" of Space, Time and Other (section 4). To speak of "genuine science" as a "presumptive end" or goal opens the doors to considerable interpretational ambiguity although it certainly would seem consonate with the attempt to articulate the feeling for the problem of genesis as the perennial task of philosophy to reclaim its "true origins" in the irreal actuality and possibility of transcendental mental living in the natural attitude. Despite the fact that Husserl appears more often at odds with Kant than in agreement with him, there is nevertheless a Kantian nuance present in this attempt to explore the "territory of pure understanding," to survey and measure out its extent, to give everything its rightful place and proper due. Yet as Kant also observed, the "territory of the understanding is an island, enclosed by nature itself within unalterable limits. It is the land of truth--enchanting name!--surrounded by a wide and stormy ocean, the native home of illusion, where many a fog bank and many a swiftly melting iceberg give the deceptive appearance of farther shores, deluding the adventurous seafarer ever anew with empty hopes, and engaging him in enterprises which he can never abandon yet is unable to carry to completion."[107] Just so "phenomenology is presented as a *beginning* science" the procedure of which is "that of an explorer journeying through an unknown part of the world, and carefully describing what is presented along his unbeaten paths, which will not always be the shortest. Such an explorer can rightfully be filled with the sure confidence that he gives utterance to what, at the time and under the circumstances, *must* be said--something which, because it is the faithful expression of something seen, will always retain its value--even though new explorations will require new descriptions with manifold improvements."[108]

It is clear, then, that whether the territory of pure understanding be the island of truth, or of the Blessed, or perchance even of Cipango beyond the pillars of Hercules and always afloat, the voyage of discovery can no more be completed than it can be abandoned. "There is no 'royal road' into phenomenology and therefore none into philosophy. There is only *one* road prescribed by phenomenology's own essence."[109] But if this Kantian nuance of our concluding reflections serves as a caveat for the investigations of Space, Time and Other, it can also provide a clue to understanding a novel aspect of Husserl's thought which we have, in part, tried to confirm. With Husserl the attempt has been made to distinguish the objectivity of the surrounding world of culture, as he called it, from the "absolutely unconditional accessibility to everyone which belongs essentially to the constitutional sense of Nature....Everyone, as a matter of a priori necessity, lives in the same Nature, a Nature moreover that, with the necessary communalization of his life and the lives of others, he has fashioned into a cultural world in his individual and communalized living and doing."[110] The clarification of the "absolutely unconditional accessibility" rests, Husserl says, "on a basis given us beforehand in the transcendental attitude: a stuctural articulation of the primordial sphere, in which we already found a world, a primordial one"[111] but likewise a quasi-objective phantom world and, in turn, pre-objective worlds.

And, at the end of the Fifth Cartesian Meditation, Husserl speaks of this clarification as comprising a "'transcendental aesthetics' in a very much broadened sense."[112]

Although the title, "transcendental aesthetics," obviously has a Kantian nuance because it includes the intuition of time and of space-time (but in Husserl's sense of the word, "intuition"), the corresponding title which one would immediately expect, "transcendental analytics," would be inconsistent with any such Kantian nuance. The reason is that for Husserl a *phenomenological* transcendental analytics comprises "the upper stories of the constitutional Apriori, which pertain to the Objective world itself and the multiplicities constituting it (at the highest level, the Apriori pertaining to the 'idealizing' and theorizing acts that ultimately constitute scientific Nature and the scientific world)."[113] As a consequence, the transcendental aesthetics and analytics are related by a complexus of interlocking structures of oriented constituting, some of which are assigned to the aesthetics while others belong to the analytics. The "cut-off point," so to speak, for the "distribution" of structures of the real, objective world is the experience of someone else which, in contrast to Husserl, must be assigned to the "upper story" of transcendental aesthetics rather than to the "first story above our 'transcendental aesthetics'."[114] Thus a transcendental phenomenological aesthetics renders a critical-epistemological account of the most basic "primitive" or quasi-objectivities constituted in an original way in primary passivity of transcendental mental living in the natural attitude. Correspondingly, transcendental phenomenological analytics renders an account of those objectivities constituted in an original way in "thinking capable of the significational function" (see above, section 10f.).[115] The novelty of this Husserlian view lies in the transcendental aesthetics being the theory basic to the transcendental analytics no matter what the extent to which "idealization" of the real, objective world is carried out, regardless of the degree of progressive formalization and symbolization on the foundation of which the objectivities of the natural sciences are produced.

But the relationship of the aesthetics and analytics is not only that of fundamentality. Indeed, although transcendental aesthetics and analytics are about mutually irreducible structures of transcendental mental living in the natural attitude they are also mutually implicative and correlative structures, the one providing the warrant for the other the validity of which, to be sure, transcends and exceeds it as something "secondarily" constituted.[116] That is to say, if our frame of reference for oriented-constitutional inquiry is the procedure of unbuilding and building-up analysis peculiar to transcendental phenomenological epoché (above, section 14f.), then the inquiry arrives, I believe, at a definite limit or boundary prescribed in advance, the seizing upon and objectivating of which require the adoption of a reflective "turn" or shift to the "intrinsically first criticism of cognition, the one in which all others are rooted...transcendental self-criticism on the part of *phenomenological cognition itself*."[117] To make that reflective turn it is necessary, first, to mark off the limit-phenomena of oriented-constitutional inquiry and, second, to indicate the way that the limit-phenomena are essential to transcendental phenomenological aesthetics and analytics. It will then be possible to provide a precise definition of the aesthetics and analytics which together form the theory of the "perception" of Space, Time and Other. The development of the reflective turn itself will have to await another occasion.

98. THE FIRST LIMIT-PHENOMENON: SOMEONE ELSE

Our starting point and frame of reference of these investigations has been and remains our own living present (see above, section 62) about which we have tried to say something true regardless of how it has been, or may have been, or will be interpreted scientifically. We disregard until further notice any interpretation of the real, objective world by scientific thinking and experiencing, and focused on a subscientific, subphilosophic facet of experience included in the perception of the spatio-temporal form of the real, objective world (see above, sections 17, 20, 36). In the language of Alfred Schütz, we retrieve a taken-for-granted, common-sense aspect of mundane reality presented and appresented as subscientifically categorialized and conceptualized, penetrated by intersubjective, socially distributed and derived meanings conferred on it by members of the community living in that world constituted in primary passivity (see above, sections 18f., 79ff.). It is in this light that the "perception of other minds" is a limit-phenomenon, and that in a two-fold way.

In the first place, unbuilding to quasi-objective phantom spaces and to prespaces discloses space perceptions that do not include in their proper essence perception of anything other than "my world," than the "world for me," in its spreadoutness but not its extensity--thus perceptions which do not necessarily include in their proper essence perception of real individual things (section 73f.). In other words, what is disclosed is the outer-most limits of "my world" of which "my subjectivity" is constitutive in primary passivity. In the second place, building-up analysis discloses that "my world" cannot be objectivated without objectivating my mental living as immediately presented and immanently correlative to other affairs "for me" immediately presented and transcendent, even if presumptive, vague and obscure. The pairing of these presentational senses founds not only syntheses of similarity and difference but also of incompatibility of immanences. It is the latter, as we said (section 78f.), that motivates the constituting of quasi-objective phantom space and spreads of phantom unities and phantom appearances as something secondarily constituted, as an "appearance" of the quasi-objective primordial world and spatial things, i.e, a space perception of "my world" in primary passivity that indeed includes in its proper essence that it be a world experienceable by someone else. It is a constitution that is equally the limit at which it is essential to the space perception of "my world" that it be space perception of "my world" "for others," for someone else; the mutual incompatibility of immanences may be said to "found" the "absolutely unconditional accessibility to everyone which belongs essentially to the constitutional sense of Nature" thus bestowing the sense, "objectivity," to the real world in the first place. Intending to and positing the identically same spatially extended primordial world "for me" and "for others" is, in other words, intending to and positing the primordial world "in itself" of which "we" intend to and posit ourselves and each other as non-self-sufficient components. Or, differently expressed, the outermost limit of "our world" as essential to the perception of "my" world is world in and for itself.

This outermost limit is what we called before a "full fledged" social relationship or consociation (section 80) included in the proper essence of space perception. It is "full fledged" in the sense that each member of the relationship intends to and posits the spatial (and temporally) extended primordial world as identically the same world intended to and posited by others.[118] And it is "full fledged" in the additional sense that each member is self-intentive and self-positing as identically the same one whom the other also intends to and posits.[119] As a consequence, the consociation essential to social relationships consists of intersubjectively (or inter-publicly) intended to and posited identities, immediately presented as transcendent to the mental living of those standing in the consociation. Precisely such a consociation, essential to space perception, is the basis in primary passivity for the taken-for-granted, common-sense aspect of mundane reality.

It is now necessary to recast this first and two-fold limit-phenomenon in the rubic of transcendental phenomenological *philosophy*. Exercising transcendental phenomenological epoché, and limiting the range of the transcendental phenomenological reduction to objectivating my own world-positing, transcendental mental living as current and actual, I see in reflection that this mental living is, first of all, immediately presented as having the sense, "one stream of mental living among others." And this noematic sense is implicitly posited by my own living with the same simple certainty that characterizes the positing and intending by other transcendental mental livings in the natural attitude. The specifically *philosophic significance* of this is that I, transcendental on-looker, can make explicit the noematic sense, "one stream of transcendental mental living among others," by considering the fact that in order to be a mundane stream of mental living is to be an immanently public stream of mental living and that any other mental living would, like mine, exercise transcendental phenomenological epoché and objectivate its world-positing transcendently inward mental living. And this constituting of a plurality of mental livings possessing the world in common is presupposed by any attempt to communicate a transcendental phenomenological conviction or opinion. In so communicating, however, I, as transcendental on-looker, address myself not as a human being to other human beings, but as transcendental phenomenologizing mental living to other transcendental phenomenologizing mental livings whom I mean as at least possibly exercising the transcendental phenomenological epoché and, as on-lookers, each observing its own transcendental stream of mental living in the natural attitude. The two-fold limit-phenomenon is that constitution, then, in which, each transcendental phenomenologizing mental living means itself as identically the same one intended to and addressed by other transcendental phenomenologizing mental livings on the basis of their world-intending and world-positing. *Strictly speaking*, then, *trasncendental intersubjectivity*, as Husserl calls it, is transcendental phenomenologizing intersubjectivity and it is that in which the real, objective world, as real world for everyone, is constituted.

In this respect *the perception of space, time and other may be defined as the full-fledged social relationship which is a consociation of transcendental, phenomenologizing mental livings each objectivating itself as mundanized, transcendental streams of mental living immanently public and publicly-transcendently inward in the natural attitude.*

In terms of the oriented constitution of the real, objective world in primary passivity this signifies that I, as transcendental on-looker, must make explicit the noematic sense, "one stream of transcendental mental living among others," and then, in the course of unbuilding procedures, proceed to exercise a further transcendental phenomenological epoché, i.e., extend the range of the initial epoché with respect to my intending to and positing of that noematic sense in consequence of which I can objectivate my own transcendental mental living as *solus ipse*. As a result, I now uncover the noematic sense, "mundane plurality of transcendental mental livings in the natural attitude as *mundanized* plurality of transcendental mental livings." In the strict sense, this is the transcendental community (Husserl: "*Wirgemeinschaft*") and presupposed by all other socialities. In exercising the transcendental phenomenological epoché, to be sure, the noematic sense, "mundanized plurality of immanently and transcendently public transcendental mental livings," *and* the noematic sense, "my own transcendental immanently public mental living as a member of the plurality of transcendental, transcendent mental livings," are parenthesized. Nonetheless, as noematic sense of my own "solipcistic" intending to and positing mental living, they are salient for my transcendental phenomenologizing mental living, "calling" to be seized upon as transcendental "clues" to *their* "solipcistic" constituting (or, as Husserl would say, their "egological" constituting).

99. THE SECOND LIMIT-PHENOMENON: MATERIAL AND FORMAL UNIVERSALS. TRANSCENDENTAL AESTHETICS AND ANALYTICS REDEFINED

The second limit-phenomenon of oriented constitution involves a different dimension of mental living to which reference has already been made in discussing Husserl's own thought as a transcendental empirical example of a motive for the self-conscious exercise of transcendental epoché (section 9), when discussing several examples drawn from the domain of actional perception in order to clarify some aspects of ideation (section 33), and in establishing the point of departure for the first step of unbuilding analysis of the real, objective world other cases of actional mental life-processes, such as collectings, were considered (section 17, b). Implicit in those discussions was the second limit-phenomenon, and to make it explicit now it is worth while developing a composite of the examples already considered.

Some pencils are seen on the table in front of me; looking at them the visual field exhibits a certain order and structurization--presented are center, foreground and background with respect to which elongated parts are prominent and, so to speak, "call" for my advertence to them. I may or may not heed that "call;" suppose that I do not and that, moreover, I am engaged in listening to the voices coming from the next room. Notwithstanding, all the while I am listening there proceeds in primary passivity the seeing of the pencils salient in the visual field. Being engaged in the listening does not eliminate the seeing nor its objects, nor need it necessarily alter the essential organization of the visual field. Were I now to advert to the pencils seen all the while, becoming actionally engaged in the seeing of them rather than in the listening, the seeing is converted into an actional seeing on the basis of which I seize upon and perhaps proceed to objectivate the seen pencils.

Adverting to what I see in front of me is, as a consequence, a condition for making thematic what I see. If it occurs for the "first time," then for the "first time" there is presented and posited, as it itself, a "This, for me;" more broadly stated: a "something" conceptualized, a "This, there in the visual field." In other words, "This" is a noematic sense presented in and through the act of adverting to, and explicitly objectivating, the visual saliency rather than in and through the act of seeing. Without adverting to that particular saliency in the visual field, thus without converting the primary passive seeing into an actional one, a seeing in which *I* am engaged, no form, "This," would be realized, *scl.* be presented as it itself, "in person."[120] Having adverted to, objectivated, something such as the pencil, I can proceed to do the same with something else, namely, another pencil: I see, seize upon and objectivate "This," then "That." Were this all that I did, however, there still would not be the presentation of two objects--one, then another thing, i.e., a mere succession of things would be presented. For there to be a presentation of a pair, or, in general, of a plurality, it is necessary to seize upon and objectivate "This" and while hold "This" in grip, so to speak, proceed to seize upon "That, different from This." Simply because "This" and "That" are presented does not signify that here is an intending to a pair. The form, "Pair," or the form, "Plurality," is realized or conferred by virtue of an actional collecting. On the basis of having objectivated first "This," and "That" from "This," it is possible to return to "This" and make it a primary thematic object; I can then proceed collectively, for instance from "This to That" whereby there is now presented in an original way the Collection ["This and That"].

The mental life-process in which the collection is constituted as it, itself, namely the act of collecting, is not *ipso facto* an objectivating of the Collection; for that objectivation another and further act is required. In other words, produced but not seized upon and objectivated in the step-by-step collecting, the Collection itself is salient and "calls" for advertence to it. Adverting to it, seizing upon and objectivating it, actualizes the *form*, "Collection." It is the act in which the form, Collection, is constituted in an original way. The phenomenological result is that, in the case of the constituting of Forms, such as "This and That," they must already be there to be objectivated. The objectivating of the collection-form in the example presupposes actionality of mental living antecendent to the objectivating, namely those actional mental life-processes in which the (particular) collection-form is constituted in an original way. In general, "forms," or "formal universals" or "formal essences" are constituted in consequence of antecendent actionality. But this is not the case with all "universal" affairs. To be sure, "material universals" or "essences" are, like "formal universals," ideal existents (in Husserl's sense, to be sure) and affairs with discernible syntaxis. But are such "material universals" such as Mental Living,

Space, Time objectivated only in consequence of antecendent actionality? It certainly seems to me not to be the case, and that here we have an objectivational distinction that establishes the second limit-phenomenon of oriented constitution.

The most convenient and simplest way to make explicit the objectivational distinction is to consider an elementary case of "pairing" (above, section 10), examined with respect to what is included in it regardless of whether or not I am engaged in the pairing intentionality. For example, suppose that I see clearly and distinctly a roughly spherical red thing, *X*, and a roughly cubical red thing, *Y*; regardless of any actionality, each thing can be seen as it, itself meant throughout the seeing as self-identical and numerically distinct from the other; moreover, still without presupposing any antecendent actionality, the things can be perceived as paired so that "automatically" the perceptual sense of each is transferred to the other: red and round are transferred to, or "overlay," red and cubical; conversely, red and cubical are transferred to red and round.[121] In so far as it is consistent with the perceptual sense, the transferred sense is imputed to the thing to which it is transferred; in so far as it is inconsistent, the transferred perceptual sense is cancelled out, resulting in what may be called "negative fulfillment" of the sense. Accordingly, in so far as the perceptual sense, "red," is concerned, there is presentation of a pair of things, *X* and *Y*, as similars. they are "similars" for the unitary visual intending to "similars" to the extent that there is seeing of them as paired owing to agreement of presented and transferred "reds." It is important to emphasize with Husserl, however, that the two moments of "red" are not constituted in an identifying synthesis as one individual; rather, as the example indicates, something non-individual is co-constituted: the non-individual Redness exemplified in each of the moments of "red" belonging to *X and Y*.

Regardless, then, of whether or not I am engaged in the seeings of cube and sphere, regardless of whether or not I make them thematic as whole or in part, something non-individual is already *implicity* intended to in the seeing of them *on the basis of agreement or "positive fulfillment" of presented and transferred perceptual and apperceptual senses*. Owing to the agreement, the non-individual (the "material universal" or "essence") Redness stands out, "calls" for advertence to it so that it may be converted into an actional seeing on the basis of which I can proceed to seize upon and objectivate the "material universal," Redness, thus making it an object on which the form, e.g., "Something at Large," is conferred. Not only the two individual physical things, *X* and *Y*, but also the "material universal" are, in this example, already meant and believed in with simple certainty regardless of whether or not there is any actionality, any act in the pregnant sense, which confers the Form in question. There are, of course, those cases of mental life-processes which cannot be converted into actional processes, which remain processes proceeding in primary passivity (above, sections 10, 17, 21). But no matter, for universally it is true of every mental life-process, as we have noted before (above, sections 62, 84ff.), that its present extent is continuously a retro-tending to itself in primary passivity, all the while protending to still more of itself to come in primary passivity. *In primary passivity*, the present extent of any mental life-process (be it actional or not, be it capable of being converted into an actional process), continuously identifies itself as the one extent previously protended to and subsequently retrotended to. This signifies that more fundamental than any actionality there is constituted, so to speak, a "point of view" with respect to which two extents of seeing (or of collecting, or of holding in grip, or of judging, and the like) are already meant and intended to as alike; owing to "overlaying" of presented and transferred senses, Seeing, eventually Perception, finally Mental Life-process at large, Other Mind, Space and Time itself, are constituted not in ideational and objectivating acts, but rather already in primary passivity. Thus in the case, for instance, of the perception of other minds, in primary passivity there is complete cancellation, negative fulfillment, of the presented and transferred senses, "previously protended to and subsequently retrotended to:" simultaneously, however, there is partial positive fulfillment of the mutually transferred, appresented sense, e.g., "seeing of the same things that I now see." Thus however we may elaborate on the earlier analysis of the perception of other minds (sections 78ff), it is clear, I believe, that perception does not, as Husserl asserts, belong "in the first story above our 'transcendental aesthetics'," that is, the story which transcends oriented constitution in primary passivity.

In short, "material universals" of whatever specificity such as Mental Living or Mental Living of Someone Else, or Physical Thing and Physical-thing Properties, Spatial Spread and Spatial Extension, or Time, are already constituted in primary passivity regardless of whether we are, or can be, engaged e.g., in the seeing of the same things in the same, public world. The second limit-phenomenon lies, then, in the mutual transfer of sense and assimilative syntheses that proceed in primary passivity. To express the limit with respect to perception of space, time and other, where only identifying syntheses are constituted in primary passivity, no "material universals" are constituted, nothing whatever is constituted "as of a kind." (See above, sections 44, 48, 52, 60, 65, 75ff.) Examples are what we called prespace-time in the narrowest meaning of the term, "prespace-time."

This second objectivational distinction has far-reaching consequences because the constituting of "material universals" proves to be quite distinct from that of "formal universals" which always require an antecendent actionality (on the basis of which, for example, formal logic, deductive systems such as "geometries" and number theories are developed). In contrast, "material universals" are constituted in an original way in assimilative and differentiative syntheses in primary passivity (mental living, space and time, social relationships and thus cultural formations of whatever sort.) On this Husserlian (but not Husserl's) view, then, a transcendental phenomenological aesthetics cannot be restricted just to real individuals constituted in primary passivity and apart from the cultural surrounding world; nor can a transcendental analytics be restricted just to actional mental life-processes such as thinkings, theorizings, conceptualizings, and their "products."

Instead a *transcendental phenomenological aesthetics* minimally must comprise the oriented constitution of "material universals" in an original way in primary passive assimilative syntheses, along with the seizing upon and objectivating of "material universals," as well as the oriented constituting of those primary passive identifying syntheses included in all assimilative syntheses in consequence of which "material universals" are constituted. Correspondingly, a transcendental phenomenological analytics minimally must comprise all the oriented constitution of "formal universals," seizing upon and objectivating of "material universals" which acquire the appearance of something "primordial" entering, with a new stratum of sense, into the secondarily constituted "formal" world and in a particular order.

100. CONCLUSION. PHILOSOPHY AND PHENOMENOLOGY

As transcendental aesthetics and analytics, phenomenological philosophy purports a comprehensive account of the oriented constitution of the protentive-retrotentive synthetic temporalization of the living, flowing intersubjective present of which the naturalness of transcendental mental living in the natural attitude consists. Not all aspects of the aesthetics have been explored with the equal emphasis they deserve. For example, both unbuilding and building-up analysis of the spatio-temporality of the real, objective world constituted in primary passivity disclose that not only are there quasi-objective perceivings and positings and sensings going on in primary passivity, but also as well willings, desirings, preferrings and valuings. These have been taken account of in so far as they figure into "motives" for building up various stages of oriented constitution (see above, sections 53ff., 65f., 71f., 75f.). But their philosophic significance is much more far-reaching for it seems that values themselves of whatever sort are constituted in primary passive "feelings," that value-hierarchies are constituted in primary passive preferrings (most primitively, e.g., of a change in the mode of presentation indirectly controlled for the sake of actualizing pleasant or unpleasant sensa correlative to controlling one or another sequence of kinaesthetic flow-processes), and that each prespatio-temporal and quasi-objective spatio-temporal storey of oriented constitution of the real, objective world includes strivings and willings peculiar to them. In short, the very milieu-worlds (Scheler) which mark and organize the real, objective world are already constituted in primary passivity. Consideration must be taken, too, of the positing that belongs to strivings, feelings, preferrings, and the like, as well as of the fact that such mental-life processes, unlike processes such as perceivings, are non-objectivating.

Thus even though the phenomenological foundations have been laid for establishing the role of striving, feelings and preferrings in the perception of space, time and other, their phenomenological analysis requires a separate study that includes still other dimensions or oriented constitution such as language in the broadest sense, cultural artifacts and social and political roles and institutions. As presented here, then, the transcendental aesthetics remains incomplete, only a fragment. Finally, the oriented constitution that figures into the transcendental analytics has been identified (sections 17, 35, 38, 41) but not developed.

By way of conclusion to the investigations traced in the preceding chapters I wish to take a first step in making explicit a *philosophical* dimension of inquiry foreshadowed by the attempt at the beginning of this book to recover the meaning of Husserl's slogan, the "will to return to the things themselves." More particularly, the dimension I have in mind is what Husserl called the "intrinsically first criticism of cognition," a self-criticism on the part of phenomenologizing itself required by the will to return to the "things themselves." Still more particularly, it is a self-criticism of that phenomenologizing which produced the transcendental aesthetics in the first place. Just here a problem arises for to seize upon and objectivate the phenomenologizing itself is to take up a stand at the other side of the limit-phenomena that circumscribe the naturalness of transcendentally pure mental living in the real, objective world. Indeed, that intrinsically first criticism does not itself seem to be a matter of oriented constitution at all. It would seem to be an inquiry *constructive rather than constitutive* in nature, more like a "transcendental dialectics" in a new and broadened meaning. The most immediate task of such intrinsically first criticism is to unfold phenomenologizing not only with respect to the work it has accomplished in producing the transcendental aesthetics, and by extension the transcendental analytics, but also with respect to its method, transcendental phenomenlogical epoché and reduction. In other words, what begins to take shape now is something akin to a "transcendental doctrine of elements." Aside from the Kantian echo of this title, of what does a transcendental doctrine of elements consist, what is its "constructive" character?

With Husserl we can note that the oriented-constitutional investigations "take on a painful and yet unavoidable relativity, a *provisionalness*, instead of the definitiveness for which we were striving: each investigation, *at its own level*, overcomes some naiveté or other, but is still accompanied by the naiveté of its level--which must then be overcome in turn by more penetrating investigations of origins[122] leading to a "formal ontology" of any possible living, flowing present of mental living in the world in the natural attitude, and that ontology eventually turns out to be a part of an "ontology" that relates to all that exists in any sense whatever. It is an important task, Husserl says, to see how these two "apriori sciences of whatever exists as worldly," are grounded in the "original sources in trancendental subjectivity:" "The original grounding of all the sciences, and of the formal ontologies of both sorts exercising in their behalf the function of a theory of science, the normative function," Husserl says, "*gives all of them unity, as branches of a constituted production from the one transcendental subjectivity.*"[123] Interpretation especially of the italicized part of this last sentence risks opening a pandora's box. The most conservative and least risky reading of Husserl's statement, it seems to me, is to note that the grounding of those ontologies is distinct in each case--in the one case, the grounding is in primary passivity; in the other case, the grounding is in the actionality of generalization and formalization. That means that the ontologies cut across the distinction between transcendental aesthetics and analytics, thus that their unity comprises transcendental phenomenology as a whole, that is, with respect to its limit-phenomena. These statements have a direct bearing on the nature of phenomenological philosophy because they are nothing other than expressions of phenomenologizing mental living theorizing in the transcendental-phenomenological unnatural attitude by making thematic to itself the work it had performed "anonymously."

"When I proceed thus," Husserl says, "reflecting and fixing my discoveries, the theoretical formations of transcendental phenomenology and this phenomenology itself, as an infinite open unity of science, are generated from myself by virtue of sources belonging to my own passivity (association) and activity, at first with a sort of naivete."[124] I speak of the "anonymity" of phenomenologizing because the purport of the transcendental aesthetics is that phenomenologizing mental living is already present in the world by being intentive to the world in primary passivity no matter whatever else it is or may be intentive to. But there is another and equally important meaning of "anonymity," for when I make thematic my phenomenologizing, when I seize upon it and express it in formal ontological statements that purport universality and validity, I do so *always and only with respect to* exercise of transcendental phenomenological epoché circumscribed by its limit-phenomena. And I do this, in other words, as disinterested transcendental on-looker so that, although as on-looker I am not constitutive of the real, objective world in primary passivity, except in so far as my mental living exemplifies any mental living whatever, I allow that oriented constitution to proceed even though I exclude the accepting of the real, objective world as existing correlative to parenthesizing the real, objective world posited (above, section 6). This signifies that, *as on-looker, I am defined with respect to the delimited or circumscribed whole of what I allow to proceed just as what proceeds in common with any actual or possible on-looker.* Another way to say the same thing is that *the time and the space-time of being in the world constituted in primary passivity delimits the whole of that toward which I am, as on-looker, disinterested, "anonymous."* The disinterestedness, the "anonymity," when made thematic comprise the philosophic-theoretic sight required by the "things themselves" to which I strive to return so as to produce critical-epistemological understanding.[125]

Accordingly, while philosophy, critical-epistemological understanding, may be said to include transcendental phenomenology, the converse is not the case. The moment I make thematic the performance of my phenomenologizing in common with other actual and possible phenomenologizings I give *philosophical* exposition to the "feeling for the problem of genesis" by taking up "*all the problems of accidental factualness, of death, of fate,* of the possibility of a '*genuine*' *human life* demanded as 'meaningful' in a particular sense--among them, therefore, the problem of the 'meaning' of history....We can say that they are the *ethico-religious* problems, but stated in the realm where everything that can have a possible sense for us must be stated."[126] In short, I must be disinterested in response to the "things themselves" to be interested in the motives and problems they raise concerning my "beginning" and "end," my life and death, my fate and history. *Philosophy is then constructive* in the sense that *its* motives and problems are built upon that towards which it is and must be disinterested, namely the whole-establishing limits of oriented constitution of the real, objective world in primary passivity, *that is, the perception of space, time and other.* It is not so much that transcendental phenomenology in the exercise of its method resolves philosophical problems, but rather that transcendental phenomenology makes the problem possible in the first place. One of the most important implications here is that while phenomenological propositions, such as those expressing the aesthetics and analytics, entail corresponding ones in philosophy, there are philosophical ones that have no correspondence in transcendental phenomenology. For example, those propositions that go to make up the transcendental doctrine of method and the transcendental doctrine of elements have no correspondence in phenomenology. Thus seizing upon and making normative the rules that establish the correctness or incorrectness of the various steps of unbuilding and building up reductions are "ratified" not in phenomenology but in philosophy. Moreover, not every problem in phenomenology is *ipso facto* a philosophical one. For example, the problems of prespatio-temporality are phenomenological, not philosophical problems. Yet to say this still does not quite express the relationship and difference of phenomenology and philosophy.

To come closer to the relationship and difference of phenomenology and philosophy it is probably best to jettison once and for all the Kantian nuance and nostalgia which have been hovering in and around this discussion. This is because, unlike Kant in the *Critique of Pure Reason* (B 735f.), not only is there a "supply of materials...just sufficiently commodius for our business on the level of experience," but we need not, as does Kant, set aside the concern with the "materials" to opt for the architectonics of the "project of pure reason." Indeed, not only is there some material left for building critical-epistemological understanding, but as a matter of fact as well as essence, all of the "materials" with which we started are left: interpublic mental living temporalized in the real, objective world, the primordially constituted "primal logos" entering into the secondarily constituted "everything else that is 'logical'." Not only does the "material" universality of mental living shape, in Kant's words, the "determination of the formal conditions of a complete system of pure reason," but it is also there to be "formed" and "formed" to start with. In short, the situation at issue, the relationship and difference of phenomenology and philosophy, is also a paradox: phenomenologizing mental living, even in primary passivity, is both a constituted moment of oriented constitution and the "origin" of that constitution, it is both relationship and difference.

Here it is not possible to further develop the nature of phenomenological philosophy. This study of the perception of space, time and other began with a metaphor, namely that of the spaces between the pickets of a fence.[127] It does not seem amiss to end this study by expressing the relationship and difference between phenomenology and philosophy with another metaphor in which phenomenologizing mental living may be likened to a story, and that same living as philosophizing may be likened to the narrator of the story. To express the metaphor we may refer to Thomas Mann's novel, *Joseph in Egypt*, for there Mann is confronted with the problem of the relation of the narrator of a story to the story itself. Should the narrator appear in the story at all, or is the telling of the story sufficient unto itself? Should the narrator be at one with the story rather than outside it, manipulating and calculating it? "But how is it with God, whom Abram brought into being and recognized? He is in the fire but He is not the fire. Thus He is at once in it and outside it. Indeed, it is one thing to be a thing, quite another to observe it. And yet there are planes and spheres where both happen at once: the narrator is in the story, yet is not the story; he is its scene but it is not his, since he is also outside it and by a turn of his nature puts himself in the position of dealing with it."[128]

NOTES

(Complete references to works cited are given in the main Bibliography)

INTRODUCTION, PART ONE

1. Karl Schuhmann, *Husserl Chronik. Denk- und Lebensweg Edmund Husserls*, pp. 32, 41.
2. Published in *Philos. Monatsheft*, 30 (1894), pp. 159-191; reprinted in Edmund Husserl, *Aufsätze und Rezensionen (1890-1910)* (=*Husserliana* XXII), pp. 92-123; English translation by R. Hudson and P. McCormick, Husserl, *Shorter Works*, pp. 126-141.
3. Schuhmann, p. 41. Some of this chapter was rewritten and included in *Ideas*, First Book, Part I. See F. Kersten, "The Originality of Gurwitsch's Theory of Intentionality," pp. 19-28.
4. Schuhmann, p. 46. See also Edmund Husserl, *Logische Untersuchungen, Erster Band: Prolegomena zur reinen Logik*, Vierte Auflage (hereinafter LU I[2]), p. XII; English translation by J.N. Findlay, Vol. 1, pp. 46f.
5. See Edmund Husserl, *Logische Untersuchungen, Erster Theil: Prolegomena zur reinen Logik* (Halle, 1900; hereinafter LUI[1]), "Vorwort." For the textual history of the composition of the *Prolegomena*, see the Introduction of Elmar Holenstein, Edmund Husserl, *Logische Untersuchungen, Erster Band: Prolegomena zur reinen Logik.* Text der 1. und der 2. Auflage (=*Husserliana* XVIII), pp. XIXff.
6. Reprinted in Husserl, *Aufsätze und Rezensionen (1890-1910)*, pp. 124-151.
7. See Schuhmann, p. 64, and p. 64, note 1.
8. A few examples are the following: Wilhelm Schapp, *Beiträge zur Phänomenologie der Wahrnehmung* (Göttingen, 1910); Theodor Conrad, *Zur Wesenslehre des psychischen Lebens und Erlebens.* Mit einem Geleitwort von H.L. Van Breda; Alexander Pfänder, "Zur Psychologie der Gesinnungen," *Jahrbuch für Philosophie und phänomenologische Forschung*, herausgegeben von Edmund Husserl, Vol. 1 (1912), Vol. III, 1916; Hedwig Conrad Martius, "Zur Ontologie und Erscheinungslehre der realen Aussenwelt," *Jahrbuch*, Vol. III (1916); Heinrich Hofmann, "Untersuchungen über

den Empfindungsbegriff," *Archiv für die gesamte Psychologie*, XXVI (1913); David Katz, *Die Erscheinungsweisen der Farben und ihre Beeinflussung durch die individuelle Erfahrung*; Herbert Leyendecker, *Zur Phänomenologie der Tauschungen*; and Adolf Reinach, "Die apriorischen Grundlagen des bürgerlichen Rechtes," *Jahrbuch*, Vol.I (1913).

9. See Edmund Husserl, *Formale und transzendentale Logik*, section 27, a.
10. Husserl, LU II[1] p. 7; LU II[2], p. 5. The translation is by J.N. Findlay, Edmund Husserl, *Logical Investigations*. Passages in square brackets are additions in the second edition. (The translation has been slightly altered for purposes of emphasis.) The introduction to the second volume was drafted about July, 1900; see Schuhmann, p. 61.
11. LU II[1], section 1, p. 4; section 2, p. 9.
12. See F. Kersten, "The Originality of Gurwitsch's Theory of Intentionality," for the nature of the changes from LU to *Ideas*, First Book. For the use of the term, 'intending to,' see below, section 10.
13. LU II[1] section 2, pp. 8f.
14. Ibid., section 7, p. 19.
15. See Edmund Husserl, *Ideen zu einer reinen Phänomenologie und phänomenologishen Philosophie, Erstes Buch: Allgemeine Einführung in die reine Phänomenologie*, *Jahrbuch*, Vol. I (1912), section 19, p. 35. (Hereinafter, *Ideas*, First Book.)
16. Despite their diversity, they give the impresson of having in common their reading only of LU, 2nd edition, or their not having read LU, Vol. II at all.
17. See Siegfried Krakauer, *From Galigari to Hitler*, Chapter 1. Husserl's own comments may be found in his letter to Arnold Metzger, 4 September, 1919, published in *Husserl: Shorter Works*, translated by E. Kohak, p. 361, col. b, and p. 362, col. a.
18. See Willy Moog, *Die Deutsche Philosophie des 20. Jahrhunderts*, Chapters V, VI.
19. Tracing of these movements here would take us too far afield; indeed, the history of the movements, for what they are, must still be written. In part, Husserl's own assessment is found in "Philosophie als strenge Wissenschaft," *Logos*, I (1910/11), pp.

294ff.; see also his comments in *Ideas*, First Book, pp. 151ff., and the comment on Meinong, p. 228.

20. See the latest edition of Herbert Spiegelberg (with the collaboration of Karl Schuhmann), *The Phenomenological Movement. A Historical Introduction*, Third Revised and Enlarged Edition, pp. 233-240.
21. See the Translator's Introduction to the *Logical Investigations* by J.N. Findlay, pp. 9ff., for a characteristic expression of this view.
22. This may also be one of the roots of the hostility toward transcendental phenomenology exemplified, at one end of the spectrum, by e.g., Fritz Heinemann in his once popular tract, *Existenzphilosophie. Lebendig oder Tod?* Chapter III; and, at the other end of the spectrum by e.g., Theodor Adorno in his *Zur Metakritik der Erkenntnistheorie*, especially Chapter IV, pp. 240ff. Much of what is chronicled under the heading of the "phenomenological movement" by Spiegelberg would seem to lie somewhere in between the two ends of the spectrum. The hostility may still be perceived, for instance, in the exchange between Calvin Schrag and Herbert Spiegelberg published in *Journal of the British Society for Phenomenology*, Vol.II (No. 3, October 1980), pp. 281f.
23. See Edmund Husserl, *Nachwort zu meinen* "Ideen ze einer reinen Phänomenologie und phänomenologischen Philosophie," *Jahrbuch für Philosophie und phänomenologische Forschung*, XI (1930), pp. 555f. For a later reflection on the nature of "theory" in the phenomenological sense, see ibid., pp. 561f.
24. Husserl, LU II[1], p. 5.
25. Ibid., see p. 18; cf. p. 4.
26. See ibid., p. 6.
27. Ibid., p. 10.
28. Ibid., p. 15.
29. In *Ideas*, First Book, "pure" also signifies purity from positing individual existence, i.e., the reduction to pure possibility; in addition, "pure" signifies *transcendentally* pure, scl. pure transcendental, as opposed to mundane, possibilities. See below, Part I, Chapter One, section 5f.
30. For psychologism, see *Prolegomena*, Chapters 3-8. It is therefore not an exaggeration to say that the idea of pure phenomenol-

ogy in the Introduction to Vol. II is presupposed by Vol. I, the *Prolegomena.*

31. LU II^1, p. 17.
32. Ibid., p. 12.
33. Ibid., pp. 4, 16. And this holds, mutatis mutandis, for psychology as well, despite the quite different relationship with phenomenology.
34. Ibid., p. 9; 2nd ed., pp. 8f.
35. See below, Part I, Chapter One, section 13.
36. The chief example would be *Ideas*, First Book.
37. Thus the literary order and background should not be confused with the epistemic order of inquiry. Some of the lectures were published in 1929 as "Vorlesungen zur Phänomenologie des inneren Zeitbewusstseins," herausgegeben von Martin Heidegger, *Jahrbuch für Philosophie und phänomenologische Forschung*, IX (1929); see also *Ding und Raum. Vorlesungen 1907*, herausgegeben von Ulrich Claesges (=*Husserliana* XVI).
38. Edmund Husserl, *Die Idee der Phänomenologie. Fünft Vorlesungen* [1907], herausgegeben von Walter Biemel (=*Husserliana* II). See in addition the discussion of Iso Kern, "The Three Ways to the Transcendental Phenomenological Reduction in the Philosophy of Edmund Husserl," p. 139 col. 2.
39. Above, pp. 6f.
40. Husserl, *Idee der Phänomenologie*, Lectures 1, 2.
41. For this interpretation of the "feeling" for the problem of phenomenological genesis, see Husserl's letter to Arnold Metzger, loc. cit., p. 361, col. a, p. 363, col a.
42. LU II^1, p. 672. The translation is mine.
43. LU II^2, 2, p. 200. The translation is mine. When, specifically, the passage was rewritten is unknown to me.
44. *Ideas*, First Book, section 19, p. 35. The translation is mine.
45. Edmund Husserl, *Cartesian Meditations*, translated by Dorion Cairns, section 5.
46. See Husserl's letter to Metzger, loc. cit., p. 361, col. 2. See also the important discussion of Husserl's slogan in Eugen Fink, "Das Problem der Phänomenologie Edmund Husserls," *Revue Internationale de Philosophie*, I (1938/39), pp. 232ff., especially 237 and 241; and below, section 53.

CHAPTER ONE

1. Edmund Husserl, *Ideen zu einer reinen Phänomenologie und phänomenologische Philosophie. Erstes Buch. Allgemeine Einführung in die reine Phänomenologie.* Edited by Karl Schumann, 2 vols. Hereinafter abbreviated *Ideas*, I; page numbers refer to the first edition, printed in the margin, Vol. I. For the literary and conceptual history of *Ideas*, I, see Karl Schuhmann, *Die Dialektik der Phänomenologie*, II: *Reine Phänomenologie und phänomenologische Philosophie*; see also his *Die Fundamentalbetrachtung der Phänomenologie*, Chapter III and Chapter VI, B.
2. See ibid., p. 6.
3. Ibid., p. 8. The translation is mine. The phenomena in question are transcendentally pure in addition to the other meanings of "pure," above, pp. 9f.
4. Ibid., section 59, pp. 141f., and p. 6. The translations are mine.
5. Ibid. The translation is mine.
6. Ibid. To speak of a "non-genuine" sense of real does not necessarily signify that it is false. See below, sections 7, 8, 91.
7. Ibid.
8. Ibid., section 1.
9. see below, section 10.
10. "Fact," of course, in the phenomenological sense; see below, Chapter Five. See also F. Kersten, "The Occasion and Novelty of Husserl's Phenomenology of Essence," in *Phenomenological Perspectives. Essays in Honor of Herbert Spiegelberg*, edited by Philip Bossert, pp. 69ff. Both terminologically and conceptually, Husserl's thinking has progressed over *Logical Investigations*, first edition, and "Philosophy as a Strict Science." But the progress also carries a number of ambiguities and difficult problems; in this connection, see Dorion Cairns, "A Letter to John Wild about Husserl," edited by Lester Embree, *Research in Phenomenology*, V (1975), pp. 159ff.
11. In this connection, see F. Kersten, "The Constancy Hypothesis in the Social Sciences," in *Life-World and Consciousness. Essays for Aron Gurwitsch*, edited by Lester E. Embree, pp. 523-527.
12. See above, pp. 17f.
13. See below, Chapter Four, for discussion of time and unity of mental life-processes.

14. See below, Chapter Two, for further discussion of the limits of the transcendental phenomenological epoché; also see *Ideas*, I, section 33. For a detailed examination of the genuine sense of "real," see F. Kersten, "Husserl's Doctrine of Noesis-Noema," in *Phenomenology: Continuation and Criticism. Essays in Memory of Dorion Cairns*, edited by F. Kersten and R. Zaner, pp. 132-138. From what has been said so far, it is clear that while I distinguish "epoché" and "reduction," I also insist with Husserl that they are strictly *correlative* terms as is stressed in section 32. For a contrary but prevalent view, with which I strongly disagree, see H. Spiegelberg, *The Phenomenological Movement. A Historical Introduction*, pp. 708-711; and "Reflections of The Phenomenological Movement," *The Journal of the British Society for Phenomenology*, XI (No. 3), October, 1980, pp. 279-280, 281-282.
15. See below, section 12. Husserl also speaks of these steps as "abstractive reductions"--although one must not take the term, "abstraction," in an empiricistic sense; see *Erfahrung und Urteil. Untersuchungen sur Genealogie der Logik*, redigiert und herausgegeben von Ludwig Landgrebe, section 12 (English translation by James S. Churchill and Karl Ameriks, pp. 51ff.); and Edmund Husserl, "Notizen zur Raumkonstitution," edited by Alfred Schutz, *Philosophy and Phenomenological Research*, I (1940), p. 28. See below, Chapter Two, for discussion of the phenomenological meaning of "abstraction."
16. The paramount advantage of the first alternative is that it establishes a critical control over the formulation of the phenomenological method. It would seem possible, however, without ignoring that vast literature, to achieve the same results by reference to the things themselves; thus the second alternative may be regarded as a beginning; the alternatives are not mutually exclusive. cf. Iso Kern, "The Three Ways to the Transcendental Reduction in the Philosophy of Edmund Husserl," p. 137, col. b, 144, col. b.
17. See above, p. 10.
18. As a consequence, it is not the case that the distinction between transcendental phenomenology and phenomenological psychology is not obvious, as M. Van De Pitte argues, "Introduction to 'Author's Preface to the English Edition of *Ideas*,'" p. 38 col. a, "because the two disciplines are precisely isomor-

phic." Husserl insists that they are not and cannot be in the text of *Ideas*, First Book; see the next note (19), and note 27.

19. Schematically, what Husserl has in mind may be represented in the following way:

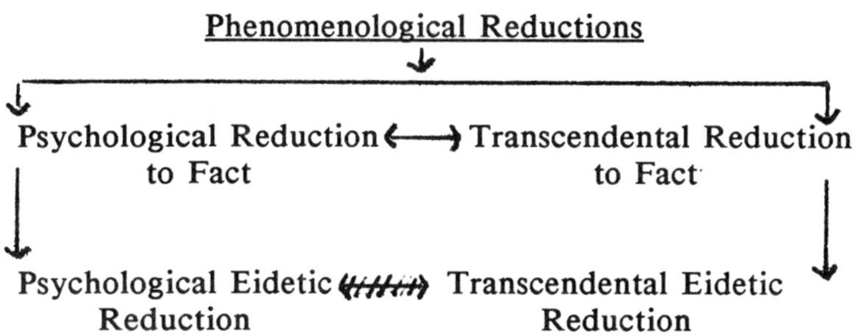

The hatched line represents Husserl's view in the "Introduction" to *Ideas*, and its impossibility; that is, while it is possible to effect the transition from the psychological reduction to fact to the transcendental reduction to fact, it is impossible, as section 33 indicates, to go from the psychological eidetic reduction to the transcendental eidetic reduction; they are not "isomorphic." The confusion that can only result when they are regarded as isomorphic is (inadvertently) illustrated by J. Claude Piguet, *De l'Esthétique à la Métaphysique*, pp. 130ff. To hold that the reductions are isomorphic would be to assert that the refrainings in question are equivalent, which is hardly the case.

20. See F. Kersten, "Introduction to Husserl's 'Origin of the Spatiality of Nature,'" p. 216; and below, Chapter Two.

21. The occasion for seeing the right order may have been Husserl's revision of parts of the text of *Ideas*, I, for Boyce Gibson's English translation (revisions which apparently Boyce Gibson did not receive); see Schuhmann's discussion in his edition of *Ideen*, Vol. 1, pp. xlvii ff.

22. See below, Chapter Five, where this reduction will be considered in detail.

23. See also Husserl's later way of expressing this situation in *Cartesian Meditations*, translated by Dorion Cairns, p. 19: "Meanwhile the world experienced in this reflectively grasped life goes on being for me (in a certain manner) 'experienced'

as before, and with just the content it has at any particular time...the only difference is that I, as reflecting philosophically, no longer keep in effect...the natural believing in existence involved in experiencing the world...."

24. See F. Kersten, "Heidegger and Transcendental Phenomenology," *The Southern Journal of Philosophy*, 11 (No. 3, Fall, 1973), pp. 213-215.
25. Eugen Fink, "Die phänomenologische Philosophie Edmund Husserls in der gegenwärtigen Kritik," pp. 102f. The translation is mine. (English translation by R.O. Elveton, *The Phenomenology of Husserl*, pp. 73-147.)
26. Dorion Cairns once gave the following example: suppose I were engaged in a reflective inquiry but failed to seize upon the intentionality of my mental life-processes and that, moreover, I were inextricably entangled in the prejudices of a sensualistic psychology so that I conclude that the mental life-processes I reflect on are but complexes of sensa; and even though I may go on to assert that some data in the complexes of sensa are clearer than others, more obscure than still others, I still do not see that they point back to intendings to them ("sensings"). Could I then, as such a psychologist, effect the transcendental phenomenological epoché? It is clear that I am in no position to do so and that I must see and seize upon the intentionality of the mental life-processes in question before I can effect the epoché. Thus reflection, although a motivating circumstance, is not a sufficient condition for exercising transcendental phenomenological epoché.
27. Dorion Cairns would seem to be in search of such a sufficient condition in his posthumously published lecture, "Philosophy as a Striving Toward Universal *Sophia*," in *Essays in Memory of Aron Gurwitsch*, edited by Lester Embree. In any case, I disagree with Rudolf Boehm, *Vom Gesichtspunkt der Phänomenologie. Husserl-Studien*, p. 168, when he says that "paradoxically the 'Considerations Fundamental to Phenomenology' reach their 'zenith' without abandoning the ground of the natural attitude" (my translation)--a statement cited with approval by Schuhmann, *Die Fundamentalbetrachtung der Phänomenologie*, p. 102. The "consideration" that is fundamental to phenomenology is nothing else but the refraining from engaging in the positing of which the naturalness of the natural attitude consists.

Boehm's paradox is only a seeming paradox: Husserl begins by trying to say something true about the real world in and of itself and finds that it only makes sense to speak of it as an object of the consciousness of it; and in making this discovery Husserl finds that he has already exercised transcendental phenomenological epoché. His task is to make this explicit in "considerations fundamental to phenomenology," i.e., he must make explicit his having abandoned the ground of the natural attitude. Van de Pitte (op.cit., p. 40, col. b) is closer to the truth of the matter when he notes that Husserl's "talk is clearly about meaning, not being; and as talk about how we view the world, rather than about the way the world is 'in itself,' it makes considerable sense. If one interprets it metaphysically, it is contradictory--the world does and does not exist independently of consciousness." What Van de Pitte does not see, in my opinion, is that the contradiction is present epistemologically as well as metaphysically, and just for that motive Husserl exercises transcendental phenomenological epoché. In this connection, see also Jan M. Brockman, *Phänomenologie und Egologie. Faktische und transzendentales Ego bei Edmund Husserl*, pp. 58ff., 63f. Brockman carefully collates the various significations of "*Einstellung*" in Husserl's writings and makes a persuasive case that "change" or "transformation" of attitude does not of itself entail the procedures of transcendental phenomenological epoché and reduction; thus the *beginning* of phenomenology is a problem (Brockman refers to Husserl, *Krisis*, pp. 139f. [English translation, pp.135f.]). Brockman then proceeds to explicate the "Cartesian" and "non-Cartesian" ways of phenomenological epoché and reduction. In this respect what I offer in sections 7ff. can only be regarded as a Husserlian, not Husserl's, procedure. See below, Chapter Four, section 35.

28. See *Formal and Transcendental Logic*, translated by Dorion Cairns, pp. 25ff.; see also Appendix II, pp. 312ff. For more detailed discussion and development of Husserl's classification as expression of the transcendental phenomenological datum, see F. Kersten, "Transcendental Phenomenology of Reason," pp. 4-9, and "Can Sartre Count?" pp. 347ff. The conclusions reached by Husserl in *Formal and Transcendental Logic*, section 4, were clearly the result of many years work; in this connec-

tion see Elmar Holenstein, *Phänomenologie der Assoziation*, especially Chapter 4; and the posthumously published manuscripts of Husserl in *Analysen zur Passiven Syntheses (1918-1926)*, herausgegeben von Margot Fleischer (=*Husserliana* XI). Some of this material will be considered below, Part Two.

29. See below, Chapter Five. From now on, the term, "mental life-processes," unless otherwise stated, means *transcendental* mental life-processes, the full expression used when emphasis is needed.
30. For Husserl's notion of "syntheses," see F. Kersten, "Can Sartre Count?", pp. 349-353; and Holenstein, Part I.
31. The transcendental ego is, then, not a transcendental phenomenological datum in the sense stated here; see F. Kersten, "Privatgesichter," in *Sozialität/Intersubjektivität*, herausgegeben von Richard Grathoff und Bernhard Waldenfels, pp. 128f.
32. See below, pp.36ff., where, with Husserl, I shall argue that in principle ego-engagement in retentions and protentions is impossible.
33. The translation, like the others from *Cartesian Meditations*, is by Dorion Cairns. See also *Ideas*, I, section 150ff.
34. *Cartesian Meditations*, p. 133.
35. Edmund Husserl, *Ding und Raum. Vorlesungen 1907*, herausgegeben von Ulrich Claesges, pp. 309f. the translation is mine.
36. Heinrich Hofmann, "Untersuchungen uber den Empfindungsbegriff," *Archiv für gesamte Psychologie*, XXVI (1913), p. 28. (p. 76 contains an important reference to Husserl's 1907 lectures cited in the preceding note.)
37. Oskar Becker, "Beiträge zur phänomenologischen Begründung der Geometrie und ihrer physikalischen Anwendungen," *Jahrbuch für Philosophie und phänomenologische Forschung*, VI (1923), pp. 446-458. I shall return to this essay in a discussion in the Second Part. For a different interpretation of Husserl's concept of "oriented constitution," see Ludwig Landgrebe, *Der Weg der Phänomenologie*, pp. 149f., 153, 156, 161.

CHAPTER TWO

1. See Edmund Husserl, *Erfahrung und Urteil*, section 11, pp. 47f., section 12, pp. 51, 56f., section 64, p. 306; see also Edmund

Husserl, "Notizen zur Raumkonstitution," p. 28: "Was wir [aber] Abstraktion nennen, ist [nichts anderes als die] Explikation von 'Momenten' auf Grund des fertig konstituierten Sinnes, in dem das Abstrahierte schon konstitutiv impliziert ist." (In this connection, see also *Ideen*, I, p. 115.) Unlike the term "*Aufbau*" in this context, so far as I know the term "*Abbau*" was not employed in the writings of Husserl published during his lifetime. But he did seem to use the term frequently in lectures and it is also found in research manuscripts (see also Dorion Cairns, *Conversations with Husserl and Fink*, edited by Richard M. Zaner, p. 65). For example, in a critical note to Lecture 32 of *Erste Philosophie* (Den Haag: Martinus Nijhoff, 1959 [*Husserliana VIII*]), herausgegeben von Rudolf Boehm, and published as Appendix II, Husserl says: "Hier *fehlt* die Beschreibung der objektiven, immerfort vorgegebenen Welt und *die Lehre von Abbau*, die zu einer reinlichen Darstellung unentbehrlich ist" (p. 312). In another of the manuscripts published in the same volume a characteristic use of the term is found: "Klammere ich die wissenschaftliche Geltungssphäre ein, so schichte ich in mir als Subjekt das Spezifische des 'Wissenschaftlers' ab und baue also mich, das konkrete Ich, ab..." (p. 268). In yet another manuscript dating from 1928 and published as an appendix to Edmund Husserl, *Analysen zur Passiven Synthesis (1918-1926)*, Husserl speaks of "Das 'Chaos' der 'Eindrücke' organisiert sich--die Eindrücke noch keine Gegenstände, Reduktionselemente, genetische Urelemente, auf die der Abbau der Intentionalität und ihre Genesis zurückführt" (p. 413). The term was also used frequently by Husserl's student Dorian Cairns when he lectured on Husserl's theory of intentionality.

2. Husserl, *Cartesianische Meditationen*, sections 44, 58, 61.
3. Ibid., section 42ff. See above, section 12f., especially pp. 44f. The context would indeed seem to be the same throughout the lectures and publications, even though the term itself may not be explicitly introduced.
4. Ibid., section 44, p. 129. (Translation, p. 98.)
5. The situation Husserl seeks to explain may be illustrated in the following way by a diagram:

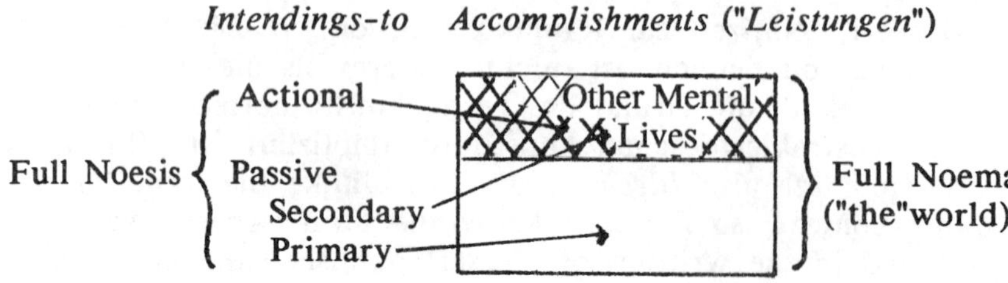

The hatched area illustrates that dimension of experience which is "screened off." See ibid., pp. 128f. (translation, pp. 98f.).

6. That is, do I first reduce the fully constituted, real objective world to the primordial quasi-objective world, *and then* to the "solipcistic" quasi-objective world? See ibid., pp. 124f. (translation, pp. 92f.).
7. See ibid., section 17.
8. Husserl, *Ideen*, Second Book, section 64.
9. This is, in fact precisely what Kant would seem to do in *De Mundi sensibilis atque intelligibilis forma et principiis*. For a discussion of Kant in this connection, see Ernst Cassirer, *Kants Leben und Werke*, pp. 105f.
10. As Husserl himself realized in eventually marking almost all of the Second Book of *Ideen* for deletion; see p. 403.
11. See especially *Cartesianische Meditationen*, section 47, pp. 134f.; and also section 45, p. 130 where Husserl would seem, for the moment, to be back on what I would consider the right track.
12. I have published my own account in "The Constancy Hypothesis in the Social Sciences," *Life-World and Consciousness. Essays for Aron Gurwitsch*, pp. 558ff.; "Transcendental Phenomenology of Reason," pp. 1-14, and "Private Faces,"pp. 167-177.
13. José Ortega y Gasset, *El Hombre y la Gente*, Chapters IV, VI; Hans Reiner, "Sinn und Recht der phänomenologischen Methode," *Edmund Husserl 1859-1959*, pp. 134-147.
14. Alfred Schütz, *Collected Papers*, Vol. III, pp. 51-83.
15. See below, Part One, Chapter Four.
16. See F. Kersten, "Can Sartre Count?" pp. 339-354.
17. Husserl, *Erfahrung und Urteil*, sections 17, 22, 23. See also in this connection, Aron Gurwitsch, "Perceptual Coherence as the Foundation of the Judgment of Predication," pp. 62-89.
18. Schütz, ibid., pp. 79ff.

19. Eugen Fink, "Die Phänomenologische Philosophie Edmund Husserls in der gegenwärtigen Kritik," pp. 322, note 1. (English translation, p. 145, note 3.)
20. See above, pp. 19f.
21. See above, pp. 29ff.
22. This is a distinction also ignored by structuralists as well as phenomenologists; see James Edie, *Speaking and Meaning*, p. 150. Schütz seems to me to be much closer to the truth of the matter in his earlier *Der Sinnhafte Aufbau der sozialen Welt. Ein Einleitung in die verstehende Soziologie*, p. 42, there Schütz says that the goal of his book is to "analyze the sense-phenomena in mundane sociality, <which> does not require the acquisition of transcendental experience beyond <mundane sociality> and therefore a further remaining in the transcendental phenomenological reduction" (the translation is mine); cf. English translation, p. 44. To explain what he means by this he refers to Husserl's "Nachwort" to *Ideen*, First Book. In the light of this text Schütz explains that he is not concerned with "constitution-phenomena in the phenomenologically reduced sphere," but is concerned instead with their "corresponding correlates in the natural attitude." This signifies for Schütz that, having formulated a transcendental phenomenological eidetic account of the consciousness of internal time, he will apply the results to the natural attitude, not as transcendentally reduced, but as psychologically phenomenologically reduced (cf. above, pp. 29ff.). He then says that in the remainder of his book we are to be phenomenological psychologists remaining in the psychological phenomenological attitude in Husserl's sense in the "Nachwort." And this reference gives the clue to what Schütz really has in mind: In the "Nachwort" Husserl restates the relationship between psychological and transcendental phenomenology by noting that there is a sense in which it can be said that they both have the "same content" and while for every true proposition of psychological phenomenology there is a corresponding true one in transcendental phenomenology, the converse is not the case. Thus Schütz, like Husserl, can methodically "leave aside," as Schütz says, "all problems in transcendental subjectivity and intersubjectivity" and develop only a psychological phenomenology and, by extension, sociology, which is not in any way inconsistent with a transcendental

phenomenology. Indeed, taking Schütz at his word in his earlier work, a psychological phenomenology ultimately only has meaning within a transcendental phenomenology (which, however, Schütz does not make explicit in his book). And because not every proposition in a transcendental phenomenology is included in a psychological one, it is possible to proceed as a psychological phenomenologist without having to maintain the transcendental phenomenological attitude (or more broadly, a psychological-*sociological* attitude can prevail without making explicit a transcendental-sociological attitude such as a community of transcendental monads in the sense of the Fifth Cartesian Meditation). Thus while an important case can be made for Schütz's abandoning his earlier position in his later writings, it seems to me that his departure into a "metaphysics" of the life-world was perhaps too hasty--as was Husserl's. For an extremely important study of this whole issue, see Helmut Wagner, "Toward an Anthropology of the Life-World: Alfred Schütz's Quest for the Ontological Justification of the Phenomenological Undertaking," pp. 239-246; and Richard Grathoff, "Das Problem der Intersubjektivität bei Aron Gurwitsch und Alfred Schütz," pp. 93ff.

23. Oscar Becker, "Beiträge zur phänomenologischen Begründung der Geometrie und ihrer physikalischen Anwendungen," pp. 508ff. For basic differences with Becker on this issue, see below, Part Two, Chapters Two and Three.
24. When used without qualification, we shall mean "primordial" quasi-objective world in this wide sense. In the next chapter it will be contrasted with a specific sense of the term.
25. In this connection, see the thorough study of Robert Sokolowski, *Husserlian Meditations. How Words Present Things*, Chapter Four.
26. Husserl, *Cartesianische Meditationen*, p. 128; translation p. 97.
27. Ibid., p. 122; translation p. 90.
28. Ibid., section 42.

CHAPTER THREE

1. See Dorion Cairns, *Conversations with Husserl and Fink*, p. 102.
2. See above, p. 61 for the wide sense of the term, "primordial."

3. See Cairns, pp. 45f.
4. The manuscript is identified as "D 3," p. 34; cited by Alwin Diemer, *Edmund Husserl. Versuch einer systematischen Darstellung seiner Phänomenologie*, p. 220, note 99.
5. Husserl, *Ideen II*, section 15b, p. 38.
6. Husserl, *Cartesian Meditations*, p. 116.
7. See above, pp. 63f.
8. Ibid.
9. See above, pp. 42f., and p. 357, note 34.
10. For the Humean dimension of consciousness at issue here, see Aron Gurwitsch, "On the Intentionality of Consciousness," *Studies*, pp. 125ff.
11. See F. Kersten, "William James and Franz Brentano," pp. 179f.
12. Husserl, *Ideas* I, sections 85, 97. For a careful analysis of the distinctions mantioned here, see Rudolf Boehm, "Les ambiguités des concepts husserliens d' 'immanence' et de 'transcendence'," pp. 482-526. See also Harmon Chapman, *Sensations and Phenomenology*, Chapters V and VII. Chapman notes, p. 150, that the term, "hyletic data," has a much wider extension than the term, "sense data:" "'Hyletic data' embraces the sensory in general, including besides the traditional sensations such sensory impressions as those of pleasure, pain, tickling, kinaesthetic sensations, and the like, also such sensory occurances as impulses, feelings, and emotions in the sphere of the will." However, the term also has a narrower signification, according to which hyletic data are sharply distinguished from other such data as listed by Chapman precisely because they can be animatingly construed as adumbrations of appearances of something physical--the same cannot be said, on Husserl's view, of kinaesthesias, impulses, sensations of pleasure and pain. It is the narrower signification of the term, "hyletic data," with which I deal here, and I believe that it is the basic one for Husserl. The whole question deserves a much longer study; see, in addition, Herman Asemissen, *Strukturanalytische Probleme der Wahrnehmung in der Phänomenologie Husserls*, section 4ff.; Karl Schuhman, *Die Fundamentalbetrachtung der Phänomenologie Zum Weltproblem in der Philosophie Edmund Husserls*, pp. 156ff.; Robert Sokolowski, *Husserlian Meditations. How Words Present Things*, sections 49, 50.
13. Husserl, *Ideas*, I, p. 173.

14. Ibid., p. 176.
15. Ibid., pp. 65, 172, 205. See also Husserl, "Vorlesungen zur Phänomenologie des inneren Zeitbewusstseins," sections 8, 42; and "Notizen zur Raumkonstitution," pp. 220ff.
16. *Ideas*, I, p. 172.
17. Aron Gurwitsch, *The Field of Consciousness*, pp. 265ff.; see also Chapman, op. cit., Chapter VII; and Sartre, *L'Etre et le néant*, pp. 26f.
18. Gurwitsch, p. 269.
19. Husserl, *Ideen, III*, p. 14. (The translation is mine.) See also *Ideas*, I, section 36, p. 65; section 85, p. 172, where Husserl's distinction between hyletic data and "strivings" ["*Triebe*"] is misleading because it tends to confuse, for example, the striving to actualize certain kinaesthetic processes with the processes themselves either actualized or to be actualized. The confusion results from what I believe is the mistaken idea that hyletic data are not only involved in doxic sensuous perceivings but also in non-doxic intentionalities as well, such as lovings, hatings, likings, dislikings, appreciatings, approvings, and so forth. Accordingly, Husserl can speak, e.g., of sensations of pleasantness, of sensations of striving. By eventually setting aside as a logical construct Husserl's idea of hyletic data, and by following instead the phenomenological procedures under discussion here, such "sensations" turn out to be non-existent. That is to say, there are no discoverable cases of hyletic data construed as adumbrations of "pleasant appearances" in and through which something liked is presented. Of course, *to assert that there are no "sensations of pleasure" signifies that pleasure is rather a posited character and not a noematic determination.* For a somewhat different translation of Husserl's 1912 manuscript, see Edmund Husserl, *Phenomenology and the Foundations of the Sciences*, translated by Ted E. Klein and William E. Pohl, pp. 12f.
20. Husserl, *Phänomenologische Psychologie*, p. 165. Mention must be made here of the earlier attempt of Sokolowski to untangle the various expressions of Husserl's concept of hyletic data in *The Formation of Husserl's Concept of Constitution*, pp. 51ff., 110ff., 139ff., 210ff. Sokolowski finds a revision of the concept of hyletic data in connection with the development of genetic phenomenology; with that, "sense data and noeses are no longer

conceived as two distinct elements; they are now seen to be one immanent reality, one inner flow of consciousness" (p. 211). But this would seem to be Husserl's view all along according to which the temporal form is the same for hyletic data and noeses; see above, p. 81. To proceed to say that hyletic data are now but one step in genetic constitution (p. 211) would signify, at the very least, that not *all* noeses and sense data are "one inner flow of consciousness" (hence they are still distinct from other noeses in other steps of genetic constitution). See also Chapman, op. cit., p. 152; Schuhmann, op. cit., pp. 128f., 77ff., and 77, note 3 where the concept of hyletic data is understood as a concept of relation, of the relation of hyle and noesis. The concept of hyletic data is introduced when we ask, not what something transcendent to consiousness is, but instead when we ask how it is constituted; thus Schuhmann: "Aside from all noetic 'formation and alteration'--to use Hegel's phrase--hyle is just an 'empty place,' i.e., simply whatever you please. No preformation is derived from it that would only need be repeated, so to speak, by noetic production. Given hyle in and for itself is only the task of positing something or other transcendent. The noetic sense-bestowal effected on it is its determinedness according to independent interpretation" (p. 158). (The translation is mine.) Hyletic data serve, then, to present something transcendent which is real and mundane (pp. 159f.) and as the basis for noetic "interpretation." It would seem as though Hume and Berkeley had finally joined hands across the Hegelian "empty place." I am not certain that Schuhmann's account answers Gurwitsch's polemical point any better than Sokolowski's.

21. See Gurwitsch, op. cit., pp. 114ff., 137ff., 271f.
22. Husserl, "Vorlesungen," p. 367.
23. See *Phänomenologische Psychologie*, section 31ff. Where Husserl elaborates this view in some detail. For instance, he says that "The stream of the subjective, thus in our sphere, the stream of subjective data of sensation, perspectives, appearances, are designated as the stream of immanently temporal objects, and themselves join together to make up a unity of temporal objectivity, therefore of the stream of mental life-processes. Everything which we single out from the immanental temporal sphere as a single, immanental temporal object, as *a* mental life-pro-

cess, exists only as flowing. Thus each datum of sensation as well as each appearance *of*, any intentional process whatever, <exists as flowing>" (p. 171; the translation is mine).

24. "Vorlesungen," section 1, p. 370 (The translation is mine.) cf. English translation, p 44. See also p. 385 for examples of the "temporal content" such as a sound or tone with its "own temporality." (English translation, pp. 43f.) --The whole issue involved here is complex; in other places in the lectures Husserl is consistent with the concept of hyletic data. Though he may, at times, seem to be unwitting about his concept, in the light of the polemics with respect to which he introduced the concept he was not uncritical. The concept of hyletic data often bothered him, and he seems to have been aware of the fact the account of the constituting of time should confirm or disconfirm what he says about hyletic data; see *Ideas*, I, pp. 162f.; also Sokolowski, *The Formation of Husserl's Concept of Constitution*, p. 140; Schuhmann, p. 77.
25. The corresponding revisions for the constituting of "internal time" will be presented below, Part II, Chapter Eight.
26. See *Ideas*, I, section 43; *Erfahrung und Urteil*, section 19.
27. See below, Part II, Introduction, where these accounts will be formulated so as to establish a literary and historical framework for specific investigations of space and time. For general overviews, see Maurice Pradine, *Philosophie de la sensation*, I, pp. 78ff.; Werner Gent, *Die Philosophie des Raumes und der Zeit*, II, pp. 206ff.; Maurice Merleau-Ponty, *Phénoménologie de la perception*, Part I, Chapters 1, 2; Part II, Chapter 2; and Heinrich Hofmann, "Untersuchungen über den Empfindungsbegriff," pp. 37-50. The latter is especially important for the development of the phenomenological "analogue" to nineteenth and twentieth century theories of space perception.
28. Rudolph Hermann Lotze, *Medicinische Psychologie, oder Physiologie der Seele*, section 30, pp. 358f., 367-368, and section 31, pp. 381f., 386f. (For a definitive statement some twenty years later, see Lotze's appendix to Carl Stumpf, *Über den psychologischen Ursprung der Raumvorstellung*, pp. 315f.) For a more recent version in this century of Lotze's theory, modified to be sure, see Stephan Witasek, *Grundlinien der Psychologie*, pp. 171f., 201f. Hermann von Helmholtz, "Die Tatsachen in der Wahrnehmung," in *Vorträge und Reden*, pp. 223ff. Oswald

Hering, "Der Raumsinn und die Bewegungen des Auges," in *Hermann's Handbuch der Physiologie*, pp. 343ff. Theodor Lipps, "Die Raumanschauung und die Augenbewegungen," *Zeitschrift für Psychologie und Physiologie der Sinnesorgane*, III, pp. 124-171. John Stuart Mill, *An Examination of Sir William Hamilton's Philosophy*, I, pp. 383ff. Alexander Bain, *The Senses and the Intellect*, pp. 183f., 366f.

29. See John Locke, *Essay Concerning Human Understanding*, Book II, Chapter XIII, sections 2f., 22f.; George Berkeley, *A Treatise Concerning the Principles of Human Knowledge*, Sections XLIV, CVXf., and *A New Theory of Vision*, Section XLVI; David Hume, *A Treatise of Human Nature*, Book I, section 11. A significant discussion of these views is found in Harmon Chapman, op. cit., Chapter IV.
30. David Katz, *Der Aufbau der Tastwelt*, sections 2f., 14f., 46ff.
31. Merleau-Ponty, op. cit., pp. 318f., 323; English translation, pp. 274ff., 278f.
32. Mill, op. cit., p. 282.
33. For this transformation of the Kantian set of problems, see Jules Vuillemin, *Physique et Métaphysique Kantiennes*, section 35f.
34. See in this connection, William James, *Principles of Psychology*, II, pp. 270f.; Stumpf, op. cit., pp. 57f.
35. For a characteristic expression of this view, see Lotze, op. cit., section 28, pp. 328f., section 31, pp. 378f.; Helmholtz, op. cit., p. 228.
36. See Aron Gurwitsch, *The Field of Consciousness*, pp. 52ff., 57ff., for an explanation of the sort of fallacy involved in the introduction of such supervenient factors. By rejecting such factors, we do nothing else than reaffirm the basis for rejecting the theory of hyletic data.
37. For a critical discussion of this view, see Ernst Cassirer, *Philosophie der symbolischen Formen*, III, Chapter Three.
38. See above, pp. 71f.
39. Husserl, "Die Welt der lebendigen Gegenwart und die Konstitution der ausserleiblishen Umwelt," pp. 339.; English translation, p. 248, col a.
40. For example, from the strivings to actualize, e.g., the liked and to prevent the actualization of the disliked sensa-sequences expected by actualizing kinaesthetic processes expected to bring

with them the preferred sense-sequences in other fields of sensa.

41. Husserl, "Notizen zur Raumkonstitution," p. 28 (the translation is mine); "Welt der lebendigen Gegenwart und die Konstitution der ausserleiblichen Umwelt," pp. 339.; English translation, p. 248.

42. Husserl, "Notizen zur Raumkonstitution," p. 28. Husserl develops this procedure in a number of "D" manuscripts; see for example, those published as *Ding and Raum*, pp. 309ff. However, Husserl would not seem to recognize the broad sense of the phrase, "setting kinaesthesia at zero."

43. See *Ding and Raum*, p. 328. What I have been considering corresponds to what Husserl there distinguishes as the "second level" of visual space constitution.

44. See ibid., pp. 309f., 329f., 371f.

45. Claesges, *Edmund Husserls Theorie der Raumkonstitution*, p. 100 (the translation is mine). See also Schuhmann, op. cit., pp. 162ff.

46. See Gurwitsch, op. cit., pp. 71ff., 87ff.

47. Cf. Claesges, pp. 86f. To be sure, Claesges, pp. 133ff., does seek to reformulate the notion of hyletic data. In so far, he says, as the constituting of the physical thing is concerned, the hyletic datum is to be regarded as an "aspect Datum" [*Aspektdatum*]; it is, accordingly, treated noetically-noematically--that is, the datum is the noetic correlate of a sensuous intending to it, hence that on which the perceiving of the physical thing is founded. However, Claesges then goes on to say that the hyletic datum is also a "local Datum" [*Stellungsdatum*], that is, a datum in the "ideal system of loci pertaining to kinaesthesias" endowed with the character of sensation and having a role in the constituting of the organism and, therefore, he states, pertaining to the "noetic side of (kinaesthetic) consciousness" (p. 135)--a version of Husserl similar to that of Sokolowski and Schuhmann mentioned earlier where the hyletic datum is both a "content" transcendent to mental life and a "content" of mental life. But by what means is it decided that the datum is one or the other "content"? Or is the datum both at the same time? For a very similar dilemma in William James's concept of "radical empiricism," see F. Kersten, "Franz Brentano and William James," loc. cit., pp. 186f.

48. For instance, the idea on Husserl's view as presented by Claesges, that something spatial can be presented without the co-presentedness of space, ibid., p. 106; or the idea that intending to kinaesthetic processes is both actional (p. 121) and passive (p. 131); or that intending to kinaesthetic processes is ipso facto intending to the fully constituted, real objective world (p. 121), so that the theme, "concrete, real objective world," need not be reduced.
49. See below, section 35.

CHAPTER FOUR

1. Above, Chapter One.
2. For a detailed account of Husserl's transcendental phenomenology of essence, including the various significations of the term, "essence" (and cognate terms), see F. Kersten, "The Novelty of Husserl's Phenomenology of Essence," in *Phenomenological Perspectives. Historical and Systematic Essays in Honor of Herbert Spiegelberg*, pp. 61-92. It is often forgotten that Husserl's discussion of "fact" and "essence" (like that of the will to return to the "things themselves") has a specific polemical context; see ibid., section 1. The polemics need not be developed here, but unless one keeps it in mind Husserl's intent and purpose is miscontrued as some sort of Platonistic metaphysics (Adorno) or as some sort of intuitional intellectualism (Levinas). See also Lothar Ely, *Die Krise des Apriori in der transzendentalen Phänomenologie Edmund Husserls*, Chapter One.
3. What follows is based primarily on Husserl, *Ideas*, First Book, section 57ff., especially section 59. See also F. Kersten, "Phenomenology, History and Myth," *Phenomenology and Social Reality. Essays in Memory of Alfred Schütz*, pp. 261ff.
4. See Kersten, "The Novelty of Husserl's Phenomenology of Essence," pp. 83ff.
5. See above, Chapter Two.
6. See Husserl, *Ideas*, section 149ff.
7. See below, Chapter Nine.
8. Husserl, *Cartesian Meditations*, section 34.
9. See above, Chapter One, and Chapter One, note 24.

10. See above, pp. 37ff.
11. See Kersten, "The Novelty of Husserl's Phenomenology of Essence," pp. 80ff.
12. See below, Chapter Nine. See above, pp. 56f.
13. A very similar view had already been worked out in 1913 by Heinrich Hofmann in his Göttingen dissertation, "Untersuchungen über den Empfindungsbegriff," pp. 52ff. It now seems clear that Hofmann's study is in part based on Husserl's 1907 lectures now published as *Ding und Raum.*
14. Becker's "transcendental deduction" will be considered below, Chapter Seven.
15. Husserl, *Cartesian Meditations*, pp. 139ff.
16. Husserl, *Formal and Transcendental Logic*, pp. 272f.
17. Ibid., p. 271.
18. I have dealt with this topic at length in "The Lifeworld Revisited," *Research in Phenomenology*, I, pp. 33-62, and "The Life-Concept and the Life-Conviction," pp. 107-128.
19. See above, pp. 7ff. In a limited way, Herbert Spiegelberg's *The Phenomenological Movement* may be regarded as providing some of the basic materials for such a history because it is Husserl's phenomenology that provides the principle of selection for inclusion of those who are to be accounted "phenomenologists." The limits lie in the lack of means to adjudicate differences among those included.
20. See above, pp. 32ff.
21. In this connection, see F. Kersten, "The Constancy Hypothesis in the Social Sciences," section 2f.
22. See Spiegelberg, ibid., pp. 92f.; and Wilhelm Szilasi, *Einführung in die Phänomenologie Husserls*, Chapters I, II.
23. See above, pp. 29ff.
24. The former would seem to be the purport of Rudolf Boehm, *Von Gesichtspunkt der Phänomenologie. Husserl Studien*, pp. 211ff.; the latter would seem to be the purport of Iso Kern, "The Three Ways to the Transcendental Phenomenological Reduction in the Philosophy of Edmund Husserl," translated by F. Elliston and P. McCormick, in *Husserl. Expositions and Appraisals*, pp. 126ff. See also Richard M. Zaner, *The Way of Phenomenology. Criticism as a Philosophical Discipline*, Chapter I, especially pp. 71ff.; Zaner actually examines various ways to phenomenology, rather than "the way" of the title of his book. Moreover, the

ways to phenomenology are not just restricted for him to Husserl whose ways are pre-figured in philosophers before him, notably Hume and Kant. This charmingly written book is a valuable guide to the philosophical problems of phenomenology.
25. Kern, ibid., pp. 127ff.
26. Ibid., pp. 134ff.
27. Ibid., p. 139.
28. Ibid., p. 126; Boehm, op. cit., p. 209.
29. Kern, p. 127; Boehm, pp. 207f.
30. Kern, p. 130.
31. Ibid.
32. Ibid., pp. 134, 130. See also Maurice Natanson, *Edmund Husserl. Philosopher of Infinite Tasks*, pp. 27ff., 55f., 65ff.
33. Kern, pp. 134f. "Abstraction," as Kern records some of Husserl's use of the term, is distinct from Husserl's other uses, such as indicated above, pp. 47ff.
34. Kern, p. 136. See above, pp. 33f. for an answer to the criticism in terms of "motives" for exercising the transcendental phenomenological epoché. As Kern states it, the problem seems to me to be a pseudo problem.
35. Kern, p. 137.
36. Kern, pp. 137f.; Boehm, p. 209.
37. Kern, pp. 139, 142.
38. See above, pp. 58ff. In addition, see the important remarks on this subject by Guido Kung, "The Phenomenological Reduction as Epoché and Explication," in *Husserl. Expositions and Appraisals*, pp. 345f.
39. See above, pp. 27ff., and p. 354, note 19.
40. See Aron Gurwitsch, "The Place of Psychology in the System of Sciences," "The Phenomenological and the Psychological Approach to Consciousness," and "Critical Study of Husserl's *Nachwort*."
41. See above, pp. 35ff., 41ff.
42. Above, pp. 32f.
43. See above, pp. 12ff.
44. In this connection, see the careful and judicious discussion of Karl Schuhmann, *Die Fundamentalbetrachtung der Phänomenologie*, pp. 101ff., especially p. 102, note 75; also Boehm, p. 168. Schuhmann rightly points out that the whole issue of the "methodic way" hinges on the issue of the "beginning" of

phenomenological reflection and refraining. But this is also the issue of the transition from the transcendental natural attitude to the transcendental unnatural attitude which Boehm expresses as the "paradox" of not leaving behind the natural attitude, or, as I expressed it before, the question concerns how consciousness can be of "two minds" about itself. See also Schuhmann, p. 16; and pp. 29ff. for a very perceptive discussion of the positing that belongs to the natural attitude.

INTRODUCTION, PART TWO

1. The image is brought to mind by the poem of Christian Morgenstern, "Der Lattenzaun," which begins: "Es war einmal ein Lattenzaun,/ mit Zwischenraum, hindurchzuschaun./ "and goes on to tell of "Ein Architekt, der dieses sah, / stand eines Abends plötzlich da-- / und nahm den Zwischenraum heraus / und baute draus ein grosses Haus." (Christian Morgenstern, *Das Mondschaf. Ein Auswahl aus den Galgenliedern*), see above, p. 57.
2. Husserl, "Foundational Investigations of the Phenomenological Origin of the Spatiality of Nature," translated by F. Kersten, in *Husserl: Shorter Works*, p. 183.
3. Maurice Merleau-Ponty, *L'Oeil et l'Esprit*, pp. 48f. For the discussion of the historical setting that follows, in addition to Werner Gent, *Die Philosophie des Raumes und der Zeit*, I have relied on Alexandre Koyré, *From the Closed World to the Infinite Universe*; Ernst Cassirer, *Das Erkenntnisproblem in der Philosophie und Wissenschaft der neuren Zeit*, Vol.II. See also F. Kersten, "The Life-Concept and The Life-Conviction," pp. 110-113. For Descartes, see *Oeuvres de Descartes*, publiées par Charles Adam & Paul Tannery, Vol. IX-2, *Principles*, II, sections 4, 11, 13, 16, 20.
4. Descartes, section 24; see also Vol. X, *Regulae ad Directionem Ingenii* p. 426 (Regula XII).
5. Descartes, *Principles*, I, section 44f.
6. See Spinoza, *Works of Spinoza*, Unabridged Elwes Translation, Vol. II: *The Ethics*, I, Propositions 11-15; Letter to Ludovicus Meyer, 20 April, 1663 (Letter XII).

7. Hobbes, *Selections*, edited by F.J.E. Woodbridge, *Elements of Philosophy*, Part II, Chapter VII.
8. See Koyré, Chapter V; for More's letters, see *Oeuvres de Descartes*, Vol. V, and Vol. V, Appendix, pp. 628ff.
9. For a summary of Law's views, see Cassirer, pp. 454ff. Law's work, now largely unknown, provides a fine presentation of Locke as well as a thorough discussion of a Lockean answer to the then-current controversies about space, time and motion.
10. See Berkeley, *De Motu* and *The Analyst*; *The Works of George Berkeley*, edited by A.A. Luce and T.E. Jessop, Vol. 4. In *The Analyst*, section 5, one finds a succinct summary of Berkeley's view.
11. See Berkeley, *A Treatise Concerning the Principles of Human Knowledge*, Introduction, section 11.
12. See Leibniz, "Lettres entre Leibniz et Clarke," in God. Guil. Leibnitii, *Opera Philosophica*, instruxit Joannes Eduardus Erdmann, Cinquieme Ecrit de Mr. Leibniz, pp. 768f. (The translation cited is by Clarke.)
13. Leonhard Euler, "Reflexions sur l'espace et le tems," in Leonhardi Euleri, *Opera Omnia*, ediderunt E. Hoppe, K. Matter, J.J. Burkhardt, Series Tertia, *Opera Physica*, Vol. II., pp. 376f. (The translation is mine.)
14. The argument was anticipated above, section 27.
15. In passing it should be mentioned that the whole issue of the post-Kantian development of so-called "geometries" of "spaces" having more than three dimensions does not seem to have refuted the idea that "complete space" has three dimensions. If one distinguishes between Euclidean geometry proper and the system-forms of Euclidean geometry, then such so-called "geometries" are not strictly geometries in the Euclidean and Kantian sense, their domains not strictly "spaces." Thus after Kant what are developed are not n-dimensional geometries, but instead n-dimensional system-forms which may well have non-spatial "interpretations;" but if they are to be system-forms of the Euclidean system, they must be "interpretable" as about space, and this space remains just 3-dimensional no matter if other system-forms are possible. It is not possible to develop this line of thought further here.
16. See above, pp. 79f., 12f.

17. See Husserl, *Ideas*, First Book, section 1, p. 7; Aron Gurwitsch, *The Field of Consciousness*, pp. 87f., 163f.
18. See above, pp. 32ff.; for still other motives, which cannot be developed in this brief review of the historical setting of the problem of space, see F. Kersten, "The Life-World Revisited," pp. 49f., pp. 58-62; and "The Life-Concept and The Life-Conviction," pp. 111-113.
19. "Abandonment" in much the same sense that Gurwitsch speaks of the "dismissal" of the constancy hypothesis by making it explicit; see Gurwitsch, pp. 90ff.; also Husserl, *Ideas*, First Book, section 60; and F. Kersten, "The Constancy Hypothesis in the Social Sciences," pp. 523f.
20. See above, pp. 16ff. See also Husserl, *Formal and Transcendental Logic*, section 96, b.

CHAPTER FIVE

1. See above, pp. 60ff., 66f., 68ff.; below, Chapter Six. For important clues to taking these further unbuilding steps, see Dorion Cairns, "The Many Senses and Denotations of the Word *Bewusstsein* ("Consciousness") in Edmund Husserl's Writings," in *Life-World and Consciousness. Essays for Aron Gurwitsch*, pp. 19-32. Husserl would seem to reserve the term, "inhomogeneous," for what I call "here-centered" quasi-objective space of phantoms, and the term, "homogeneous," for what I call "centerless" quasi-objective space of phantoms. See, e.g., Edmund Husserl, *Zur Phänomenologie der Intersubjektivität. Texte aus dem Nachlass, Zweiter Teil: 1921-1928*, pp. 538ff., 549ff.
2. See above, pp. 70f.
3. See above, pp. 71f., 81ff.
4. Oskar Becker, "Beiträge zur phänomenologischen Begründung der Geometrie und ihrer physikalischen Anwendungen," p. 452. The translation is mine.
5. Ibid. See also Husserl, op. cit., pp. 543f., 548f.
6. Ibid.
7. Thomas Reid, *Inquiry into the Human Mind*, published in *The Works of Thomas Reid*, Chapter V, section 2. See also David Katz, *Der Aufbau der Tastwelt*, section 9.

8. See Edmund Husserl, Mss D 12 III (1931), "Zur Konstitution der Tastwelt. - Die haptischen Kinästhesen," cited by Claesges, *Edmund Husserls Theorie der Raumkonstitution*, pp. 106f.: "Das sagt, das Tastfeld hat (ungleich den visuellen Feld) die Grundeigentümlichkeit, dass zwar alle seine Daten im *Aussereinander* sind, dass aber zwei Daten und ihre Stellen im Aussereinander sich trotzdem decken können, dass sie zwar dabei in gewisser Weise verschmelsen, aber doch sich nicht wirklich mischen. Die Lokalitäten bleiben getrennt und um in Aussereinander als stetig durchlaufbarem von einem zum andern stetig zu kommen, muss man eben die im Lokalfeld vorgezeichneten Wege einschlagen, um so von einem Datum hier zu einem anderen dort zu kommen. Die lokal verschiedenen Daten 'beruhen' sich, 'decken' sich, ohne sich doch zu verdecken. Das ist ein phänomenologisch ganz einzigartiges Vorkommnis, keineswegs zu vermengen mit dem Sichverdecken von visuellen Schatten, die als Schatten schon nicht mehr Empfindungsdaten sind." For identifying and assimilative syntheses, see above, pp. 37f.
9. Cf. Claesges, op.cit., p. 106. Merleau-Ponty, *Phénoménologie de la perception*, pp. 362f. (English translation, pp. 315f.)
10. This is, to be sure, a characteristic shared by other prespatial sensum-fields, such as the visual and auditory fields. But what is unique in the case of the prespatial tactual field is the presentedness of being "closed," yet without a "boundary;" as it were, nothing "passes it by." See Husserl, *Zur Phänomenologie der Intersubjektivität*, p. 549; and especially Katz, *Aufbau der Tastwelt*, sections 7f.
11. This represents an important consequence mentioned earlier: it is possible to discriminate within the unbuilding reduction to tactual prespace a substratum where no "universals" are constituted in primary passivity, namely "material universals," because such "universals" necessarily require as founding substratum primary passive assimilative syntheses; see above, pp. 58f.; below, Chapter Nine; and F. Kersten, "Can Sartre Count?" pp. 349f., and "The Lifeworld Revisited," pp. 56f. Katz notes that perhaps an anomaly of touch, when compared to sight, is that touch informs us only of specific surfaces and shapes, but not the "generality," Surface and Shape; see *Aufbau der Tastwelt*, sections 10, 2.

12. This is not to be confused with "covering" and "uncovering" in the visual prespatial spread where, among other things, "perspectives" are involved.
13. Like many of the other terms and phrases used, they are not strictly accurate because implicitly they refer to higher substrata of constitution.
14. Cf. Husserl, "Die Welt der lebendigen Gegenwart," p. 341. Cf. also Stumpf, *Über den psychologischen Ursprung der Raumvorstellung*, pp. 383ff. The adjective, "closed," may not be the best but it is difficult to find a better one.
15. Cf. Katz, op. cit., p. 42; cf. also Husserl, "Notizen zur Raumkonstitution," pp. 37f. "Ground" and "background" are used in analogy with the visual presentedness of "figure-ground."
16. Katz, op. cit.
17. See below, Chapter Nine.
18. See Katz, ibid.
19. See Hedwig Conrad-Martius, "Zur Ontologie und Erscheinungslehre der realen Aussenwelt," pp. 428, 482f., 524ff.
20. Here, too, with Scheler I find "resistance" presented apart from any correlation of sensing the tactual spread correlative to kinaesthetic flow-sequences; see Scheler, "Die Idole der Selbsterkenntnis," in Scheler, *Vom Umsturz der Werte*, p. 258. (English translation, p. 56.)
21. See above, p. 127, and the statement of Becker cited there.
22. See above, pp. 82ff. In part, the anomaly lies in the fact that it is not a quasi-objective organism in the sense that it is an organ of perceiving, nor an organ which, by moving, in turn moves other phantom things. Nor is this visible thing something on or in which, at a higher stratum of constituting, a sensum-field is located as in the case of the quasi-objective tactual organism.
23. For a phenomenological account of sight as an analogue of touch, see Jean Paul Sartre, *L'Imaginaire*, pp. 103ff. See also Katz, *Aufbau der Tastwelt*, sections 8ff., 46, who explores the many analogies of touch and sight, especially between film and plane colors and two- and three-dimensional tactual surfaces. A modified version of Sartre's analysis is introduced below, pp. 174ff. For a criticism of Sartre's account of the analogue of sight and touch, understood as the last remnant of Cartesianism, see Maurice Merleau-Ponty, *L'Oeil et l'Esprit*, pp. 32ff

(English translation, pp. 163ff.). Taken as an analogue of touch, or even as reducible to touch, on the Cartesian view sight is then likewise a coming into contact with things, a "con-frontation;" the Cartesian view is nicely summed by Merleau-Ponty with Descartes' assertion that the blind "voient des mains" (ibid., p. 37; the reference is to Adam et Tannery, eds., *Oeuvres de Descartes*, *Discours de Méthode*, Vol. VI, p. 83). Descartes' text makes clear that "seeing with the hands" in particular, and touch in general, is the model for vision. In contradistinction, Merleau-Ponty suggests that it is basically visibility and tactuality which characterizes the human organism, that visibility is an "indwelling" rather than a confrontational mode of being in the world. Certainly it would seem, in this light, that the Cartesian idea of space perception sketched above, section 37, and its resulting paradox (pp. 111f.) presuppose that touch is the model not only for sight, but of all the other senses as well. The phenomenological account developed in this chapter is much closer to that of Merleau-Ponty, although not without qualifications; see below, Chapter Eight, pp. 239ff. The implications are far-reaching because it would follow, for instance, that if the Cartesian (and by extension, the Kantian) view is rejected the human organism is not subject to the laws of mechanics as conceived by Descartes--yet another reason for taking issue with the assumption that the spatio-temporal form of the real, objective world presented in sub-scientific experience is apperceived in the "naturalistic conception" as the self-same spatio-temporal form presented in scientific experiencing (see above, pp. 120ff.).

24. See Ernst Mach, *The Analysis of Sensations*, translated by C. M. Williams, p. 135, note 2. Mach proceeds to compare this experience with the difficulty encountered by people, blind from birth, in fixing and accommodating their sight after their eyes have been operated on. The context is not relevant, although the example is.
25. H.H. Price, *Perception*, p. 28. I disagree with the conclusion Price draws from his example, namely that "the sense-datum, the color expanse which we sense, *is* flat and vertical." It is only at a higher level of oriented constitution, where the hillside can be brought into the perceptual near-sphere, that these spreads are apprehended as "distortions" of the visual

appearances of the hillside. Also see above, p. 759, and the example drawn from Gurwitsch.

26. We therefore have the situation where anomalies of visual perceptual experience are constituted at higher stories or levels, such as that of centerless quasi-objective phantom space, rather than at the level of prespatial sensum-fields, as in the case of touch. This raises again the whole question of the constituting of anomalies; because the perceptual anomalies are, in any case, originarily constituted at different stories, they do not acquire the "intersensorial" character of "normal" perceptual appearances, i.e., they do not become "appearances" of the next higher stratum of oriented constitution; they may be regarded, rather, as limiting cases of oriented constitution.
27. This determinational sense of the quasi-objective spatial field, the "appearance" which prespatial curvature acquires at the next higher level, will be made thematic below, Chapter Seven, sections 64ff.
28. See above, pp. 83f.; below, pp. 158ff. It should be noted that at this substratum of constitution I find no objectivational difference between so-called surface and film colors.
29. The phrase, "no longer," is used in a constitutive rather than temporal meaning.
30. See Becker, op. cit., pp. 463f.
31. See Merleau-Ponty, *Phénoménologie de la Perception*, p. 368 (English translation, pp. 318f.); and "Le Doute de Cezanne," in *Sens et Non-sens*, p. 23 (English translation, p. 14); see also above, p. 80, and below, p. 143.
32. See Becker, p. 465. In contrast to "fading" or "diminishing," "annihilation" can only be objectivated with respect to something at a higher stratum of constituting than the substratum of the prespatial visual field. With respect to the latter, nothing is objectivated as sensed as leaving or entering the field when all kinaesthesias are "set at zero."
33. See the similar account by Max Scheler, *Der Formalismus in der Ethik und die materiale Wertethik*, pp. 76f. (English translation, pp. 56f.). What I find is somewhat at odds with Husserl in the "Vorlesungen zur Phänomenologie des inneren Zeitbewusstseins," section 1, p. 370, as well as with Becker, op. cit., pp. 448, 493f. Following Husserl, Becker, as noted before, believes that the prespatial visual field, and prespatial spreads of all sorts,

are "two-dimensional." In a certain way what I find is much closer to Carl Stumpf, op. cit., pp. 154, 169f., 176f., 180f., 282f.; Stumpf argues for the primitive "three-dimensionality" of each of the radically heterogeneous prespatial fields, while for Husserl and Becker depth is objectivated only when there is operative a functional correlation between kinaesthetic and visual sensa. See Becker, op. cit., pp. 470f., 485, 489; also Husserl, *Ding und Raum*, pp. 310f. For an indication of Husserl's reaction to Stumpf's idea about space perception, see the valuable introduction of Claesges to *Ding und Raum*, pp. XXIIIff., especially note 3, p. XXIII, and note 1, p. XXVI. The reason why, I believe, Husserl and Becker suggest that visual prespace is "two-dimensional" is because they both adhere to the theory of hyletic data and thus can find no such data construed or construable as adumbrative of quasi-objective or objective visual appearances in and through which things are *seen at a distance*. Not holding such a theory, Stumpf arrives at quite different results. Indeed, so long as the theory of hyletic data is adhered to, depth will always be a supervenient factor in the analysis of space perception. See above, pp. 80f., and below, Chapter Seven, section 70f.

34. Becker, op. cit., p. 495.
35. In the case of "covering" and "uncovering," what remains as an anomaly, indeed, a perceptual abnormality, at a higher story of oriented constitution, is the case of the clouds moving mentioned earlier, p. 80; see also Husserl, *Ding und Raum*, p. 260, who, however, deals with it in terms of "empty space." (See below, p. 147 for the meaning of "empty prespace.") It is worth emphasizing again that at the lowest substratum of oriented constitution of space it makes no sense to speak of anomalies of any sort.
36. The implication is crucial for it signifies that constituting of the organism presupposes the constituting of "sub-organismic" prespatial fields. Also, see below section 71.
37. The example is adapted from José Ortega y Gasset, "On the Point of View in the Arts," in José Ortega y Gasset, *The Dehumanization of Art and Other Essays*, pp. 110ff.
38. A parallel example will be considered below, Chapter Six, pp. 174ff.

39. At the lowest constitutive substratum of prespace, there is no objectivatable distinction between the "cloud moving" and other kinds of change in the sensum-field.
40. See above, pp. 137f. As will be seen in Chapter Seven below, this ubiquitous visual prespatial "centricity" is the basis for founding "centerless" or "homogeneous" quasi-objective visual space.
41. At the end of Chapter Seven the consequences of this view will be reconsidered for the constitution of "geometries."
42. See above, pp. 115f.
43. Husserl, *Ding und Raum*, section 31, pp. 104f.
44. See Alfred Schütz, "Fragments on the Phenomenology of Music," edited with and Introduction by F. Kersten, *Music and Man*, VI (1976), section 22.
45. Husserl, op. cit., p. 83.
46. Involved, for instance, is the difficulty of accounting for why some hyletic data, e.g., sounds, are not constitutive of space in the same way and to the same extent as other hyletic data (e.g., visual, tactual data).
47. In this connection, see the still intriguing and important study of P.T. Young, "Auditory localization with accoustical transposition of the ears," *Journal of Experimental Psychology*, pp. 399-429.
48. Ibid., p. 424. See also Hans Wallach, "On Sound Localization" in *Readings in Perception*, pp. 476f.
49. In this connection, see Heinz Werner, *Grundfragen der Intensitäts-psychologie*, pp. 69f.; cf. also Husserl, op. cit., p. 104; and Herbert Spiegelberg, "Über das Wesen der Idee," *Jahrbuch für Philosophie und phän. Forschung*, Vol. II, pp. 32ff., 40ff. (The assimilative synthesis most typical of constituting of sounds "inside one another" distinguished by Spiegelberg is what he calls "differentiated similarity" [*differenzierte Ähnlichkeit*].)
50. Edith Stein, "Beiträge zur philosophischen Begründung der Psychologie und der Geisteswissenschaften," *Jahrbuch für Philosophie und phän. Forschung*, Vol. 5, p. 10. See also the detailed phenomenological descriptions of silence by Don Ihde, *Listening and Voice. A Phenomenology of Sound*, pp. 185ff.
51. See Wallach, op. cit., loc. cit.
52. See Spiegelberg, op. cit., pp. 40f.

53. See above, section 17ff.
54. Husserl, *Cartesian Meditations*, p. 21. See above, sections 7, 8.
55. See above, section 20, pp. 63f.
56. Husserl, op. cit., p. 97.
57. See Husserl, op. cit., section 61; and *Formal and Transcendental Logic*, section 6, and Conclusion; see above, section 17, b, section 33; below Chapter Nine, pp. 290f.
58. See Suzanne Bachelard, *A Study of Husserl's Formal and Transcendental Logic*, translated by Lester E. Embree, pp. 65ff. Cf. Jean Cavaillés, *Sur la Logique et la Théorie de la Science*, pp. 54ff.
59. See Husserl, *Vorlesungen zur Phänomenologie des inneren Zeitbewusstseins*, p. 370. The refraining from positing my own mental living as having a locus in real, objective time will be examined in detail below, Chapter Eight. For the moment, the parallelism of the "exclusion" of "objective time" and "objective space" is understood in terms of the preceding citations from the *Cartesian Meditations*; also see above, section 7.
60. See George Berkeley, *Of the Principles of Human Knowledge*, Part I, section 1.
61. Grasquet, *Cézanne* (Paris: Bernheim Jeume, 1926), p. 81. Cited by Maurice Merleau-Ponty, *Phenomenologie de la perception*, p. 368; English translation, p. 318.
62. See above, pp. 79f., 121ff.
63. Above, sections 25, 27.
64. Husserl, *Ding und Raum*, p. 534. The translation is mine. See also pp. 550ff. Probably one reason for this double meaning of kinaesthesia can be found in Dorion Cairns' *Conversations with Husserl*, 5 January 1932, where Husserl is reported to say, in effect, that what he means by "kinaesthesia is not the bodily sensations accompanying movement or muscular tension, or the inner sensations, but rather something volitional or quasi-volitional <scl. volition in primary passivity?> that remains when one abstracts from such sensations." And in the conversation of 13 January 1932 Husserl distinguishes between a "hyletic stream" and a kinaesthetic striving, thus collapsing the difference between the hyletic and the kinaesthetic into the difference between sensation and striving, e.g., to prolong or diminish the sensation. It is this collapsing of differences that gives rise to the ambiguity in the meaning of "kinaesthesia."

65. Claesges, op. cit., section 22.
66. See above, p. 368, note 47.
67. Claesges, pp. 109f.
68. See above, pp. 81ff.

CHAPTER SIX

1. Husserl, "Notizen zur Raumkonstitution," p. 35. The translation is mine.
2. Ibid., p. 35, note 7. The translation is mine.
3. See above, sections 24ff. for discussion of the correct procedures to be employed in reducing the theme of inquiry.
4. See above, pp. 72ff., 77f.
5. Dorion Cairns, "The Many Senses and Denotations of the Word *Bewusstsein* ("Consciousness") in Edmund Husserl's Writings," p. 28.
6. Ibid., p. 29.
7. See above, p. 74. For general consideration of the first three sorts of sensa distinguished, see Moritz Geiger, "Beiträge zur Phänomenologie des aesthetischen Genusses," *Jahrbuch für Philosophie und phänomenologische Forschung*, Vol. I, Part II, pp. 678, 680f.; see also Price, *Perception*, p. 25, note 2; and Scheler, *Der Formalismus in der Ethik und die materiale Wertethik*, pp. 341ff. (English translation, pp. 328ff.) Also see Husserl, *Ideen*, II, section 39.
8. See Husserl, *Logische Untersuchungen*, II, 1, pp. 392f. (English translation, pp. 572f.)
9. See above, pp. 76f.
10. See above, pp. 156, 158f.
11. See above, p. 162f.
12. See above, pp. 128f.
13. Ibid. for what follows, see Katz, *Aufbau der Tastwelt*, section 10ff.
14. In this connection, see *Ideen*, III, pp. 118f.; Geiger, op. cit., p. 679; Conrad-Martius, op. cit., pp. 401f., 412f.; Claesges, op.cit., p. 98; Merleau-Ponty, op. cit., p. 364 (translation, p. 315).
15. Cf. Husserl, *Ideen*, II, pp. 57f.; *Ideen*, III, pp. 119ff.
16. See Husserl, *Zur Phänomenologie der Intersubjektivität, Zweiter Teil*, pp. 518ff.

17. See above, pp. 64f.
18. See Husserl, op. cit., pp. 516ff. See also Merleau-Ponty, *Phénoménologie de la perception*, p. 365: "Tactual experience...adheres to the surface of our organism; it cannot become unfolded before us, and never quite becomes an object. As the subject of touch, I cannot flatter myself that I am, correspondingly, everywhere and nowhere. I cannot forget here that it is through my body that I go to the world, and tactual experience occurs 'ahead' of me, and is not centered in me. It is not I who touch, but my organism...." (The translation is mine.) Cf. English translation, p. 315. At issue here is the oriented constituting of the unique Hereness peculiar to the quasi-objective and objective organism. See below, Chapter Seven.
19. See Husserl, *Phänomenologische Psychologie*, section 34, p. 173; and Husserl, *Analysen zur Passiven Synthesis*, p. 414: "Die Sinnesfelder miteinander sind chaotisch verbunden. Nur qualitativ haben wir bisher ein Chaos. Aber die Sinnesphären brauchen sich heran auch nicht um einander zu kümmern, und insofern doch Chaos. Aber jedes Sinnesfeld ist eine harmonische kosmische Einheit. Man kann auch mit diesem Chaos anfangen und dann fragen: Wenn in der urimpressionalen Gegenwart solche Ordnung schon herrscht, innere Vereinhetlichung, dann gehört zu jeder die zeitliche Synthese...." See ibid., p. 415 for examples of the "temporal syntheses." (This whole manuscript written around 1926 is quite interesting because Husserl finds no need to introduce the idea of adumbration of hyletic data. On the other hand, the theme, "real, objective world," or "primal impressional present," does not seem to be reduced.)
20. The meaning of "identity" in this context will be considered in the next chapter. For the moment it is necessary to note that "identity" is always "*meant* (and posited) identity," i.e., that which is constituted in passive syntheses of identification and difference and in assimilative syntheses of perfect likeness. In this connection, see Aron Gurwitsch, "Phenomenology of Thematics and of the Pure Ego," *Studies in Phenomenology and Psychology*, pp. 237ff.
21. Jean Paul Sartre, *L'Imaginaire. Psychologie phénoménologique de l'imagination*, p. 102. (The translation is mine.)
22. Husserl, *Cartesian Meditations*, section 29; cf. Sartre, op. cit., p. 103.

23. Above, p. 164. In other words, it is an intentive process that would otherwise have as foundation a sensing as substratum of its proper essence.
24. Cf. Young, "Auditory Localization with accoustical transposition of the ears," pp. 416f., 424f.; and Wallach, "On Sound Localization," pp. 484ff.
25. Cf. Young, p. 416.
26. Cf. ibid., p. 424
27. Cf. ibid., pp. 425f.
28. Cf. Kurt Goldstein, "Uber induzierte Tonusveränderungen," pp. 294ff. Goldstein makes the point that "localization," in our terms, dependent on sensing of the self-same kinaesthetic sensa-sequences, is functionally related to movements of the eyes both in the case of "normal" individuals as well as those suffering from brain lesions; more particularly, "localization" of sound is not merely a function of the ear but rather is an "accomplishment" of the whole organism (thus involving oculomotor and cephalic kinaesthesias regularly accompanying both sensing of visual and auditory sensa).
29. There is, therefore, no single sensum-field that is more basic than any other.
30. See below, Chapter Seven. To be sure, the sense in which the constituting of the anomaly, "organism," is "completed" is a rather Pickwickian one. Nietzsche's dictum that "man is an unfinished animal" holds true within the frame of the oriented constitution of the real, objective world as well as within its many substrata.
31. In this connection, see the passages referred to in note 19, and see above, p. 164, and pp. 64f., 72.
32. Op. cit., section 34, p. 173. (The translation is mine.)
33. Ibid., section 30, p. 165. (The translation is mine.)
34. See Husserl, *Zur Phänomenologie der Intersubjektivität* Zweiter Teil, pp. 516ff.
35. See above, p. 127.
36. Cf. above, pp. 127f.

CHAPTER SEVEN

1. Husserl, *Vorlesungen zur Phänomenologie des inneren Zeitbewusstseins*, section 7ff.
2. See Husserl, op. cit., and Aron Gurwitsch, "Phenomenology of Thematics and of the Pure Ego," pp. 276ff. Also see above, p. 11f.; below, pp. 272f., 275ff.
3. For more details, see F. Kersten, "Can Sartre Ccount?" pp. 344f., 347, 351; see above, pp. 36ff., 58f., 155ff. For the "same," "one and the same," "the identical," and similar locutions, see above, p. 383, note 20.
4. See above, pp. 56f., 182.
5. See above, pp. 179f. Shortly cases will be considered where the analogue cannot be converted into, e.g., a genuine seeing.
6. See Husserl, *Zur Phänomenologie der Intersubjektivität, Zweiter Teil: 1921-1928*, pp. 517ff., 554f.
7. Ulrich Claesges, *Edmund Husserl's Theorie der Raumkonstitution*, p. 103. The translation is mine.
8. Husserl, Mss. D 12 III (1931), p. 13. Cited by Claesges, p. 103. The translation is mine.
9. At this level of oriented constitution, there is already the basis for the anomaly of the so-called "phantom limb;" see Merleau-Ponty, *Phénoménologie de la Perception*, pp. 90ff., 115ff.; English translation, pp. 76ff.
10. See Husserl, "Notizen zur Raumkonstitution," p. 224.
11. Merleau-Ponty, p. 365.
12. See Husserl, "Welt der lebendigen Gegenwart," p. 340; English translation, p. 249.
13. Ibid.
14. Ibid: "they can be seen like pictures in a picture-book."
15. See Husserl, *Ideen*, II, section 37, pp. 147f.; also Conrad Martius, "Zur Ontologie und Erscheinungslehre der realen Aussenwelt," pp. 400, 464.
16. Merleau-Ponty, p. 363; English translation, p. 314.
17. Ibid.
18. See above, pp. 111f., 119ff.; below, pp. 214ff. I have already sketched the problem of access to world in "The Constancy Hypothesis in the Social Sciences," pp. 555ff.
19. Heinrich Hofmann, "Untersuchungen über den Empfindungsbegreiff," pp. 64ff. This long-neglected work is important not

only in it own right but also because of its influence on recent Spanish philosophy through a famous review by José Ortega y Gasset published in 1913 (reprinted in *Obras Completas*, Vol. I, pp. 245-261). It is also significant in still another way: Hofmann had heard Husserl's lectures on "Physical Thing and Space" of 1907, and his essay provides important information on how to read them; see Hofmann, p. 76.

20. Oskar Becker, "Beiträge zur phänomenologischen Begründung der Geometrie und ihrer physikalischen Anwendungen," p. 456.
21. Becker, p. 470. The translation is mine.
22. See Hofmann, pp. 56f., 67.; Becker, p. 455, and p. 456, footnote 1. Hofmann and Becker use the term at times as equivalent to "*Sehding*" ("sight thing"), and Hofmann also distinguishes it from the "*visuelles Sinnending*" ("visual thing of the senses") or phantom. As the term is used here, skiagraph refers to the visual or visual-tactual phantom unity just and only just as it is presented in here-centered quasi-objective space; what Hofmann and Becker call "sight thing" is what is here referred to as the changing visual aspects of the phantom unity presented as rotating--scl. the series ("cyclical series") of rotational appearances of the phantom unity which, as it rotates, does not essentially change its distance from the here-centered phantom organism. As Becker notes, the term, "skiagraph," along with the other terms used here, present difficulties for which there is no easy solution.
23. See above, pp. 198f. Left out of account here are the analyses of "deception" and "illusion" peculiar to quasi-objective perceivings of the here-centered and centerless phantom worlds; in this connection, see Herbert Leyendecker, *Zur Phänomenologie der Täuschungen*, pp. 137ff.
24. See Husserl, "Notizen zur Raumkonstitution," p. 217.
25. For an analysis of the phantom ground as an "earth basis," see Husserl, "Grundlegende Untersuchungen zum phänomenologischen Ursprung der Räumlichkeit der Nature," pp. 309f. See above, pp. 64f., 108f.
26. I have disregarded the cases of phantom unities in flight, riding or being carried, jumping off the phantom ground, and the like. See "Notizen zur Raumkonsitution," pp. 26f.
27. In this connection, see Adolf Reinach, "Über das Wesen der Bewegung," in Adolf Reinach, *Gesammelte Schriften*, pp. 446f.

28. See Hofmann, pp. 62ff.; Becker, pp. 485ff., 505f.
29. See Husserl, *Ding und Raum*, Chapter 9; Claesges, section 14f.
30. See Noel Mouloud, "Le principe spatial d'individuation: Fondement phénoménologique et signification géometrique," *Revue de Métaphysique et de Morale*, pp. 61f.
31. See above, sections 27, 38-41.
32. See above, pp. 107ff., 121f.
33. See above, pp. 95ff. For a review and discussion of the principal accounts of depth perception on the background of which Husserl worked in the first decade of the century, see Hofmann, pp. 104ff.
34. The immediate setting of the problem for Becker is found in the *Prolegomena* to the *Logical Investigations*: "If we use the term 'space' of the familiar type of order of the world of phenomena, talk of 'spaces' for which, e.g., the axiom of parallels does not hold, is naturally senseless. It is just as senseless to speak of differing geometries, when 'geometry' names the science of the space of the world phenomena. But if we mean by 'space' the categorical form of world-space, and, correlatively, by 'geometry' the categorical theoretic form of geometry in the ordinary sense, then space falls under a genus, which we can bound by laws, of pure, categorially determinate manifolds, in regard to which it is natural to speak of 'space' in a yet more extended sense. Just so, geometric theory falls under a corresponding genus of theoretically interrelated theory-forms determined in a purely categorial fashion, which in a correspondingly extended sense can be called 'geometries' of these 'spatial' manifolds. At any rate, the theory of n-dimensional spaces realizes a theoretically closed piece of the theory of theory in the sense defined above. The theory of Euclidean manifold of three dimensions is an ultimate single particularity of something ideal in this legally interconnected series of *a priori*, purely categorial theoretic forms (formal deductive systems). This manifold itself is related to 'our' space, i.e., space in the ordinary sense, as its pure categorial form, the ideal genus of which the latter represents, so to say a single particularity of something individual rather than a specific difference." The translation is by J.N. Findlay, Husserl, *Logical Investigations*, Vol. I, pp. 242f. In his *Krisis der europäischen Wissenschaften*, section 9, Husserl considers the

historical form of the problem relating "space" in the ordinary sense and "space" in the sense of the "categorial form of world-space." Cf. also H.H. Grunwalt, *Die Phänomenologie Husserls mit Berüchtsichtigung auf Galileo*, pp. 52ff.

35. Becker, p. 493.
36. Ibid., pp. 482ff.
37. Ibid., pp. 477ff.
38. Ibid., p. 428.
39. Ibid., pp. 428f. The translation is mine.
40. Ibid., p. 493. "Actual" space is not ipso facto a Euclidean manifold; see Hofmann, pp. 130f. On the other hand, no numerical distinction is made between "actual" space and e.g., "seen space," as in Hering, in Hermanns *Handbuch der Physiologie* (Leipzig, 1879), pp. 343ff. (cited by Hofmann, p. 102). Rather "actual space" signifies uniformity and homogeneity of the spatial spread with respect to three dimensions.
41. Claesges, pp. 87ff.; Mouloud, pp. 271ff.
42. See above, pp. 82f.
43. Becker, p. 494.
44. Ibid., p. 495. The translation is mine.
45. Ibid., the translation is mine.
46. See above, pp. 136ff.
47. Becker, p. 496. The translation is mine.
48. Ibid., pp. 496ff.
49. Ibid., p. 497. The translation is mine.
50. In addition to Becker, pp. 497ff., see Mouloud, pp. 53ff., 62ff., 273, ff.; and Elisabeth Ströker, *Philosophische Untersuchungen zum Raum*, Part II, Section II; Hans Voss, *Transzendenz und Raumanschauung*, section 11. See above, pp. 95f., 121f., and p. 373, note 15.
51. In this connection, see the "Einleitung" of Claesges to *Ding und Raum*, pp. XXII and XXII, note 4.
52. See above, pp. 73ff., 173f., 186ff.
53. See above, pp. 79ff., 84f.
54. Becker, p. 496. The translation is mine. See also ibid., pp. 439ff.
55. See above, sections 21, 37, 53.
56. See above, pp. 118f. The "intervention" is similar to that found in the theories of consciousness formulated by von Ehrenfels, Benussi, Meinong; see Aron Gurwitsch, *The Field of Consciousness*, pp. 57ff. The resemblance of Becker's transcendental

deduction with that of Kant (see above, p.119) would seem to run deeper than he probably realized: the criterion of dimensionality expresses an idea similar to what Kant calls "incomplete space," i.e., where prespaces are limits of other spaces (e.g., quasi-objective and objective ones). And, like Kant, for Becker the mind must be such that "complete space" is given as an immensity of hierarchical stratifications, all of which are "spaces." It also seems to me that Becker rather blindly follows Husserl's whole-part theory which, as has been shown elsewhere, is unacceptable (see F. Kersten, "The Originality of Gurwitsch's Theory of Intentionality," *Research in Phenomenology*, pp. 20ff., especially p. 24).

57. See above, p. 75.
58. Gurwitsch, pp. 271f.
59. See above, pp. 139ff.
60. See Hofmann, pp. 115f., who provides an example similar to that of Gurwitsch: I see a glass of clear water; the surface boundaries are distinguishable when I am "close" to the glass and are indistinct when I am "farther away" from the glass; these phenomenological variations in distinctness are independent of changes in direction, size, etc.; ibid., p. 124. Still another example (p. 115) given by Hofmann is that of going to the window and looking out at the wall of the building across the street; when I retreat from the window farther and farther back into the room away from the window, the distinctness and localization of depth diminishes. To be sure, neither Gurwitsch nor Hofmann have reduced their phenomenological theme to prespace. For Husserl's view c. 1907, see *Ding und Raum*, pp. 261f.; Husserl begins with the example of two balls standing out on an empty background such as the sky: "Between them is empty space." "Moreover, one can also say that there is always empty space beween two objectively 'distant' point-areas, i.e., being constituted in discrete spreadness." And although we cannot see empty space, we can, he says, phantasy a possible Betweenness as empty "but continuously fulfillable space." "Bodies are seen, and the Betweenness is seized upon along with what is seen--Betweenness that, in phantasy, can be filled out corporeally in such or such a way. <Space> is thus already co-seen." See also the 1909 manuscript published as Appendix

VII, ibid., pp. 361f., which also refers to the view of Hofmann just mentioned.

61. See above, pp. 140f.
62. Above, pp. 143f.
63. Cf. Hofmann, pp. 125ff.
64. See above, pp. 144ff.
65. See above, pp. 133f.
66. See above, 151f. Cf. Hofmann, pp. 126ff., 132ff., for a similar view argued against Lipps, Hering, Stumpf and Külpe.
67. See above, pp. 179, 191f.
68. See above, pp. 198f.
69. See above, pp. 144f.
70. Claesges, pp. 87f.
71. Because of the way in which I understand a "transcendental aesthetics" and "transcendental analytics," the whole setting of the problem, as I see it, is different from that of Claesges; see below, Chapter Nine. Perhaps the whole distinction between "transcendental phenomenology" and "mundane ontology" ought to be jettisoned; it seems superfluous. See above, section 35; also sections 12, 17 b, and 34.
72. For the problem of idealization in Euclidean geometry, see F. Kersten, "Phenomenology, History and Myth," in *Phenomenology and Social Reality. Essays in Memory of Alfred Schütz*, pp. 262ff.
73. Mouloud, pp. 271ff.
74. Ibid., p. 273.
75. Ibid. The translation is mine.
76. Ibid., p. 274.
77. Ibid. The translation is mine. This, too, has a strong Kantian flavor, especially of Kant's reconciliation of the various approaches to the problem of space; see above, pp. 118f.
78. See ibid., p. 53.
79. Husserl, *Ideas*, First Book, section 150.
80. Ibid. For the relationship of eidetic necessity and contingency, see ibid., section 9; and above, sections 31, 32.
81. Ibid.
82. Ibid., section 151.

CHAPTER EIGHT

1. See above, section 68.
2. Hedwig Conrad-Martius, "Realontologie," *Jahrbuch für Philosophie und phänomenologische Forschung*, VI, section 72f.
3. Cf. Husserl, Mss. D13 I (1921), p. 15: "<If we discriminate [*Abstrahieren*] in the entire oculomotoric system all aspect Data, all> 'stuff of perception,' <then> we find the order of kinaesthetic sequences and 'empty' fields...empty space is constituted in this combination." (The glosses are by Claesges, who cited this passage, *Edmund Husserls Theorie der Raumkonstitution*, p. 82. The translation is mine.)
4. Thus Husserl's assertion in D 13 I, p. 20 (cited by Claesges, op. cit., p. 82): "Empty space is the potentiality of objects (phantoms)." "Empty 'space'" is then no longer, as in 1907, when Husserl was arguing against Heinrich Hofmann, a *phantasied* space that can be filled out with phantoms and physical things.
5. See above, section 69.
6. For a general statement of the idea of apperception from Husserl's middle period, see Edmund Husserl, *Analysen zur Passiven Synthesis aus Vorlesungens- und Forschungsmanuskripten 1918-1926* pp. 336ff., especially p. 337, note 1.
7. Husserl, *Cartesian Meditations*, pp. 46, 47, 78, 79. For a later account of apperception in primary passivity, especially as it applies to the various elements of oriented constitution, see Husserl, "Die Welt der lebendigen Gegenwart und die Konstitution der Ausserleiblichen Umwelt," pp. 337f. (English translation by Frederick A. Elliston and Lenore Langsdorf in *Husserl: Shorter Works*, pp. 247f.); Elmar Holenstein, *Phänomenologie der Assoziation*, section 26ff. Also see above, section 13.
8. See above, section 18ff.
9. Husserl, *Ideen, Zweites Buch*, section 18b., p. 64.
10. See Husserl, ibid., section 15ff.; and *Ideas, First Book*, section 150; *Cartesian Meditations*, section 61.
11. See above, pp. 70ff.; also the discussion of Becker's "transcendental deduction," pp. 342ff. What Becker says about the prespaces being "parts" of quasi-objective space and objective space would rather seem to apply to phantom and primordial spaces--though, to be sure, now in a quite altered meaning because it is no longer a question of the acquisition at a

higher level of another dimension by the homogeneous spatial continuum. The term, "part," is admittedly not the best term, but will serve for the moment.

12. See above, pp. 35ff., 39ff., See also F. Kersten, "Can Sartre Count?" pp. 345ff.
13. For more detailed discussion, see Aron Gurwitsch, "Perceptual Coherence as the Foundation of the Judgement of Predication," pp. 78ff.; and Giuseppina Moneta, *On Identity. A Study in Genetic Phenomenology*, pp. 46ff.
14. See Husserl's statement from the *Prolegomena* cited above, note 34 p. 387; see above, pp. 54, 87ff., where the difference between formal and material universals is stated.
15. A brief comparison with the Kantian idea of "causality" may be instructive here. Both Husserl and Kant agree that anything experienceable as a "world" of necessity has a "causal" structure of some sort, and both agree, in addition, that the "causal" structure of the experienced "world" derives from something they both call "transcendental consciousness." Finally, they both agree that human beings are in some sense "free." But they disagree about the nature of the idea of "transcendental consciousness" and about the way in which the "causal" structure of a "world" is derived from "transcendental consciousness." (See in this connection, Aron Gurwitsch, "The Kantian and Husserlian Conceptions of Consciousness," *Studies in Phenomenology and Psychology*, pp. 150ff.) In contrast to Kant, only Husserl would seem to distinguish between the physicist's "causality" and that of the daily world, the former being an "idealization" of a specific kind of the latter and which is mathematically "exact." Moreover, because Husserl finds the causal style of the daily world to be "inexact," there is a limited freedom compatible with "determinism," i.e., with that degree of "determinism" that really exists in the phenomenal world. (See Max Scheler, "Zur Phänomenologie und Metaphysik der Freiheit," *Schriften aus dem Nachlass*, Vol. I, pp. 155-174.) Thus there is no question of freedom in the physicist's world because there is no mundanization of transcendental natural living in that "world." A "thing-in-itself" therefore need not be postulated to "save the appearance" of freedom.
16. See above, pp. 143f. See also Husserl, "Notizen zur Raumkonstitution," pp. 218ff.

17. David Katz, *Die Erscheinungweisen der Farben*, pp. 8f.
18. Ibid., p. 12.
19. See above, pp. 204ff.
20. See Wilhelm Schapp, *Beiträge zur Phänomenologie der Wahrnehmung*, pp. 86f.
21. Katz, op. cit., p. 13.
22. Maurice Merleau-Ponty, *Phenomenology of Perception*, translated by Colin Smith, p. 311.
23. Schapp, pp. 78, 96f.
24. Ibid., p. 97, note 1.
25. See above, pp. 146f. for the constituting of "resistance" in visual prespace.
26. Schapp, p. 80.
27. Ibid., pp. 114, 116. The translation is mine. See also ibid., pp. 86f. for a different case of alteration of what is inherent in the presentation of physical thing.
28. Husserl, Notizen zur Raumkonstitution," p. 218. The translation is mine.
29. Ibid.
30. Husserl, "Die Welt der lebendigen Gegenwart," p. 337. English translation p. 247.
31. See Husserl, *Cartesian Meditations*, pp. 116f.: "As reflexively related to itself, my animate bodily organism (in my primordial sphere) has the central 'Here' as its mode of givenness; every other body, and accordingly the 'other's' body, has the mode 'There.' This orientation, 'There,' can be freely changed by virtue of my kinaesthesias. Thus, in my primordial sphere, the *one spatial 'Nature'* is constituted throughout the change in orientations, and constituted moreover with an intentional relatedness to my animate organism as functioning perceptually. Now the fact that my bodily organism can be (and is) apprehended as a *natural body existing and moveable in space like any other* is manifestly connected with the possibility expressed in the words: By free modification of my kinaesthesias, particularly those of locomotion, I can change my position in such a manner that I convert any There into a Here--that is to say, I could occupy any spatial locus with my organism. This implies that, perceiving from there, I should see the same physical things, only in correspondingly different modes of appearance, such as pertain to my being there. It

implies, then, that not only the systems of appearance that pertain to my current perceiving 'from here,' but other quite determinate systems, corresponding to the change of position that puts me 'there,' belong constitutively to each physical thing. And the same in the case of every other 'There.'" The translation is by Dorion Cairns. Cf. also section 17. in addition, see *Ideas*, First Book, section 53, p. 103, and section 150, p. 315; *Ideen*, Zweites Buch, section 41a, pp. 158f.

32. See above, p. 238.
33. Alfred Schütz, "Husserl and the Social Sciences," *Edmund Husserl, 1859-1959* (The Hague: Martinus Nijhoff, 1959), pp. 95f.
34. See above, pp. 197ff., 259ff.; also David Katz, *Der Aufbau der Tastwelt*, p. 255. See also the discussion by Ernst Cassirer, *Philosophie der symbolischen Formen*, Dritter Teil: *Phänomenologie der Erkenntnis*, p. 151.
35. See above, pp. 79ff., 108ff.; Katz, section 48, for a review of the various conceptions of the function of touch in perception in the history of philosophy and psychology.
36. See Katz, section 46, especially the remarks on Gelb and Goldstein and on Wittmann.
37. See above, pp. 179ff.; Katz, section 46.
38. Cassirer, op. cit., p. 151. Strictly speaking, then, phenomenologically it is necessary to deny the traditional idea that tactual experience has an epistemic primacy. Indeed, it would seem to follow from what has been developed that it makes little sense to speak of "primacy" with respect to any of the senses; certainly one can distinguish different roles of touch or vision at different levels of oriented constitution. But that does not of necessity signify that one sense has a primacy over another. Thus it is not a question, for instance, of vision acquiring a primacy over touch in the case of quasi-objective primordial space; the question is rather of the way in which quasi-objective primordial space is an "intersubjective" quasi-objective space.
39. Cassirer, p. 151.
40. Ibid. The translation is mine.
41. See above, sections 8, 24.
42. See Erwin Strauss, "The Upright Posture," *Psychiatric Quarterly*, XXVI (1952). To be sure, the objectivation of the constituting of "upright posture" as an essentially necessary feature of the

quasi-objective bodily organism pertains to a higher level of oriented constitution with respect to the phenomenon of expression that Strauss analyzes. At its most constitutively primitive, the upright posture is the secondarily constituted appearance of the full phantom organism "under circumstances." Although important, it is not possible to examine this feature of constitution of the organism further here.

43. In contrast to the modes of original presentation of the quasi-objective phantom organism; see above, pp. 201ff.
44. See above, pp. 36f., 186f.
45. Husserl, *Cartesian Meditations*, section 18, p. 43. (The translation is slightly altered.)
46. See above, pp. 216ff., 226f.
47. See above, pp. 30f.
48. See *Cartesian Meditations*, section 61, pp. 141ff.
49. Ibid., p. 117. The emphasis is mine.
50. See above, pp. 145f.
51. "Private" or "inward" are words that therefore can only be employed in the second and never in the first person which, in turn, signifies that presupposed is an archontic belief that may be called a "life-conviction" (see F. Kersten, "The Life Concept and Life Conviction," pp. 115f.) I do not, for example, speak to the piece of paper but rather to you, to what, for me, is not only transcendent to my mental life-processes but which is also presented immediately as private, as inward. On the basis of this assumptive life-conviction I perceive a piece of paper public as much for me as for you, a piece of paper public for *us*--a publicity that *we* share, that we have in common. I always address a privacy and never a public something, yet I can only address a privacy *in public*.
52. See Rene Toulemont, *L'Essence de la Société selon Husserl*, pp. 97ff., 311ff.; Aron Gurwitsch, *Human Encounters in the Social World*, sections 18, 21, 22, 29; and Husserl, *Erste Philosophie (1923/24)*, Zweiter Teil, pp. 505f.
53. Cf. *Cartesian Meditations*, section 35f.
54. See above, pp. 29ff.
55. This situation has been examined in much greater detail in F. Kersten, "The Constancy Hypothesis in the Social Sciences," pp. 560ff.
56. Ibid.

57. See ibid., p. 561; and F. Kersten, "Loneliness and Solitude," *Humanitas*, X (1974), pp. 305f. If not the first, Kulpe, *Die Realisierung* (Leipzig: Hirzel Verlag, 1920), II, pp. 191ff., was one of the first to call attention to the phenomenon of incompatibility by showing that the so-called argument by analogy for the "existence of other minds" is fallacious precisely because it smuggles in the fourth premise of compatibility. See also José Ortega y Gasset, *El Hombre y la Gente*, Chapter 6, pp. 152ff. for important supplementations to the analysis of the phenomenon of incompatibility especially with respect to the equally basic phenomenon of sexuality.
58. Husserl, *Cartesian Meditations*, section 42; the translation is by Dorion Cairns. See above, section 21.
59. Cf. *Cartesian Meditations*, section 44; see above, pp. 28f.
60. Ibid., section 42, p. 90. The translation is by Dorion Cairns.
61. See above, section 19, pp. 58ff.
62. See above, pp. 235ff. It should be added here that the expressions, "organism of my mental living," or "organism of my mental life-processes," or "physical component of a psychophysical thing," and similar expressions, are admittedly ambiguous. No "mind-body" dualism is purported, however; the datum is, strictly, "consciousness of my organism," a set of intendings to somatic states of various sorts.
63. See F. Kersten, "The Constancy Hypothesis in the Social Sciences," pp. 561f.
64. See Max Scheler, *Der Formalismus in der Ethik*, pp. 158ff. (English translation, pp. 139ff.); and Alfred Schütz, "The Problem of Transcendental Intersubjectivity in Husserl," pp. 79ff.
65. Nor need one introduce another *deus ex machina*, the theory of the "organization of consciousness" such as found in Aron Gurwitsch, "An Apparent Paradox in Leibnizianism," *Social Research*, 33 (1966), pp. 47ff.; see also Aron Gurwitsch, *Leibniz. Philosophie des Panlogismus*, pp. 450ff.
66. See above, p. 214. This is also the foundation for the (secondarily) constituted "milieu-worlds" of the real, objective world. See below, Chapter Nine.
67. Husserl, "Foundational Investigations of the Phenomenological Origin of the Spatiality of Nature," p. 224.
68. Ibid., p. 225.
69. Ibid., p. 227.

70. It is in this context that Husserl introduces a version of the Cartesian problem (above, p. 111): if space defines the real, what then is space--if earth defines motion, what then is earth neither in motion nor at rest? See "Foundational Investigations," p. 225.
71. Husserl, p. 226.
72. Ibid.
73. Ibid., p. 227; see above, pp. 261f.
74. Although we cannot develop it here, it must be noted that this is equally the intentive foundation for the constituting of "history;" see ibid., p. 230. In any case, it is in this constitutive setting that the distinction between physical thing-body and animate body-organism is fixed and consistent, stable, i.e., "substantive." See Husserl, *Ding und Raum*, pp. 343ff.
75. Husserl, "Foundational Investigations," pp. 227f.
76. Ibid., p. 227.
77. Husserl, "Vorlesungen zur Phänomenologie des inneren Zeitbewusstseins," section 1. (English translation, pp. 22f.) See above, sections 4, 9, 62.
78. Husserl, p. 368. The translation is mine. Cf. also ibid., pp. 373f.
79. Ibid., sections 1, 2. See above, sections 4, 53.
80. See above, sections 8, 24, 77.
81. Above, p. 250.
82. Husserl, sections 34, 36. In this connection, see also the study of Yvonne Picard, "Les temps chez Husserl et chez Heidegger," *Deucalion*, I (1946), Part I. (A student of Jean Wahl, Picard was killed in a German concentration camp in 1941.) For a discussion of the polemic that lies behind the beginning of Husserl's 1905 lectures, see the fragment of Henri Dussort's Introduction to his translation of the lectures (*Leçons pour une Phénoménologie de la Conscience Intime du Temps*, pp. 185ff.) and the Introduction by Rudolf Boehm to Edmund Husserl, *Zur Phänomenologie des Inneren Zeitbewusstseines* (*1893-1917*), pp. XXXff.
83. See above, section 25, and the further remarks there on the 1905 lectures.
84. See above, p. 187; cf. Husserl, "Vorlesungen," section 35, p. 329; section 39, pp. 436f.; Appendix VI, p. 465 (English translation, pp. 99, 108f., 149ff.) The intending previously protended to and subsequently retrotended to is precisely what is signified by the phrase, "living present." See also *Ideas*, First Book, section

81, where Husserl describes the same thing as the stream or flow of mental life-processes as temporally extended and flowing in world-time; the account is somewhat misleading, however, because the non-genuine time of what is transcendent to mental living is not sharply distinguished from the genuine time of mental living itself.

85. Husserl, "Vorlesungen," section 10. See also Aron Gurwitsch, "Phenomenology of Thematics and of the Pure Ego," pp. 274ff.
86. See Gurwitsch, p. 277.
87. Husserl, *Cartesian Meditations*, p. 43.
88. Ibid., section 39, p. 81.
89. See Husserl, Mss. C 3 II, p. 7; cited by Gerd Brand, *Welt, Ich und Zeit*, p. 75.
90. Husserl, *Formal and Transcendental Logic*, section 96, pp. 238f. The translation is by Dorion Cairns, and I have added the emphasis. For what immediately follows, see *Ideas*, First Book, section 81, and *Experience and Judgement*, sections 42, a, and 64, c.
91. Thus Aron Gurwitsch, "On the Intentionality of Consciousness," p. 137. See above, pp. 187ff. and the discussion of the constituting of the temporal determination, "as-present."
92. Husserl, *Formal and Transcendental Logic*, p. 241. The translation is by Dorion Cairns.
93. Husserl, *Cartesian Meditations*, p. 147. The translation is by Dorion Cairns.
94. See Husserl, *Nachwort zu meinen* "Ideen zu einer reinen Phänomenologie und phänomenologischen Philosophie," p. 562: I find myself "as transcendental Ego-self, concrete in my own transcendental conscious living" and this signifies that transcendental "fellow subjects become legitimated in my <own> transcendental living within a transcendental community of a plurality of <transcendental> subjects [*Wirgemeinschaft*] which is likewise legitimated <in my own transcendental living>. Transcendental intersubjectivity <in the natural attitude> is accordingly that transcendental intersubjectivity in which the real world, as Objective, as <the real world> for 'everyone,' becomes constituted." The translation is mine, and the glosses express what Husserl says before and after this passage. See also *Formal and Transcendental Logic*, p. 248.

95. Gurwitsch, "On the Intentionality of Consciousness," p. 138. Neither noematic unity--broadly speaking, mental living itself or world--can, as identities, be reduced to the time-flow of mental living; It must also be noted that the mental living which, as intentive to the one pole of identity is ipso facto intentive to the other pole, is not itself extended and flowing in time. To speak of several "flows of time," as Husserl tends to do in the 1905 lectures, is misleading. Only one flow of time is presented and posited as temporally extended, flowing contemporaneously with the multi-storied real, objective world transcendent to it. To be sure, with respect to itself, mental living is not contemporary with anything. See F. Kersten, "Can Sartre Count?" pp. 351f.
96. When used without qualification, "time" is time of mental life-processes. Unless otherwise specified, "intending" is an auditory sensing of a tone-sensum.
97. For a rather different reading of these texts of Husserl, see Klaus Held, *Lebendige Gegenwart. Die Frage nach der Seinsweise des transzendentalen Ich bei Edmund Husserl, entwichelt am Leitfaden der Zeitproblematik*, pp. 80ff., 134ff. In the light of what is said in the text, it rather seems to me that regardless of actional or passive character of any given mental life-process or extent of any given mental life-process with respect to its presenting and positing, it is always self-presenting and self-positing in an identifying synthesis in primary passivity; no antecendent actionality, functioning, anonymous or otherwise need be presupposed. On phenomenological grounds, it is, after all, impossible to speak of the same intending as recurring. What the diagram represents is the constituting of something identical in a multiplicity of nonrecurring auditory sensings: at "t_1" there is intending to the tone-sensum E_n and at "t_2" an intending to E_2...as identical with E; and so forth. Were the two intendings, auditory sensings, not distinct, were the one simply a recurrence of the other, there would be no intending to the identity of E_n and E_2 at "t_2". Similarly, unless a current extent of the sensing were self-differentiating--the self-same, previously protended to and subsequently retrotended to--there would be no intending to identity itself of mental living. What is said about the sensing holds equally for mental living as a whole. Mental living, and any given extent of

mental living is constituted in self-positing syntheses of identification and difference as an identity presented and posited throughout a multiplicity of protendings and retrotendings to itself, and each partial extent of mental living is constituted as identical and different from every other partial extent if not in specific kind then at least always in tense. On this basis, mental life-processes and their partial extents are protended to and retrotended to as paired in primary passivity, and that includes actional as well as passive processes, and no antecendent actionality, no antecendent engagement of the ego in mental living is required for the pairing, hence for the original constituting of processes as of a certain kind (e.g., actional processes of judging; see above, pp. 36ff.). Finally, it is equally gratuitous, it seems to me, to posit or postulate an ultimate "*nunc stans*" of "time-consciousness," for mental living is always presented and posited as flowing through a "stationary" or "unflowing" frame in such a way that the relation of any partial extent to each part of the frame continually changes (the tense changing from "will be later" to "will be soon" etc.). The "unflowing frame," however, is not a *nunc stans* because that part of the frame which is the past was never present, although that segment or extent of mental living which is past was present and became past, and so forth. Moreover, unless the flow of mental living and its relations to the frame were not as described then mental living would be nothing but continuous recurrence: that would be the *nunc stans* and, it seems to me, just the sort of postulate Husserl's analysis of time avoids. Mental living is rather constituted as "self-identity" and "non-recurrence."

CHAPTER NINE

1. See above, pp. 28, 95.
2. See above, pp. 32f., 107f.
3. See above, pp.108ff., 121ff.
4. This situation has made itself felt in recent discussions of so-called "medical ethics;" see F. Kersten, "The Life Concept and Life Conviction: A Phenomenological Sketch," pp. 110ff.

5. This leads to the substitution of concepts of scientific thinking for those of subscientific thinking and experiencing; see in this connection, F. Kersten, "The Constancy Hypothesis in the Social Sciences," pp. 524ff.
6. See above, p. 109.
7. See above, pp. 155ff.
8. See F. Kersten, "The Life Concept and Life Conception," pp. 112f.
9. In this connection, see Claesges, *Edmund Husserls Theorie der Raumkonstitution*, section 9, pp. 45ff.; Elisabeth Ströker, *Philosophische Untersuchungen zum Raum*, especially III. Abschnitt.
10. See above, pp. 43ff., 71.
11. See above, pp. 55, 89ff.
12. See above, p. 89.
13. See above, pp. 92f.
14. See above, p. 93f.
15. See above, p. 92. For a detailed review of Husserl's discussion, see Claesges, sections 5-10.
16. Claesges, p. 34. The translation is mine.
17. See *Ideas*, I, sections 23f.; 12.
18. Ibid., section 15.
19. Ibid.
20. Ibid., section 72.
21. Ibid., section 73.
22. The analogy has its origin in the Third Logical Investigation, section 3ff., and especially section 25, p. 289 (second edition): "Die Zerstückung der quasi-räumlichen Ausbreitung eines visuallen, unverändert dauernden, aber in Abstraktion von dem zeitlichen Moment betrachteten Inhalts bestimmt auch eine Zerstückung dieses Inhalts selbst."
23. See *Ideas*, I, section 88.
24. See *Logical Investigations*, English translation, Vol. II, p. 485. The German text reads (second edition, pp. 288f.): "die Zerstückung eines unselbständigen Moments bedingt eine Zerstückung des konkreten Ganzen, indem die sich ausschliessenden Stücke, ohne selbst in ein Fundierungsverhältnis zueinander zu treten, neue Momente an sich ziehen, durch die sie nun einzeln zu Stücken des Ganzen suppliert werden." The translation has been altered slightly to make it consistent with the terminology of Husserl used here.

25. *Ideas*, I, section 49.
26. Ibid., section 50.
27. Ibid., section 53; see above, pp. 12ff., 55f., 111.
28. Primarily in Carl Stumpf, *Über den psychologischen Ursprung der Raumvorstellung*, pp. 106ff.
29. Gurwitsch, Phenomenology of Thematics," p. 262.
30. Ibid., pp. 266ff. It should be noted in this connection that Gurwitsch avoids Husserl's reformulations in *Ideas*, I, sections 12 and 15, and that he does not consider the distinction between discrete and non-discrete parts.
31. See above, pp. 15f.
32. To be sure, Gurwitsch wants to understand something different by "ego" than does Husserl; see Gurwitsch, "A non-egological Concept of Consciousness," *Studies in Phenomenology and Psychology*, pp. 287ff. For a more Husserlian account of the "ego," see Robert W. Jordan, "Being and Time: Some Aspects of the Ego's Involvement in his Mental Life," in *Phenomenology: Continuation and Criticism*, pp. 105ff.; for a more conservative view, see F. Kersten, "Privatgesichter," pp. 128f.
33. In *Ideas*, I, section 15, an *individuum* is a material "universal" exemplified by a reality--in contrast to the Third Logical Investigation where an individuum is a reality and not an essence as here. In contradistinction to an individuum, a concretum is exemplified by a material ideality--i.e., a material "universal" exemplified by an ideality which, in turn, is exemplified by a reality. This is further clarified in Husserl, *Erfahrung und Urteil*, section 84. In a note added to one of his copies of *Ideas*, First Book, Husserl notes that the concepts of *individuum* and *concretum* in section 15 are "somewhat modified in contrast to" what he said in the *Logical Investigations*; this is, of course, a gross understatement.
34. This would seem to be the gist of *Ideas*, Second Book, section
64. In this connection, see Ludwig Landgrebe, "Seinsregionen und regionale Ontologien in Husserls Phänomenologie," in Ludwig Landgrebe, *Der Weg der Phänomenologie*, pp. 143ff.
35. Landgrebe, p. 148.
36. For another version of this criticism, see F. Kersten, "Heidegger and Transcendental Phenomenology," pp. 211ff.
37. Landgrebe, p. 153. The translation is mine. For Landgrebe's role

in the preparation of *Ideas*, Second Book, see *Husserliana IV*, pp. xviiif.

38. Landgrebe, p. 153; Husserl, *Ideas*, II, p. 208. See also Alfred Schütz, "Husserl's Ideas, Volume II," in *Collected Papers*, edited by I. Schütz (The Hague: Martinus Nijhoff, 1966), pp. 27ff.; and Paul Ricoeur, "Husserl's Ideas II: Analyses and Problems," in Paul Ricoeur, *Husserl. An Analysis of His Phenomenology*, pp. 76ff.
39. See above, pp. 298ff.
40. Husserl, *Ideas*, II, p. 208.
41. Ibid., section 2. See Landgrebe, p. 154.
42. Ibid., section 49ff.
43. Ibid., pp. 297f.
44. Ibid., pp. 299, 301f. The translations are mine.
45. See the "Einleitung des Herausgebers," *Husserliana IV*, pp. xviiiff.
46. Landgrebe, p. 157.
47. Ibid., p. 158. The translation is mine.
48. Husserl, *Ideas*, I, sections 78, 79.
49. See above, section 4, p. 14.
50. See above, pp. 27f.
51. See above, pp. 19f., 24f., 28f., 35f.
52. Descartes, *Philosophical Essays*, translated with an Introduction and Notes by Laurence J. Lafleur (Indianapolis: Bobbs-Merrill Company, 1964), pp. 100, 102. See above. section 38.
53. Gottfried Wilhelm von Leibniz, *Monadology and Other Philosophical Essays*, translated by Paul Schrecker and Anne Martin Schrecker, p. 6.
54. See above, section 39.
55. David Hume, *A Treatise of Human Nature*, edited by L.A. Selby-Bigge, p. 65.
56. Ibid., p. 66.
57. Immanuel Kant, *Kritik der reinen Vernunft*, B 629f. The translation is by Norman Kemp Smith, *Immanuel Kant's Critique of Pure Reason*, p. 506f.
58. Ibid., B 626; translation, p. 504. See also Eugen Fink, *Alles und Nichts, Ein Umweg zur Philosophie*, pp. 140ff.; Roman Ingarden, *Der Streit um die Existenz der Welt*, p. 61. See above, section 40f.

59. Ingarden, pp. 60f.; see also Roman Ingarden, *Das Literarische Kunstwerk*, section 15. In addition, see H.J. Paton, *Kant's Metaphysic of Experience*, II, pp. 358ff.
60. See above, pp. 27f.
61. Jules Vuillemin, *Physique et Métaphysique Kantiennes*, p. 343. See also pp. 345ff. Exactly what, by implication, are the "postulates of the epoché" Vuillemin does not say.
62. For a review of Brentano's ideas in this connection, see F. Kersten, "Franz Brentano and William James," pp. 177ff.
63. Husserl, *Logische Untersuchungen*,[2] II, 1, p. 428; II, 2, pp. 139f. English translation, pp. 598f., 780ff.
64. Ibid., II, 1, p. 430; translation, p. 599.
65. That is to say, Kant, like Brentano, holds that there is but one generic feature of mental living; accordingly, as Gurwitsch has shown, there is likewise but one meaning of "objectivity" in Kant. See Aron Gurwitsch, "La Conception de la conscience chez Kant et chez Husserl," pp. 73ff.
66. See above, p. 287.
67. Husserl, *Logische Untersuchungen*, II, 1, p. 437. The translation is mine.
68. See *Ideas*, I, section 104.
69. See Husserl, *Logische Untersuchungen*, II, 2, section 1ff.; Adolf Reinach, "Zur Theorie des negativen Urteils," *Gesammelte Schriften*, pp. 57ff.
70. See Husserl, *Logische Untersuchungen*, II, 2, section 43; see above, pp. 89ff., and F. Kersten, "Can Sartre Count?" pp. 343f.
71. See above, p. 28.
72. See above, p. 286.
73. See above, pp. 15f.
74. *Ideas*, I, section 90, p. 185.
75. See above, pp. 14ff., 33ff.
76. *Ideas*, I p. 186.
77. Ibid.
78. Kant, *Kritik der reinen Vernunft*, B 627; translation, p. 505.
79. Ibid., B 629; translation, p. 506.
80. *Ideas*, I, p. 187.
81. Ibid., p. 188.
82. Ibid., section 102, p. 214; cf. section 103ff.
83. Rudolf Boehm, "Zum Begriff des 'Absoluten' bei Husserl," p. 221.

Translation in R.O. Elveton, *The Phenomenology of Husserl*, p. 179.

84. Although the situation is more complex, it nonetheless seems to me to boil down to the following: for Kant and, I suggest also for Brentano, the real, objective world constituted in sub-scientific experiencing and thinking is itself "merely" presented, and only posited as existent in scientific experiencing and thinking. It is then an easy step to *substitute* the world posited for that presented.
85. If the seeings alternate back and forth then they are constituted in what Husserl calls "discontinuous syntheses of identification;" see *Ideas*, I, section 118.
86. See Husserl, *Logische Untersuchungen*, Zweiter Band, Zweiter Teil, (*Husserliana* XIX/2), pp. 851f. for the distinction between spread and extension.
87. See *Ideas*, I, section 44ff.
88. See above, pp. 302f.
89. Boehm, translation, p. 180.
90. *Ideas*, I, section 46.
91. Cf. Thomas Hobbes, *The Elements of Philosophy*, Part II, Chapter VII. Also see above, pp. 120ff.
92. *Ideas*, I, p. 110.
93. Boehm, translation, pp. 181, 182.
94. See *Ideas*, I, section 103ff.
95. Note that in *Ideas* Husserl also speaks of "potential" or "inactional" positing, e.g., in section 113, which must be extended to include, in later writings, primary passive positings and its modes in addition to the primal, unmodalized "passive Doxa" in which all world-experience is said to be rooted; see *Erfahrung und Urteil*, sections 7ff.
96. Or to express the situation as Husserl does in *Logische Untersuchungen*, II, 1^2 pp. 400ff., especially pp. 435f. (translation, pp. 578ff., 604f.), the distinction is between the object which is intended to and the object as it is intended to; see also Gurwitsch, *The Field of Consciousness*, pp. 185ff.
97. Nor is this yet a case of predication; see Gurwitsch, "Perceptual Coherence as the Foundation of the Judgment of Predication," pp. 84ff.
98. See above, pp. 27f.
99. Above, pp. 30f., 34f.

100. Above, pp. 28f.
101. Husserl, *Formal and Transcendental Logic*, pp. 272f. Translation by Cairns.
102. See above, p. 70.
103. *Formal and Transcendental Logic*, p. 273. See above, sections 20, 34, 78f., 80.
104. See above, p. 325.
105. Boehm, translation, pp. 196f.
106. Some indication is given above, p. 226, in the discussion of Becker; see also Becker, "Beiträge zur phänomenologischen Begründung," pp. 419ff., 457ff. See also above, pp. 60ff., 119, 154f., for some indication of the scope and nature of the problems that arise at this juncture.
107. Kant, *Kritik der reinen Vernunft*, B 294f. Translation, p. 257.
108. Husserl, *Ideas*, I, p. 235.
109. Ibid.
110. Husserl, *Cartesian Meditations*, p. 132.
111. Ibid., p. 142.
112. Ibid., p. 146.
113. Ibid.
114. Ibid. Also see above, sections 17, 18. In *Formal and Transcendental Logic* Husserl gives a yet broader scope to his formulation of transcendental aesthetics and analytics. He proceeds to disclose what he calls the "logos of the aesthetic world" and the "logos in a higher sense" that belongs to the real, objective world and to science, an "analytic logos." The same "distribution" of dimensions or oriented constitution is to be found as in the *Cartesian Meditations*, but now set in a specific historical context, namely the context of modern science. This context emerges when Husserl concludes a discussion of the transformation of traditional logic into transcendental phenomenological logic by observing that all the material *a priori* disciplines the nature of which he developed under the heading of "logic" are developed in the pregnant sense of the term, "logic." Transcendental logic or, equivalently stated, ontology, elaborates the universal Apriori of a possible world. Such elaboration gives rise at various stages to the problem of a "world-logic" or "ontology of the world" ("*mundane Ontologie*"). As the study of the underpinings of that structure we find, again in a new sense of the term,

transcendental aesthetics which "deals with the eidetic problem of any possible world *as a world given in 'pure experience'* and thus precedes all science in the 'higher' sense; accordingly it undertakes the eidetic description of the all-embracing Apriori, without which no Objects could appear unitarily in mere experience, prior to categorial actions..., and therefore without which the unity of a Nature, the unity of a world, as a passively synthesized unity, could not become constituted at all." (Translation by Dorion Cairns, p. 292.) To be a genuine science, this "logos of the aesthetic world" requires its own appropriate constitutional investigations that eventually lead to "logics" in another sense "among which the highest and most inclusive would be the *logic* of the absolute *science*, the logic of transcendental-phenomenological *philosophy* itself." (Ibid., p. 291) In another connection, I have tried to formulate the same situation in a rather different way; cf. "Phenomenology, History and Myth," pp. 262ff.

115. See above, section 10f.

116. See F. Kersten, "The Life-World Revisited," pp. 48ff. See also Grathoff, "Das Problem der Intersubjektivität bei Aron Gurwitsch und Alfred Schütz," pp. 96ff. and the distinction between the life-world and the "work-world;" the former pertains constitutively to the aesthetics and the latter to the analytics (see ibid., pp. 104ff. for an inventory of those sorts of actional intendings included in the analytics). To express the relationship between them in transcendental terms, we may say that the work-world is the "appearance" of something primordial, the life-world, thus something secondarily constituted. The work-world then includes in its essence the life-world, but the converse in not necessarily the case.

117. *Formal and Transcendental Logic*, p. 289.

118. See ibid., section 95; *Cartesian Meditations*, section 55. Husserl tends to use the term, "*Ichgemeinschaft*" (equivalently, "*Allgemeinschaft*"), "community of egos," to designate something similar to the constituting of what Tönnies called *Gemeinbesitz*, the possessing in common of a shared world in common. See Gurwitsch, *Encounters in the Social World*, pp. 117ff., 125ff.

119. See Husserl, *Formal and Transcendental Logic*, section 95; *Ideas*, I, section 135.

120. See Husserl, *Erfahrung und Urteil*, sections 17ff., 34, 40ff., 61ff.
121. See above, pp. 38ff.
122. Husserl, *Formal and Transcendental Logic*, pp. 270f.
123. Ibid., p. 272.
124. Ibid., p. 275
125. Above, pp. 17f.
126. *Cartesian Meditations*, P. 156.
127. Above, p. 106.
128. Thomas Mann, *Joseph in Egypt*, translated by H.T. Lowe-Porter, Vol. I, pp. 180f. See also Dorion Cairns "Philosophy as a Striving Toward Universal *Sophia* in the Integral Sense, pp. 27-43, for an account of the meaning of "philosophy" itself as "the objective correlate, a critically justified objective sense" of a critically justified striving to return to the things themselves as an ideal to be approximated not only in one's own life but in that of one's predecessors and one's successors. Cf. ibid., pp. 41ff., and p. 42, note 8.

LIST OF WORKS CITED

Adorno, Theodor. *Zur Metakritik der Erkenntnistheorie*. Stuttgart: Kohlhammer, 1956.

Asemissen, Herman. *Strukturanalytische Probleme der Wahrnehmung in der Phänomenologie Husserls*. Köln: Kölner Universitäts-Verlag, 1957.

Bachelard, Suzanne. *A Study of Husserl's Formal and Transcendental Logic*. Translated by Lester E. Embree. Evanston: Northwestern University Press, 1968.

Bain, Alexander. *The Senses and the Intellect*. London: Longmans, Green & Co., 1868.

Becker, Oskar. "Beiträge zur phänomenologischen Begründung der Geometrie und ihrer physikalischen Anwendungen." *Jahrbuch für Philosophie und phänomenologische Forschung*, VI (1923), pp. 385-560.

Berkeley, George. *The Works of George Berkeley*. Edited by A.A. Luce and T.E. Jessop. London: Thomas Nelson and Sons, 1951.

Boehm, Rudolf. "Les ambiguités des concepts husserliens d''immanence' et de 'transcendence'." *Revue Philosophique de la France et de L'Étranger*, CXLIX (1959), pp. 481-526.

_______. "Zum Begriff des 'Absoluten' bei Husserl." *Zeitschrift für Philosophische Forschung*, XIII (1959), pp. 214-242.

_______. "The Concept of the 'Absolute' in Husserl." Translated by R.O. Elveton. *The Phenomenology of Husserl. Selected Critical Readings*. Edited, Translated, and with an Introduction by R.O. Elveton. Chicago: Quandrangle Books, 1970, pp. 174-203.

_______. *Vom Gesichtspunkt der Phänomenologie. Husserl-Studien*. Den Haag: Martinus Nijhoff, 1968.

_______. Einleitung des Herausgebers, Edmund Husserl, *Zur Phänomenologie des inneren Zeitbewusstesins (1893-1917)*. Den Haag: Martinus Nijhoff, 1966.

Brand, Gerd. *Welt, Ich und Zeit*. Den Haag: Martinus Nijhoff, 1955.

Brockman, Jan M. *Phänomenologie und Egologie. Faktische und transzendentales Ego bei Edmund Husserl*. Den Haag: Martinus Nijhoff, 1963.

Cairns, Dorion. *Conversations With Husserl and Fink*. Edited by Richard M. Zaner. The Hague: Martinus Nijhoff, 1976.

_______. "The Many Senses and Denotations of the Word *Bewusstsein* ("Consciousness") in Edmund Husserl's Writings." *Life-World and Consciousness. Essays for Aron Gurwitsch*. Edited by Lester E. Embree. Evanston: Northwestern University Press, 1972, pp. 19-31.

_______. "Philosophy as a Striving Toward Universal *Sophia*." *Essays in Memory of Aron Gurwitsch*. Edited by Lester E. Embree. Washington, D.C.: The University Press of America, 1983, pp. 28-43.

Cassirer, Ernst. *Das Erkenntnisproblem in der Philosophie und Wissenschaft der neueren Zeit*. 3 Vols. Berlin: Bruno Cassirer, 1922.

_______. *Philosophie der symbolischen Formen*. 3 Vols. Oxford: Bruno Cassirer, 1954.

Chapman, Harmon. *Sensations and Phenomenology*. Bloomington: Indiana University Press, 1966.

Claesges, Ulrich. *Edmund Husserls Theorie der Raumkonstitution*. Den Haag: Martinus Nijhoff, 1964.

Conrad, Theodor. *Zur Wesenslehre des psychischen Lebens und Erlebens*. Mit einem Geleitwort von H. L. Van Breda. Den Haag: Martinus Nijhoff, 1968.

Conrad-Martius, Hedwig. "Realontologie." *Jahrbuch für Philosophie und phänomenologische Forschung*, VI (1923), pp. 159-333.

_______. "Zur Ontologie und Erscheinungslehre der realen Aussenwelt. Verbunden mit einer Kritik positivistischer Theorien." *Jahrbuch für Philosophie und phänomenologische Forschung*, III (1916), pp. 345-542.

Descartes, Rene. *Philosophical Essays*. Translated with an Introduction and Notes by Laurence J. Lafleur. Indianapolis: Bobbs-Merrill Co., 1964.

Descartes, Rene. *Oeuvres de Descartes*. Publiées par Charles Adam et Paul Tannery. Paris: Vrin, 1964ff.

Diemer, Alwin. *Edmund Husserl*. Meisenheim am Glan: Verlag Anton Hain K.G., 1956.

Dussort, Henri. "Note d'Henri Dussort sur la partie polemique des Leçons sur le temps." Edmund Husserl, *Leçons Pour Une Phé-*

noménologie de la Conscience Intime du temps. Traduit de L'Allemand par Henri Dussort. Preface de Gerard Granel. Paris: Presses Universitaires de France, 1964, pp. 185-192.

Edie, James. *Speaking and Meaning.* Evanston: Northwestern University Press, 1977.

Ely, Lothar. *Die Krise des Apriori in der transzendentalen Phänomenologie Edmund Husserls.* Den Haag: Martinus Nijhoff, 1962.

Euler, Leonard. "Reflexions sur L'espace et le tems." *Opera Omnia.* Ediderunt E. Hoppet, K. Matter, J.J. Burkhardt. *Series Tertis: Opera Physica. Miscellanea epistolae*, Volumen Secundem, pp. 376-383. Leipzig: B.G. Teubner, 1952.

Fink, Eugen. *Alles und Nichts. Ein Umweg zur Philosophie.* Den Haag: Martinus Nijhoff, 1959.

_______. "Die phänomenologische Philosophie Edmund Husserls in der gegenwärtigen Kritik." *Kant-Studien*, 38 (1933), pp. 319-383. English translation by R.O. Elveton, "The Phenomenological Philosophy of Edmund Husserl and Contemporary Criticism," in *The Phenomenology of Husserl.* Selected Critical Readings. Edited, Translated and with an Introduction by R.O. Elveton. Chicago: Quadrangle Books, 1970, pp. 73-147.

_______. "Das Problem der Phänomenologie Edmund Husserls." *Revue Internationale de Philosophie*, I (1938/39), pp. 226-270. English translation by Robert M. Harlan, "The Problem of the Phenomenology of Edmund Husserl," in *Apriori and World*, edited and translated by W. McKenna, R.M. Harlan and L.E. Winters. The Hague: Martinus Nijhoff, 1981, pp. 21-55.

Geiger, Moritz "Beiträge zur Phänomenologie des aesthetischen Genusses." *Jahrbuch für Philosophie und phänomenologische Forschung*, I (1913), pp. 567-684.

Gent, Werner. *Die Philosophie des Raumes and der Zeit.* Hildesheim: Georg Olms Verlagsbuchhandlung, 1962. 2nd Edition.

Goldstein, Kurt. "Über induzierte Tonusveränderung." *Klin. Wochschriften*, 4 (1925).

Grasquet, Bernhard. *Cézanne.* Paris: Bernheim Jeune, 1926.

Grathoff, Richard. "Das problem der Intersubjektivität bei Aron Gurwitsch und Alfred Schütz." In *Sozialität und Intersubjektivität. Phänomenologische Perspektiven der Sozialwissenschaften.* Herausgegeben von Richard Grathoff und Bernard Waldenfels. München: Wilhem Fink Verlag, 1983, pp. 87-120.

Grunwalt, H.H. *Die Phänomenologie Husserls mit Berüchtsichtung auf Galileo*. Berlin: Emil Ebering, 1927.

Gurwitsch, Aron. "An Apparent Paradox in Leibnizianism." *Social Research*, 33 (1966), pp. 49-64.

________. *The Field of Consciousness*. Pittsburgh: Duquesne University, 1964.

________. *Human Encounters in the Social World*. Edited by Alexander Metraux. Translated By F. Kersten. Pittsburgh: Duquesne university Press, 1977.

________. *Leibniz. Philosophie des Panlogismus*. Berlin and New York: Walter de Gruyter, 1974.

________. "Perceptual Coherence as the Foundation of the Judgment of Prediction." *Phenomenology: Continuation and Criticism. Essays in Memory of Dorion Cairns*. Edited by F. Kersten and R. Zaner. The Hague: Martinus Nijhoff, 1973, pp. 62-89.

________. *Studies in Phenomenology and Psychology*. Evanston: Northwestern University Press, 1966. ("Critical Study of Husserl's *Nachwort*." pp. 107-115; "A Non-egological Concept of Consciousness," pp. 287-300; "Phenomenology of Thematics and of the Pure Ego," pp. 175-286; "The Phenomological and the Psychological Approach to Consciousness," pp.89-106; "The Kantian and Husserlian Concetions of Consciousness," pp. 148-174; "The Place of Psychology in the system of Sciences," pp. 56-68; "On the Intentionality of Consciousness," pp. 124-140.)

________. "La Conception de la conscience chez Kant et chez Husserl." *Bulletin de la Société Francaise de Philosophie*, 54 (1960), pp. 65-96.

Heinemann, Fritz. *Existenzphilosophie. Lebendig oder Tod*? Stuttgart: Kohlhammer, 1954.

Held, Klaus. *Lebendige Gegenwart. Die Frage nach der Seinsweise des transzendentalen Ich bei Edmund Husserl, Entwickelt an Leitfaden der Zeitproblematik*. Den Haag: Martinus Nijhoff, 1966.

Helmholtz, Hermann von. "Die Thatsachen in der Wahrnehmung,": *Vorträge und Reden*, Zweiter Band. Braunschweig: Friedrich Vieweg und Sohn, 1896, pp. 213-248.

Hering, Oswald. "Der Raumsinn und die Bewegungen des Auges." *Hermanns Handbuch der Physiologie*. Leipzig: S. Hirzel, 1879.

Hofmann, Heinrich. "Untersuchungen uber den Empfindungsbegriff." *Archiv für die gesamte Psychologie*, XXVI (1913), pp. 1-135.

Holenstein, Elmar. *Phänomenologie der Assoziation. Zu Struktur und Funktion eines Grundprinzips der passiven Genesis bei Edmund Husserl.* Den Haag: Martinus Nijhoff, 1972.

Hume, David. *A Treatise of Human Nature.* Edited by L.A. Selby-Bigge. Oxford: The Clarendon Press, 1951.

Husserl, Edmund. *Analysen zur Passiven Synthesis (1918-1926).* Herausgegeben von Margot Fleischer. Den Haag: Martinus Nijhoff, 1966.

________. *Aufsätze und Rezension (1890-1910). Mit Ergänzenden Texten.* herausgegeben von Bernard Rang. Den Haag: Martinus Nijhoff, 1979.

________. *Cartesianische Meditationen und Pariser Vorträge.* Herausgegeben und eingeleitet von Dr. S. Strasser. Den Haag: Martinus Nijhoff, 1950.

________. *Cartesian Meditations.* Translated by Dorion Cairns. The Hague: Martinus Nijhoff, 1960.

________. *Ding und Raum. Vorlesungen 1907.* Herausgegeben von Ulrich Claesges. Den Haag: Martinus Nijhoff, 1977.

________. *Erfahrung und Urteil. Untersuchung zur Genealogie der Logik.* Redigiert und herausgegeben von Ludwig Landgrebe. Hamburg: Classen Verlag, 1954.

________. *Experience and Judgment. Investigations in a Genealogy of Logic.* Revised and Edited by Ludwig Landgrebe. Translated by James S. Churchill and Karl Ameriks. Afterword by Lothar Eley. Evanston: Northwestern University Press, 1973.

________. *Erste Philosophie (1923/24).* Zweiter Teil. *Theorie der Phänomenologischen Reduktion.* Herausgegeben von Rudolf Boehm. Den Haag: Martinus Nijhoff, 1959.

________. *Formale und transzendentalen Logik.* Halle: Max Niemeyer, 1929.

________. *Formal and Transcendental Logic.* Translated by Dorion Cairns. The Hague: Martinus Nijhoff, 1969.

________. "Grundlegende Untersuchungen zum phänomenologischen Ursprung der Räumlichkeit." *Philosophical Essays in Memory of Edmund Husserl. Edited by Marvin Farber. Cambridge, Mass.: Harvard University Press, 1940, pp. 305-326.*

________. *Die Idee der Phänomenologie. Fünfte Vorlesungen.* herausgegeben von Walter Biemel. Den Haag: Maritnus Nijhoff, 1950.

________. *Ideen zu einer reinen Phänomenologie und phänomenologischen Philosophie.* Erstes Buch: *Allgemeine Einführung in die reine Phänomenologie*. Neu herausgegeben von Karl Schuhmann. 1. Halbband, Text der 1.-3. Auflage. Den Haag: Martinus Nijhoff, 1976.

_______. *Ideas Pertaining to a Pure Phenomenology and to a Phenomenological Philosophy*. First Book: *General Introduction to a Pure Phenomenology*. Translated by F. Kersten. The Hague: Martinus Nijhoff, 1982.

_______. *Ideen zu einer reinen Phänomenologie und phänomenologischen Philosophie*. Zweites Buch: *Phänomenologische Untersuchungen zur Konstitution*. Herausgegeben von Marly Biemel. Den Haag: Martinus Nijhoff, 1952.

_______. *Ideen zu einer reinen Phänomenologie und phänomenologischen Philosophie*. Drittes Buch: *Die Phänomenologie und die Fundamente der Wissenschaften*. Herausgegeben von Marly Biemel. Den Haag: Martinus Nijhoff, 1952.

_______. *Ideas Pertaining to a Pure Phenomenology and to a Phenomenological Philosophy*. Third Book: *Phenomenology and the Foundations of the Sciences*. Translated by Ted E. Klein and William Pohl. The Hague: Martinus Nijhoff, 1980.

_______. *Krisis der europäischen Wissenschaften und die transzendentalen Phänomenologie. Eine Einleitung in die phänomenologische Philosophie*. Herausgegeben von Walter Biemel. Den Haag: Martinus Nijhoff, 1954.

_______. *The Crisis of European Sciences and Transcendental Phenomenology. An Introduction to Phenomenological Philosophy*. Translated with an Introduction by David Carr. Evanston: Northwestern University Press, 1970.

_______. *Logische Untersuchungen*. Halle: Max Niemeyer, 1900/01.

_______. *Logische Untersuchungen*. Vierte Auflage. Halle: Max Niemeyer, 1928.

_______. *Logische Untersuchungen*. Erster band, *Prolegomena zur Reinen Logik*. Text der 1. and 2. Auflage, herausgegeben von Elmar Holenstein. Den Haag: Martinus Nijhoff, 1955.

_______. *Logische Untersuchungen*, Zeiter Band, Zweiter Teil. Herausgegeben von Ursula Panzer. Dan Haag: Martinus Nijhoff, 1984.

_______. *Logical Investigations*. Translated by J. N. Findlay. London: Routledge & Kegan Paul, 1970.

_______. "Nachwort zu meinen 'Ideen zu einer reinen Phänomenologie und phänomenologischen Philosophie." *Jahrbuch für Philosophie und Phänomenologische Forschung*, XI (1930), pp. 549-570.

_______. "Author's Preface to the English Edition," *Ideas: General Introduction to Pure Phenomenology*, translated by W. R. Boyce Gibson. London: George Allen & Unwin Ltd., 1931.

_______. "Notizen zur Raumkonstitution." Edited by Alfred Schutz. *Philosophy and Phenomenological Research*, I (1940), pp. 21-37, 217-226.

_______. *Phänomenologische Psychologie. Vorlesungen Sommersemester 1925*. Herausgegeben von Walter Biemel. Den Haag: Martinus Nijhoff, 1962.

_______. *Husserl: Shorter works*. Edited by F. Elliston and P. McCormack. Notre Dame: University of Notre Dame Press, 1982. ("The World of the Living Present and the Constitution of the Surrounding World External to the Organism," translated by Fredrick A. Elliston, pp. 238-250; "Foundational Investigations of the Phenomenological Origin of the Spatiality of Nature," translated by F. Kersten, pp. 222-233; "Psychological Studies for Elementary Logic," translated by P. Hudson and P. McCormick, pp. 126-141; "Letter to Arnold Metzger of September 4th, 1919, "translated by Erazim Kohak, pp. 360-364.)

_______. "Vorlesungen zur Phänomenologie des inneren Zeitbewusstseins." Herausgegeben von Martin Heidegger. *Jahrbuch für Philosophie und phänomenologische Forschung*, IX (1929), pp. 367-496.

_______. *Zur Phänomenologie des inneren Zeitbewusstseins*. Herausgegeben von Rudolf Boehm. Den Haag: Martinus Nijhoff, 1966.

_______. *The Phenomenology of Internal Time Consciousness*. Translated by James S. Churchill. Bloomington: Indiana University Press, 1964.

_______. *Zur Phänomenologie des Intersubjektivität*. Texte aus dem Nachlass, Zweiter Teil: 1921-1928. herausgegeben von Iso Kern. Den Haag: Martinus Nijhoff, 1973.

_______. "Die Welt der lebendige Gegenwart und die Konstitution der ausserleiblichen Umwelt." Edited by Alfred Schütz. *Philosophy and Phenomenological Research*, VI (1946), pp. 323-343.

Ihde, Don *Listening and Voice. A Phenomenology of Sound*. Athens, Ohio: Ohio University Press, 1976.

Ingarden, Roman. *Das Literarische Kunstwerk*. Tubingen: Max Niemeyer, 1964. English translation, *The Literary Work of Art*, trans. George. G. Grabowicz. Evanston: Northwestern University Press, 1973.

_______. *Der Streit um die Existenz der Welt.* Tubingen: Max Niemeyer, 1964.

James, William. *The Principles of Psychology.* New York: Dover Publications, 1950.

Jordan, Robert W. "Being and Time: Some Aspects of the Ego's Involvement in His Mental Life." *Phenomenology: Continuation and Criticism. Essays in Memory of Dorion Cairns.* Edited by F. Kersten and R. Zaner. The Hague: Martinus Nijhoff, 1973, pp. 105-113.

Kant, Immanuel. *Critique of Pure Reason.* Translated by Norman Kemp Smith. London: Macmillan, 1953.

_______. *Kritik der reinen Vernunft.* Nach der ersten und zweiten Original-Ausgabe neu herausgegeben von Raymond Schmidt. Hamburg: Verlag von Felix Meiner.

_______. *De mundi sensibilis atque intelligibilis forma et principiis.* Hamburg: Verlag von Felix Meiner, 1958.

Katz, David. *Die Erscheinungsweisen der Farben und ihre Beeinflussung durch die individuelle Erfahrung.* Leipzig: Johann Ambrosius Barth, 1911.

_______. *Der Aufbau der Tastwelt.* Leipzig: Johann Ambrosius Barth, 1925.

Kersten, Fred. "Can Sartre Count?" *Philosophy and Phenomenological Research*, XXXIV (1974), pp. 339-354.

_______. "The Constancy Hypothesis in the Social Sciences." *Life-World and Consciousness. Essays for Aron Gurwitsch.* Edited by Lester E. Embree. Evanston: Northwestern University Press, 1972, pp. 521-563.

_______. "Heidegger and Transcendental Phenomenology." *The Southern Journal of Philosophy*, 11 (1973), pp. 202-215.

_______. "Husserl's Doctrine of Noesis-Noema." *Phenomenology: Continuation and Criticism. Essays in Memory of Dorion Cairns.* Edited by F. Kersten and R. Zaner. The Hague: Martinus Nijhoff, 1973, pp. 114-141.

_______. "The Life-Concept and the Life-Conviction." *Journal of Philosophy and Medicine*, 3 (1978), pp. 107-128.

_______. "Loneliness and Solitude." *Humanitas*, X (1974), pp. 301-312.

_______. "The Occasion and Novelty of Husserl's Phenomenology of Essence." *Phenomenological Perspectives. Essays in Honor of*

Herbert Spiegelberg. Edited by Philip Bossart. The Hague: Martinus Nijhoff, 1975, pp. 61-92.

_______. "The Originality of Gurwitsch's Theory of Intentionality." *Research in Phenomenology*, V (1975), pp. 19-28.

_______. "Phenomenology, History and Myth." *Phenomenology and Social Reality. Essays in Memory of Alfred Schutz*. Edited by Maurice Natanson. The Hague: Martinus Nijhoff, 1970, pp. 234-269.

________. "Private Faces." *Research in Phenomenology*, XII (1983), pp. 167-177.

Koyré, Alexandre. *From the Closed World to the Infinite Universe*. New York: Harper Torchbooks, 1957.

Krakauer, Siegfried. *From Galigari to Hitler*. Princeton: Princeton University Press, 1970.

Külpe, Oswald. *Die Realisierung. Ein Beitrag zur Grundlegung der Realwissenschaften*. Leipzig: Verlag von S. Hirzel, 1920.

Küng, Guido. "The Phenomenological Reduction as Epoché and Explication." *Husserl: Expositions and Appraisals*. Edited by Frederick Elliston and Peter McCormick. Foreward by Paul Ricoeur. Notre Dame: University of Notre Dame press, 1977, pp. 338-349.

Landgrebe, Ludwig. *Der Weg der Phänomenologie*. Gutersloh: Gutersloher Verlagshaus Gerd Mohn, 1963.

Leibniz, Wilhelm Gottfried. *Monadology and Other Philosophical Essays*. Translated by Paul Schreker and Anne Martin Schreker. With an Introduction and Notes by Paul Schreker. Indianapolis: Bobbs-Merrill, 1966.

_______. *Opera Philosophica*. Instruxit Joannes Eduardus Erdmann. Berolini: Sumtibus G. Eichleri, 1840.

Leyendecker, Herbert. *Zur Phänomenologie der Täuschungen*. Halle: Max Niemeyer, 1913.

Lipps, Theodor. "Die Raumanschauungen und die Augenbewegungen." *Zeitschrift für Psychologie und Physiologie der Sinnesorgane*, III (n.d.), pp. 123-171.

Locke, John. *Essay Cconcerning Human Understanding*. Edited with an Introduction Critical Apparatus and Glossary by Peter H. Nidditch. Oxford: At The Clarendon Press, 1975.

Lotze, Rudolf Hermann. *Medicinische Psychologie, oder Physiologie der Seele*. Leipzig: S. Hirzel, 1852.

Mach, Ernst. *The Analysis of Sensations*. Translated by C.N. Williams. New York: Dover Publications, 1959.

Mann, Thomas. *Joseph in Egypt*. Translated by H.-T. Lowe-Porter. New York: Alfred Knopf, 1938.

Merleau-Ponty, Maurice. *L'Oeil et l'esprit*. Paris: Gallimard, 1964.

_______. *Phénoménologie de la perception*. Paris: Gallimard, 1945.

_______. *Phenomenology of Perception*. Translated by Colin Smith. New York: The Humanities Press, 1962.

_______. *Sens et Non-sens*. Paris: Nagel, 1948.

_______. *Sense and Non-Sense*. Translated with a Preface by Hubert L. Dreyfus & Patricia Allen Dreyfus. Evanston: Northwestern University Press, 1964.

Mill, John Stuart. *An Examination of Sir William Hamilton's Philosophy*. New York: Henry Holt, 1884.

Moneta, Giuseppina C. *On Identity. A Study in Genetic Phenomenology*. The Hague: Martinus Nijhoff, 1976.

Moog, Willy. *Die Deutsche Philosophie des 20. Jahrhunderts*. Stuttgart: Enke, 1922.

Morgenstern, Christian. *Das Mondschaf. Eine Auswahl aus den Galgenliedern*. Wiesbaden: Insel-Verlag, 1953.

Mouloud, Noel. "Le Principe spatial d'individuation: Fondement phénoménologique et signification géometrique." *Revue de Métaphysique et de Morale*, 1956, pp. 53-73, 259-282.

Natanson, Maurice. *Edmund Husserl. Philosopher of Infinite Tasks*. Evanston: Northwestern University Press, 1973.

Ortega y Gasset, José. *The Dehumanization of Art and Other Essays*. Translated by Paul Snodgrass and Joseph Frank. Princeton: Princeton University Press, 1958.

_______. *El Hombre y la Gente*. Madrid: Revista de Occidente, 1957.

_______. "Sobre el concepto de sensacion" [1913] *Obras Completas*. Madrid: Revista de Occidente, 1950, Vol. 1, pp. 245-261.

Paton, H.J. *Kant's Metaphysics of Experience*. London: Allen & Unwin, 1956.

Pfänder, Alexander. "Zur Psychologie der Gesinnungen." *Jahrbuch für Philosophie und phänomenologische Forschung*, I (1912), pp. 325-404; III (1916), pp. 1-125.

Picard, Yvonne. "Le Temps chez Husserl et chez Heidegger." *Deucalion*, I (1946/47), pp. 95-124.

Piguet, J.-Claude. *De L'Ésthétique à la Métaphysique*. La Haye: Martinus Nijhoff, 1959.

Pradine, Maurice. *Philosophie de la Sensation*. Strassbourg: publications de l'Universite de Strassbourg, 1928.

Price, H.H. *Perception*. London: Methuen, 1961.

Reid, Thomas. *Inquiry into the Human Mind*. Preface, Notes and Supplementary Dissertations by Sir William Hamilton. Edinburgh: Maclachlan and Stewart, 1852.

Reinach, Adolf. *Gesammelte Schriften*. Herausgegeben von seinen Schulern. Halle: Max Niemeyer, 1921. ("Zur Theorie des negativen Urteils," pp. 56-120; "Die apriorischen Grundlagen des bürgerlichen Rechtes," pp. 166-350; "Über das Wesen der Bewegung," pp. 406-461.)

Reiner, Hans. "Sinn und Recht der phänomenologischen Methode." *Edmund Husserl 1859-1959*. La Haye: Martinus Nijhoff, 1959, pp. 134-147.

Ricoeur, Paul. *Husserl. An Analysis of His Phenomenology*. Translated by Edward G. Ballard and Lester E. Embree. Evanston: Northwestern University Press, 1967.

Sartre, Jean Paul. *L'Etre et le néant*. Paris: Gallimard, 1948.

________. *L'Imaginaire. Psychologie phénoménologique de la imagination*. Paris: Gallimard, 1948.

Schapp, Wilhelm. *Beiträge zur Phänomenologie der Wahrnehmung*. Gottingen, 1910.

Scheler, Max. *Der Formalismus in dr Ethik und dei materiale Wertethik. Neuer Versuch der Grundlegung eines Ethischen Personalismus*. Vierte durchgesehene Auflage herausgegeben mit einem neuen Sachregister von Maria Scheler. Bern: Francke Verlag, 1954.

________. *Formalism in Ethics and Non-formal Ethics of Values*. Translated by Manfred S. Fringes & Roger L. Funk. Evanston: Northwestern University Press, 1973.

________. *Selected Philosophical Essays*. Translated with an Introduction by David Lachterman. Evanston: Northwestern University Press, 1973.

________. *Von Umsturz der Werte. Abhandlungen und Aufsätze*. Herausgegeben von Maria Scheler. Bern: Francke Verlag, 1955.

________. "Zur Phänomenologie und Metaphysik der Freiheit." *Schriften aus dem Nachlass*. Band 1. Zweite, durchgesehene und erweiterte Auflage mit einem Anhang. Herausgegeben von Maria Scheler. Bern: Francke Verlag, 1957, pp. 155-178.

Schuhmann, Karl. *Die Dialektik der Phänomenologie.* Den Haag: Martinus Nijhoff, 1973.

_______. *Die Fundamentalbetrachtung der Phänomenologie.* Den Haag: Martinus Nijhoff, 1971.

_______. *Husserl Chronik. Denk- und Lebensweg Edmund Husserls.* Den Haag: Martinus Nijhoff, 1977.

Schütz, Alfred. "Fragments on the Phenomenology of Music." Edited with an Introduction by F. Kersten. *Music and Man*, VI (1976), pp. 6-71.

_______. *Collected Papers*, Vol. III. Edited by I. Schutz with an Introduction by Aron Gurwitsch. The Hague: Martinus Nijhoff, 1966. ("The Problem of Transcendental Intersubjectivity in Husserl,"pp. 51-83; "Edmund Husserl's Ideas, Volume II," pp. 15-39.)

_______. "Husserl and the Social Sciences." *Edmund Husserl 1859-1959*. The Hague: Martinus Nijhoff, 1959, pp. 86-98.

_______. *Der Sinnhafte Aufbau der sozialen Welt. Eine Einleitung in die verstehende Soziologie.* Wien: Springer Verlag, 1960.

_______. *The Phenomenology of the Social World.* Translated by George Walsh and Frederick Lehnert. Evanston: Northwestern University Press, 1967.

Sokolowski, Robert. *Husserlian Meditations. How Words Present Things.* Evanston: Northwestern University Press, 1974.

_______. *The Formation of Husserl's Concept of Constitution.* The Hague: Martinus Nijhoff, 1964

Spiegelberg, Herbert. *The Phenomenological Movement. A Historical Introduction.* Third Revised and Enlarged Edition. With the Collaboration of Karl Schuhmann. The Hague: Martinus Nijhoff, 1982.

_______. "Reflections on the Phenomenological Movement." *Journal of the British Society for Phenomenology*, XI (1980).

_______. "Über das Wesen der Idee. Eine ontologische Untersuchung." *Jahrbuch für Philosophie und phänomenologische Forschung*, XI (1930), pp. 1-238.

Spinoza, Baruch. *Works of Spinoza.* Unabridged Elwes Translation. New York: Dover Publications, 1955.

Stein, Edith. "Beiträge zur philosophischen Begründung der Psychologie und der Geisteswissenschaften." *Jahrbuch für Philosophie und phänomenologische Forschung*, V (1926), pp. 1-284.

Strauss, Erwin. "The Upright Posture." *Essays in Phenomenology*. Edited by Maurice Natanson. The Hague: Martinus Nijhoff, 1966, pp. 164-192.

Ströker, Elisabeth. *Philosophische Untersuchung zum Raum*. Frankfurt am Main: Vittorio Klostermann, 1965. English translation, *Investigations in Philosophy of Space*, trans. Algis Mickunas. Athens: Ohio University Press, 1987.

Stumpf, Carl. *Über den psychologischen Ursprung der Raumvorstellung*. Leipzig: S. Hirzel, 1873.

Szilasi, Wilhelm. *Einführung in die Phänomenologie Husserls*. Tubingen: Max Niemeyer, 1959.

Toulemont, Rene. *L'Essence de la Société selon Husserl*. Paris: Presses Universitaires de France, 1962.

Van de Pitte, M. "Introduction to 'Author's Preface to the English Edition of *Ideas*'." *Husserl. Shorter Works*. Edited by F. Elliston and P. McCormick. Notre Dame: Notre Dame University Press, 1981, pp. 36-41.

Voss, Hans. *Transzendenz und Raumanschauung*. Frankfurt am Main: Vittorio Klosterman, 1940.

Vuillemin, Jules. *Physique et Métaphysique Kantiennes*. Paris: Presses Universitaires de France, 1955.

Wagner, Helmut. "Toward an Anthropology of the Life-World: Alfred Schütz's Quest for the Ontological Justification of the Phenomenological Undertaking." *Human Studies*, 6 (1983), pp. 239-246.

Wallach, Hans. "On Sound Localization." *Readings in Perception*. edited by Beardslee and Wertheimer. New York: Van nostrand Company, 1958, pp. 476-483.

Werner, Heinz. *Grundfragen der Intensitätspsychologie*. Leipzig: Verlag von Johann Ambrosius Barth, 1929.

Witasek, Stephan. *Grundlinien der Psychologie*. Leipzig: Durr'schen Buchhandlung, 1908.

Young, P.T. "Auditory Localization with accoustical transposition of the ears." *Journal of Experimental Psychology*, 11 (1928), pp. 399-429.

Zaner, Richard M. *The Way of Phenomenology. Criticism as a Philosophical Discipline*. New York: Pegasus, 1970.

INDEX OF NAMES

INDEX OF TOPICS